Greenhouse Gas Emission and Mitigation in Agriculture

Greenhouse Gas Emission and Mitigation in Agriculture

Hudson Keech

SYRAWOOD
PUBLISHING HOUSE
New York

Published by Syrawood Publishing House,
750 Third Avenue, 9th Floor,
New York, NY 10017, USA
www.syrawoodpublishinghouse.com

Greenhouse Gas Emission and Mitigation in Agriculture
Hudson Keech

International Standard Book Number: 978-1-64740-351-5 (Hardback)

Cataloging-in-Publication Data

Greenhouse gas emission and mitigation in agriculture / Hudson Keech.
p. cm.
Includes bibliographical references and index.
ISBN 978-1-64740-351-5
1. Agriculture--Environmental aspects. 2. Greenhouse gases--Environmental aspects.
3. Greenhouse gas mitigation. 4. Land use, Rural--Environmental aspects.
5. Crops and climate. I. Keech, Hudson.

S589.75 .G74 2023
630--dc23

TABLE OF CONTENTS

PREFACE

Greenhouse gases, or GHGs, are compound gases with the property of absorbing heat or longwave radiation in the atmosphere. They lead to the greenhouse effect, global warming and climate change. Nuclear tools and techniques help the researchers in monitoring greenhouse gas emissions and understanding environmental changes to oceans, mountains and their ecosystems. They also play an important role in developing ways to adapt to food and water shortages exacerbated by changing weather patterns. In agriculture, nuclear and isotopic techniques are used to reduce emissions by detecting the amount of nitrous oxide a plant can consume. Isotopes are also used to determine the status and movement of water in soil. They can be used to assess the status and movement of soil, which in turn, helps in understanding how strong crops can be produced under changing conditions. This book outlines the varied aspects of greenhouse gas emission and mitigation in agriculture using nuclear methods. It aims to serve as a resource guide for scholars and researchers alike.

Various studies have approached the subject by analyzing it with a single perspective, but the present book provides diverse methodologies and techniques to address this field. This book contains theories and applications needed for understanding the subject from different perspectives. The aim is to keep the readers informed about the progresses in the field; therefore, the contributions were carefully examined to compile novel researches by specialists from across the globe.

Indeed, the job of the editor is the most crucial and challenging in compiling all chapters into a single book. In the end, I would extend my sincere thanks to the chapter authors for their profound work. I am also thankful for the support provided by my family and colleagues during the compilation of this book.

Hudson Keech

1

Agriculture and GHGs

M. Zaman, K. Kleineidam, L. Bakken, J. Berendt, C. Bracken, K. Butterbach-Bahl, Z. Cai, S. X. Chang, T. Clough, K. Dawar, W. X. Ding, P. Dörsch, M. dos Reis Martins, C. Eckhardt, S. Fiedler, T. Frosch, J. Goopy, C.-M. Görres, A. Gupta, S. Henjes, M. E. G. Hofmann, M. A. Horn, M. M. R. Jahangir, A. Jansen-Willems, K. Lenhart, L. Heng, D. Lewicka-Szczebak, G. Lucic, L. Merbold, J. Mohn, L. Molstad, G. Moser, P. Murphy, A. Sanz-Cobena, M. Šimek, S. Urquiaga, R. Well, N. Wrage-Mönnig, S. Zaman, J. Zhang, and C. Müller

Abstract The rapidly changing global climate due to increased emission of anthropogenic greenhouse gases (GHGs) is leading to an increased occurrence of extreme weather events such as droughts, floods, and heatwaves. The three major GHGs are carbon dioxide (CO_2), methane (CH_4), and nitrous oxide (N_2O). The major natural sources of CO_2 include ocean–atmosphere exchange, respiration of animals, soils

M. Zaman (✉) · L. Heng
Soil and Water Management & Crop Nutrition (SWMCN) Section, Joint FAO/IAEA Division of Nuclear Techniques in Food and Agriculture, International Atomic Energy Agency (IAEA), Vienna, Austria
e-mail: m.zaman@iaea.org; zamanm_99@yahoo.com

K. Kleineidam · C. Eckhardt · A. Jansen-Willems · G. Moser · C. Müller
Institute of Plant Ecology, Justus Liebig University Giessen, Giessen, Germany

L. Bakken
Norwegian University of Life Sciences (NMBU), Aas, Norway

J. Berendt · S. Fiedler · N. Wrage-Mönnig
University of Rostock, Rostock, Germany

C. Bracken
School of Agriculture and Food Science and Earth Institute, University College Dublin, Dublin, Ireland

K. Butterbach-Bahl
Institute of Meteorology and Climate Research, Atmospheric Environmental Research (IMK-IFU), Karlsruhe Institute of Technology, Karlsruhe, Germany

Z. Cai
School of Geography Sciences, Nanjing Normal University, Jiangsu, China

S. X. Chang
Department of Renewable Resources, University of Alberta, Edmonton, AB T6G 2E3, Canada

T. Clough
Department of Soil & Physical Sciences, Faculty of Agriculture & Life Sciences, Lincoln University, Lincoln, New Zealand

(microbial respiration) and plants, and volcanic eruption; while the anthropogenic sources include burning of fossil fuel (coal, natural gas, and oil), deforestation, and the cultivation of land that increases the decomposition of soil organic matter and crop and animal residues. Natural sources of CH_4 emission include wetlands, termite activities, and oceans. Paddy fields used for rice production, livestock production systems (enteric emission from ruminants), landfills, and the production and use of fossil fuels are the main anthropogenic sources of CH_4. Nitrous oxide, in addition to being a major GHG, is also an ozone-depleting gas. N_2O is emitted by natural processes from oceans and terrestrial ecosystems. Anthropogenic N_2O emissions occur mostly through agricultural and other land-use activities and are associated with the intensification of agricultural and other human activities such as increased use of synthetic fertiliser (119.4 million tonnes of N worldwide in 2019), inefficient use of irrigation water, deposition of animal excreta (urine and dung) from grazing animals, excessive and inefficient application of farm effluents and animal manure to croplands and pastures, and management practices that enhance soil organic N mineralisation and C decomposition. Agriculture could act as a source and a sink of GHGs. Besides direct sources, GHGs also come from various indirect sources, including upstream and downstream emissions in agricultural systems and ammonia (NH_3) deposition from fertiliser and animal manure.

K. Dawar
Department of Soil and environmental Sciences, University of Agriculture, Peshawar, Pakistan

W. X. Ding
Institute of Soil Science, Chinese Academy of Sciences, Nanjing, China

P. Dörsch · L. Molstad
Faculty of Environmental Sciences and Natural Resource Management,
Norwegian University of Life Sciences (NMBU), Aas, Norway

T. Frosch
Leibniz Institute of Photonic Technology, Technical University
Darmstadt, Darmstadt, Germany

J. Goopy
International Livestock Research Institute (ILRI), Nairobi, Kenya

C.-M. Görres
Department of Soil Science and Plant Nutrition/Department of Applied Ecology, Hochschule
Geisenheim University, Geisenheim, Germany

A. Gupta
Independent Consultant India, Mumbai, India

S. Henjes · M. A. Horn
Institute of Microbiology, Leibniz University Hannover, Hannover, Germany

M. E. G. Hofmann
Picarro B.V., 's-Hertogenbosch, The Netherlands

M. M. R. Jahangir
Department of Soil Science, Bangladesh Agricultural University, Mymensingh, Bangladesh

Keywords Climate change · GHG · N_2O · CO_2 · NH_3 · Animals

1.1 Introduction

The global climate is changing rapidly. This leads to the increasing occurrence of extreme weather events such as droughts and floods. The major cause of these events is the rising temperature in the Earth's atmosphere, which is driven by increasing emissions of climate-relevant greenhouse gases (GHGs) that trap heat in the atmosphere. Major GHGs include carbon dioxide (CO_2), methane (CH_4), and nitrous oxide (N_2O) (Fig. 1.1).

Carbon dioxide is the major GHG responsible for the increasing greenhouse effect of the atmosphere. Key natural sources of CO_2 include ocean–atmosphere exchange, respiration of animals, soils (microbial respiration) and plants, and volcanic eruption. Major anthropogenic sources of CO_2 include burning of fossil fuel (coal, natural gas, and oil), deforestation, and the cultivation of land that increases the decomposition of soil organic matter and crop and animal residues (Xu and Shang 2016).

Aside from CO_2, CH_4 is a major GHG, which is emitted by natural and anthropogenic processes. Natural sources of CH_4 emission include wetlands, termite activities, and ocean. Paddy fields used for rice production, livestock production systems

J. Zhang
School of Geography, Nanjing Normal University, Nanjing, China

K. Lenhart
Bingen University of Applied Sciences, Berlinstr. 109, Bingen 55411, Germany

D. Lewicka-Szczebak
Laboratory of Isotope Geology and Geoecology, Institute of Geological Sciences, University of Wrocław, Wrocław, Poland

G. Lucic
Picarro Inc., Santa Clara, CA, USA

L. Merbold
Mazingira Centre, International Livestock Research Institute (ILRI), Nairobi, Kenya

J. Mohn
Laboratory for Air Pollution & Environmental Technology, Empa Dübendorf, Dübendorf, Switzerland

P. Murphy
Environment & Sustainable Resource Management Section, School of Agriculture & Food Science, and UCD Earth Institute, University College, Dublin, Ireland

A. Sanz-Cobena
Research Center for the Management of Environmental and Agricultural Risks (CEIGRAM), ETSIAAB, Universidad Politécnica de Madrid, Madrid, Spain

M. Šimek
Institute of Soil Biology, Biology Centre of the Czech Academy of Sciences, and Faculty of Science, University of South Bohemia, České Budějovice, Czech Republic

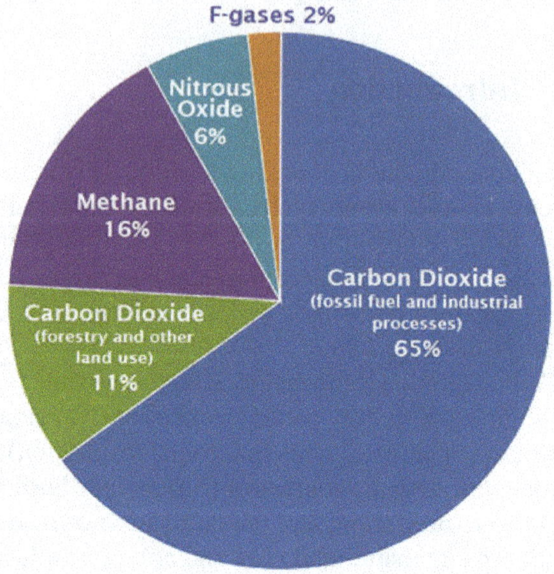

Fig. 1.1 Major greenhouse gas emissions and contributions by various sectors (IPCC 2014a, b)

(enteric emission from ruminants), landfills, and production and use of fossil fuels are the main anthropogenic sources of CH_4. Furthermore, CH_4 can be produced by anaerobic mineralization by methanogenic archaea in both natural and man-made systems. Also, plants have been shown to emit CH_4.

The third major GHG is N_2O (Zaman et al. 2012). Besides being a major GHG, N_2O is a major ozone-depleting gas (Ravishankara et al. 2009). Oceans and soils under natural vegetation are non-anthropogenic sources of N_2O. However, at a global scale, the emission of N_2O is mostly caused by, or related to, anthropogenic agricultural and other land-use activities. The atmospheric concentration of N_2O has increased by more than 20% from ~271 ppb to 331 ppb since the industrial era (ca. 1750) to 2018 (WMO 2019). Over the last decade, the rate of N_2O increase was equal to 0.95 ppb yr^{-1} (IPCC 2013b; WMO 2019) with an increasing trend (Makowski 2019; Thompson et al. 2019). In 2006, the total anthropogenic source of N_2O was 6.9 Tg N_2O-N. Out of these direct emissions, agricultural sources dominated (4.1 Tg N_2O-N), while indirect emissions accounted for 0.6 (with a range of 0.1–2.9) Tg N_2O-N (IPCC 2013a). Such a large N_2O emission is attributed to various factors including the intensification of agricultural and other human activities, increased use of synthetic fertiliser (119.4 million tonnes of N worldwide in 2019), inefficient

M. dos Reis Martins · S. Urquiaga
EMBRAPA Agrobiologia Seropédica, Brazilian Agricultural Research Corporation, Seropedica, RJ, Brazil

R. Well
Thünen Institute of Climate-Smart Agriculture, Braunschweig, Germany
S. Zaman
University of Canterbury, Christchurch, New Zealand

use of irrigation water, deposition of animal excreta (urine and dung) from grazing animals, excessive and inefficient application of farm effluents and animal manure to croplands and pastures, and management practices that enhance soil organic N mineralisation and C decomposition. These activities affect the N cycle. The N cycle is rather complex, and it has even been disrupted due to increased N inputs and intensification of agriculture. Sources of increased N inputs, in particular reactive N, into the N cycle stem, for instance, from the Haber–Bosch process (Erisman et al. 2011), thereby transforming the N cycle into the so-called N cascade, which is characterised by the release of reactive N forms into the environment with various consequences (Sutton et al. 2011). There are still many uncertainties concerning the N cycle. For example, the role of individual factors controlling the occurrence and rate of the key N transformation processes, such as denitrification and nitrification, is uncertain (Butterbach-Bahl et al. 2013; Müller and Clough 2014; Smith 2017). Nitrification results in ammonium (NH_4^+) being converted to nitrate (NO_3^-) under aerobic conditions, while denitrification is the reduction of NO_3^- to N_2 under anaerobic conditions. Nitrifier denitrification occupies the niche between nitrification and denitrification and occurs as oxygen concentrations approach an anaerobic status. Under these conditions, nitrifiers actually convert nitrite into N_2O and N_2 instead of nitrate (Wrage-Mönnig et al. 2018).

The three GHGs (CO_2, CH_4, and N_2O), derived from various sectors, play a major role in regulating Earth's temperature (Fig. 1.1). Without GHGs in the atmosphere, the average global soil surface temperature would be ~19 °C, compared to the present values of 14 °C (Hossain 2019).

Recent data from the UN Intergovernmental Panel on Climate Change (IPCC) clearly show that anthropogenic emissions of GHGs are at the highest in history (IPCC 2014a). Since 1990, Earth's average surface air temperature has increased by about 0.8 °C, with much of the emission increases taking place since the mid-1970s (Fig. 1.2).

The global warming potential (GWP) of a GHG relates to the amount of heat trapped by a certain mass of a gas to the amount of heat trapped by a similar mass of CO_2 calculated over a 100-year time horizon (IPCC 2016). For example, the GWP of N_2O is 265–298, which means if the same masses of N_2O and CO_2 were emitted into the atmosphere, N_2O would trap 265–298 times more heat than CO_2 over a 100-year time period (Table 1.1) (IPCC 2016).

The Kyoto Protocol was negotiated by parties to the United Nations Framework Convention on Climate Change (UNFCCC) in an effort to stabilise the continued increase in atmospheric GHG concentrations. The Kyoto protocol outlines GHG reduction targets for participating countries. The signatories to the protocol must develop and report on their annual national inventory of anthropogenic GHG emissions. Guidelines on how to construct inventories were prescribed by the IPCC (2014b). Country-specific emission data can be considered but that requires accurate inventory data based on precise GHG measurements.

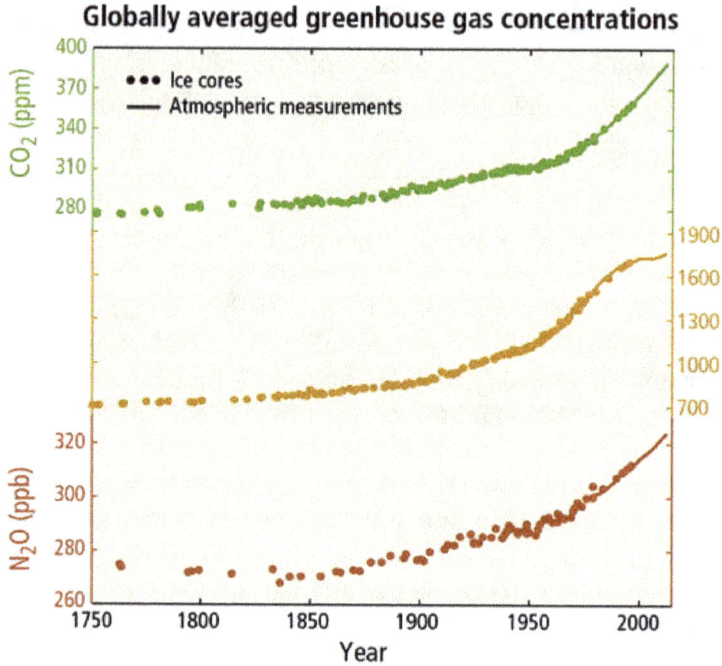

Fig. 1.2 Recent anthropogenic emissions of GHGs (IPCC 2014a; WMO 2019)

Table 1.1 Global warming potential (GWP) and atmospheric lifespan of various GHGs (IPCC 2016)

Greenhouse gas	Chemical formula	Global warming potential (GWP) for a 100-year time horizon	Lifespan (years)
Nitrous oxide	N_2O	265–310	114
Methane	CH_4	21–28	12
Carbon dioxide	CO_2	1	Variable
Water vapour	H_2O	n/a	Variable
Ozone	O_3	n/a	Hours to days

1.2 Impact of Ammonia on GHG Emissions

Besides direct sources, GHGs also come from different indirect sources such as upstream and downstream in agricultural systems (Plate 1.1).

Ammonia (NH_3) itself has no direct greenhouse effect. It is a gas with a relatively short residence time in the atmosphere (2–10 days) compared to some GHGs, such as CO_2 (3–4 years), CH_4 (12 years), and N_2O (114 years). However, after NH_3 is emitted to the atmosphere and reacts with acids, it forms salts. These salts then return to The Earth's surface and act as a N source source for N_2O emissions, similar to a fertiliser-N application. When the soil is submitted to conditions near the optimum for

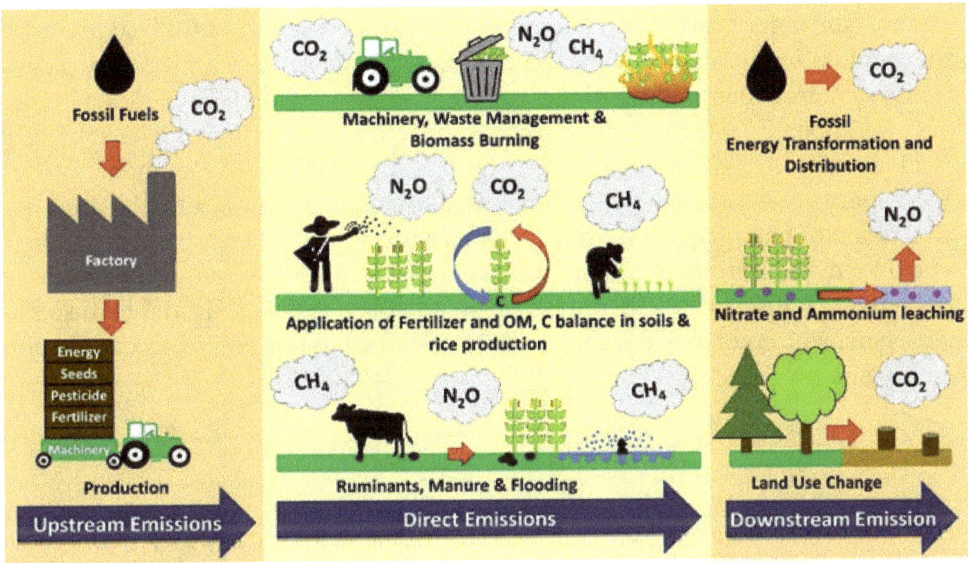

Plate 1.1 Schematic representation of direct (from the cropping system) and indirect (upstream and downstream) GHG emissions from crop production. As an example, ammonia emitted from a cropping system will be deposited and potentially oxidised to nitrate, which can further be denitrified, thus enhancing the risk for N_2O emission (Aguilera Fernández 2016)

urease activity (e.g. pH close to neutrality, soil moisture near field capacity, temperature >30 °C), the N losses through NH_3 volatilisation from urea-based fertilisers applied on soil surface can be as high as 50% (Martins et al. 2017; Rochette et al. 2013). Therefore, the measurement of NH_3 emission is important to estimate indirect N_2O emissions derived from soil amendments, such as urea-based fertilisers, green manures, animal excreta, or ammonium-based fertilisers in alkaline soils. A default emission factor defined by IPCC, known as EF_4, can be applied for the estimation of indirect N_2O emissions derived from volatilisation of NH_3 and other nitrogen oxides (NO_x) (IPCC 2006). The mean value of EF_4, considering the N volatilisation and consequent re-deposition, is 0.01 kg N_2O-N per kg N volatilised as NH_3 with an uncertainty ranging from 0.002 to 0.05 (IPCC 2006). Therefore, management options that reduce NH_3 volatilisation from soils are considered mitigation practices because they reduce indirect N_2O emissions (IPCC 2014a; Lam et al. 2017).

1.3 Aim of the Book

In 1992, the International Atomic Energy Agency (IAEA) published a "Manual on measurement of CH_4 and N_2O emissions from agriculture" (IAEA 1992). Since the publication of the manual, progress has been made in analytical techniques. The progress includes advances in automation technologies as well as the theoretical understanding of how soil microbial processes affect CH_4 and N_2O emissions and the factors influencing those microbial processes.

Hence, the aim of this book is to provide an updated account of the state-of-the-art techniques to measure direct GHG emissions (Plate 1.1), as a necessary step to propose and assess any mitigation strategy. The focus is on CH_4 and N_2O emissions. Additionally, information on techniques to measure indirect GHG sources is provided. Indirect GHG sources in this book include volatilised ammonia (NH_3). NH_3 is a reactive N gas highly affecting the environment through eutrophication and acidification of natural ecosystems as well as human health due to the promotion of particulate matter formation (Sanz Cobena et al. 2014). Moreover, in line with the 1992 IAEA Tecdoc document, the hands-on approach is also followed here so that researchers, who want to use the techniques described in this book, can easily apply them to their own work.

References

Aguilera Fernández EM (2016) The influence of management practices on the greenhouse gas balance of Mediterranean cropping systems. Identifying the climate change mitigation potential through quantitative review and life cycle assessment. Departmento de Geografía, Historia y filosofía laboratorio de historia de los agroecosistemas. Universidad Pablo de Olavide, Sevilla, pp 428

Butterbach-Bahl K, Baggs EM, Dannenmann M, Kiese R, Zechmeister-Boltenstern S (2013) Nitrous oxide emissions from soils, how well do we understand the processes and their controls. Philos T R Soc London B 368:16–21

Erisman JW, Galloway J, Seitzinger S, Bleeker A, Butterbach-Bahl K (2011) Reactive nitrogen in the environment and its effect on climate change. Curr Opin Environ Sustain 3:281–295

Hossain F (2019) Sustainable design and build building, energy, roads, bridges, water and sewer systems. Butterworth-Heinemann, Elsevier, p 462

IAEA (1992) Manual on measurement of methane and nitrous oxide emissions from agriculture. International Atomic Energy Agency IAEA, Vienna, p 90

IPCC (2006) 2006 IPCC guidelines for national greenhouse gas inventories Chapter 11 N_2O emissions from managed soils, and CO_2 emissions from lime and urea application, 2006 IPCC guidelines for national greenhouse gas inventories

IPCC (2013a) Climate change 2013: The physical science basis. In: Stocker TF, Qin D, Plattner GK, Tignor M, Allen SK, Boschung J, Nauels A, Xia Y, Bex V, Midgley PM (eds) Contribution of working group I to the fifth assessment report of the intergovernmental panel on climate change. United Kingdom and New York, pp 1535

IPCC (2013b) Working group I contribution to the IPCC fifth assessment report climate change 2013: The physical science basis summary for policymakers. Intergovernmental Panel for Climate Change, pp 36

IPCC (2014a) Climate change 2014 synthesis report: Contribution of working groups I, II and III to the fifth assessment report of the intergovernmental panel on climate change. In: Team TCW, Pachauri RK, Meyer LA (eds) A report of the intergovernmental panel on climate change. Geneva, Switzerland, pp 151

IPCC (2014b) Supplement to the 2006 IPCC guidelines for national greenhouse gas inventories: Wetlands, In: Hirashi T, Krug T, Tanabe K, Srivastava N, Baasansuren J, Fukuda M, Troxler TG (eds) IPCC task force on national greenhouse gas inventories. Switzerland, pp 354

IPCC (2016) Global warming potential values on climate change 2013: The physical science basis. In: Contribution of working group I fifth assessment report of the intergovernmental panel on climate change, pp 4

Lam SK, Suter H, Mosier AR, Chen D (2017) Using nitrification inhibitors to mitigate agricultural N_2O emission: a double-edged sword? Glob Chang Biol 23:485–489

Makowski D (2019) N_2O increasing faster than expected. Nat Clim Chang 9:909–910

Martins MR, Sant'Anna SAC, Zaman M, Santos RC, Monteiro RC, Alves BJR, Jantalia CP, Boddey RM, Urquiaga S (2017) Strategies for the use of urease and nitrification inhibitors with urea: Impact on N_2O and NH_3 emissions, fertilizer-^{15}N recovery and maize yield in a tropical soil. Agric Ecosy Environ 247:54–62

Müller C, Clough TJ (2014) Advances in understanding nitrogen flows and transformations: gaps and research pathways. J Agric Sci Cambridge 152:S34–S44

Ravishankara AR, Daniel JS, Portmann RW (2009) Nitrous oxide (N_2O): the dominant ozone-depleting substance emitted in the 21st century. Science 326:123–125

Rochette P, Angers DA, Chantigny MH, Gasser MO, MacDonald JD, Pelster DE, Bertrand N (2013) NH_3 volatilization, soil NH_4^+ concentration and soil pH following subsurface banding of urea at increasing rates. Can J Soil Sci 93:261–268

Sanz Cobena A, Lassaletta L, Estellés F, Del Prado A, Guardia G, Abalos D, Aguilera Fernández EM, Pardo G, Vallejo A, Sutton MA, Garnier J, Billen G (2014) Yield-scaled mitigation of ammonia emission from N fertilization: the Spanish case. Environ Res Lett 9(125012): 125005

Smith KA (2017) Changing views of nitrous oxide emissions from agricultural soil: key controlling processes and assessment at different spatial scales. Eur J Soil Sci 68:137–155

Sutton MA, Oenema O, Erisman JW, Leip A, van Grinsven H, Winiwarter W (2011) Too much of a good thing. Nature 472:159–161

Thompson RL, Lassaletta L, Patra PK, Wilson C, Wells KC, Gressent A, Koffi EN, Chipperfield MP, Winiwarter W, Davidson EA, Tian H, Canadell JG (2019) Acceleration of global N_2O emissions seen from two decades of atmospheric inversion. Nat Clim Chang 9:993–998

WMO (2019) The state of greenhouse gases in the atmosphere based on global observations through 2018. WMO Greenhouse Gas Bulletin 15:1–8

Wrage-Mönnig N, Horn MA, Well R, Müller C, Velthof G, Oenema O (2018) The role of nitrifier denitrification in the production of nitrous oxide revisited. Soil Biol Biochem 123:A3–A16

Xu M, Shang H (2016) Contribution of soil respiration to the global carbon equation. J Plant Physio 203:16–28

Zaman M, Nguyen ML, Šimek M, Nawaz S, Khan MJ, Babar MN, and Zaman S (2012) Emissions of nitrous oxide (N_2O) and di-nitrogen (N_2) from agricultural landscape, sources, sinks, and factors affecting N_2O and N_2 ratios. In: Guoxiang Liu (ed) Greenhouse gases-emission, measurement and management. Intech, Croatia, pp 1–32

2

Techniques for Measuring Greenhouse Gases from Agricultural Sites

M. Zaman, K. Kleineidam, L. Bakken, J. Berendt, C. Bracken, K. Butterbach-Bahl, Z. Cai, S. X. Chang, T. Clough, K. Dawar, W. X. Ding, P. Dörsch, M. dos Reis Martins, C. Eckhardt, S. Fiedler, T. Frosch, J. Goopy, C.-M. Görres, A. Gupta, S. Henjes, M. E. G. Hofmann, M. A. Horn, M. M. R. Jahangir, A. Jansen-Willems, K. Lenhart, L. Heng, D. Lewicka-Szczebak, G. Lucic, L. Merbold, J. Mohn, L. Molstad, G. Moser, P. Murphy, A. Sanz-Cobena, M. Šimek, S. Urquiaga, R. Well, N. Wrage-Mönnig, S. Zaman, J. Zhang, and C. Müller

Abstract Several approaches exist for measuring greenhouse gases (GHGs), mainly CO_2, N_2O, and CH_4, from soil surfaces. The principle methods that are used to measure GHG from agricultural sites are chamber-based techniques. Both open and closed chamber techniques are in use; however, the majority of field applications use closed chambers. The advantages and disadvantages of different chamber techniques

---

M. Zaman (✉) · L. Heng
Soil and Water Management & Crop Nutrition (SWMCN) Section, Joint FAO/IAEA Division of Nuclear Techniques in Food and Agriculture, International Atomic Energy Agency (IAEA), Vienna, Austria
e-mail: m.zaman@iaea.org; zamanm_99@yahoo.com

K. Kleineidam · C. Eckhardt · A. Jansen-Willems · G. Moser · C. Müller
Institute of Plant Ecology, Justus Liebig University Giessen, Giessen, Germany

L. Bakken
Norwegian University of Life Sciences (NMBU), Aas, Norway

J. Berendt · S. Fiedler · N. Wrage-Mönnig
University of Rostock, Rostock, Germany

C. Bracken
School of Agriculture and Food Science and Earth Institute, University College Dublin, Dublin, Ireland

K. Butterbach-Bahl
Institute of Meteorology and Climate Research, Atmospheric Environmental Research (IMK-IFU), Karlsruhe Institute of Technology, Karlsruhe, Germany

Z. Cai
School of Geography Sciences, Nanjing Normal University, Jiangsu, China

S. X. Chang
Department of Renewable Resources, University of Alberta, Edmonton, AB T6G 2E3, Canada

and the principal steps of operation are described. An important part of determining the quality of the flux measurements is the storage and the transportation of the gas samples from the field to the laboratory where the analyses are carried out. Traditionally, analyses of GHGs are carried out via gas chromatographs (GCs). In recent years, optical analysers are becoming increasingly available; these are user-friendly machines and they provide a cost-effective alternative to GCs. Another technique which is still under development, but provides a potentially superior method, is Raman spectroscopy. Not only the GHGs, but also N_2, can potentially be analysed if the precision of these techniques is increased in future development. An important part of this chapter deals with the analyses of the gas concentrations, the calculation of fluxes, and the required safety measures. Since non-upland agricultural lands (i.e. flooded paddy soils) are steadily increasing, a section is devoted to the specificities of GHG measurements in these ecosystems. Specialised techniques are also required for GHG measurements in aquatic systems (i.e. rivers), which are often affected by the transfer of nutrients from agricultural fields and therefore are an important indirect source of emission of GHGs. A simple, robust, and more precise method of ammonia (NH_3) emission measurement is also described.

Keywords GHG · Chamber-based technique · Raman spectroscopy · Ammonia emission · Optical spectroscopy

---

T. Clough
Department of Soil & Physical Sciences, Faculty of Agriculture & Life Sciences, Lincoln University, Lincoln, New Zealand

K. Dawar
Department of Soil and environmental Sciences, University of Agriculture, Peshawar, Pakistan

W. X. Ding
Institute of Soil Science, Chinese Academy of Sciences, Nanjing, China

P. Dörsch · L. Molstad
Faculty of Environmental Sciences and Natural Resource Management, Norwegian University of Life Sciences (NMBU), Aas, Norway

T. Frosch
Leibniz Institute of Photonic Technology, Technical University Darmstadt, Darmstadt, Germany

J. Goopy
International Livestock Research Institute (ILRI), Nairobi, Kenya

C.-M. Görres
Department of Soil Science and Plant Nutrition/Department of Applied Ecology, Hochschule Geisenheim University, Geisenheim, Germany

A. Gupta
Independent Consultant India, Mumbai, India

S. Henjes · M. A. Horn
Institute of Microbiology, Leibniz University Hannover, Hannover, Germany

2.1 Introduction

Given the complexity of emissions, process-based models are not able to accurately estimate daily fluxes or the variations in fluxes due to variations in management practices. This limits our understanding of the factors affecting greenhouse gas (GHG) emissions and eventually restricts the development of agricultural management options that minimise GHG emissions. The Intergovernmental Panel on Climate Change (IPCC) requires local data based on field studies. Estimation of emission factors (EF) along with quantification of EF-associated physical, chemical, and biological processes that produce CH_4 and N_2O is required for field-scale GHG measurements. Field measurement of GHG is the basis of GHG flux estimates and a means of evaluating potential countermeasures for reducing emissions (Minamikawa et al. 2015). Several approaches exist for measuring GHG fluxes from soil surfaces. The two most important approaches are chamber-based methods and micrometeorological techniques (Denmead 2008; Oertel et al. 2012) (Chap. 4); more sophisticated approaches include space and airborne measurements.

Micrometeorological techniques usually integrate much larger surface areas in comparison to chamber-based techniques, thereby substantially reducing spatial variability problems that are inherent to chamber-based methods (Mosier 1990). Micrometeorological techniques are often more expensive, require special analytical instruments, and need knowledge and expertise that are largely not available in most developing countries. There is also a third technique that could be used for gas flux estimation–the measurement of gas concentration in different layers of soil.

M. E. G. Hofmann
Picarro B.V., 's-Hertogenbosch, The Netherlands

M. M. R. Jahangir
Department of Soil Science, Bangladesh Agricultural University, Mymensingh, Bangladesh

D. Lewicka-Szczebak
Laboratory of Isotope Geology and Geoecology, Institute of Geological Sciences, University of Wrocław, Wrocław, Poland

G. Lucic
Picarro Inc., Santa Clara, CA, USA

L. Merbold
Mazingira Centre, International Livestock Research Institute (ILRI), Nairobi, Kenya

J. Mohn
Laboratory for Air Pollution & Environmental Technology, Empa Dübendorf, Dübendorf, Switzerland

P. Murphy
Environment & Sustainable Resource Management Section, School of Agriculture & Food Science, and UCD Earth Institute, University College, Dublin, Ireland

A. Sanz-Cobena
Research Center for the Management of Environmental and Agricultural Risks (CEIGRAM), ETSIAAB, Universidad Politécnica de Madrid, Madrid, Spain

Information about gas concentration in a soil profile can also be used for gas flux prediction (Chirinda et al. 2014; Kammann et al. 2001). However, this technique requires additional information on soil physical and chemical properties, including hydraulic characteristics, to calculate GHG fluxes based on gas diffusion in the soil matrix (Diel et al. 2019). Therefore, in most instances, chamber-based methods have been used to study GHG fluxes from agricultural soils. Nonetheless, in combination with chamber-based techniques, soil profile techniques can provide valuable additional information to explain and analyse GHG emissions from the soil surface (Müller et al. 2004).

2.2 Chamber-Based Methods

The great majority of GHG emission studies published in the past three decades have used chamber-based techniques–in particular, non-flow-through, non-steady-state chambers (Rochette 2011). These methods have been described in detail in several excellent review papers (De Klein and Harvey 2012; Hutchinson and Livingston 1993; Hutchinson and Rochette 2003; Mosier 1989). The following text mainly comes from these reviews that precisely address the topic and represent a comprehensive overview of the method. In addition to information from the literature, experimental data and our own experience with field gas flux determination will also be presented.

Mosier (1990) characterised three basic chamber-based techniques: open soil chambers (open dynamic chambers) that use flow-through air circulation, closed soil chambers with closed-loop air circulation (closed dynamic chambers), or no air circulation (static closed chambers).

M. Šimek
Institute of Soil Biology, Biology Centre of the Czech Academy of Sciences, and Faculty of Science, University of South Bohemia, České Budějovice, Czech Republic

M. dos Reis Martins · S. Urquiaga
EMBRAPA Agrobiologia Seropédica, Brazilian Agricultural Research Corporation, Seropedica, RJ, Brazil

R. Well
Thünen Institute of Climate-Smart Agriculture, Braunschweig, Germany

S. Zaman
University of Canterbury, Christchurch, New Zealand

J. Zhang
School of Geography, Nanjing Normal University, Nanjing, China

K. Lenhart
Bingen University of Applied Sciences, Berlinstr. 109, Bingen 55411, Germany

2.2.1 *Advantages and Disadvantages of Closed Chamber-Based Methods*

The closed chamber technique has several advantages (Mosier 1990; Oertel et al. 2016), including the following:

- Closed chambers are simple and inexpensive to construct from various materials in different designs, shapes, and sizes, which makes it easier to find the type best suited for a given task.
- Operation of chambers and the measurement are simple, and therefore the method provides an opportunity to measure GHG from different locations at different times with the same equipment and personnel.
- Closed chambers can measure very low rates of GHG fluxes for pasture, cropland, rice paddy, wetland, drain, and ditches in a short period of time (from 30 min to 1 h), without the need for electrical supply.

The closed chamber technique, however, has some limitations:

- Increasing gas concentrations in the enclosed headspace leads to a decrease in the concentration gradient and therefore a reduction in gas diffusion, causing non-linear fluxes between the soil and the air. However, suppression of fluxes due to increased gas concentration in the headspace can be minimised by reducing the enclosure period.
- Closed chambers alter (or even eliminate) fluctuations of atmospheric pressure; however, special vents can equilibrate air pressure inside and outside of the chamber (Hutchinson and Mosier 1981; Mosier 1989).
- Temperature changes either in the soil or in the atmosphere within the chamber can occur. Such temperature differences within and outside the chamber are able to be reduced by covering the chamber with a reflective and insulating material (Mosier 1990; Šimek et al. 2014).

In summary, closed-chamber methods represent an inexpensive and easy to use technique, suitable for the determination of GHG fluxes between the soil and the atmosphere from a wide range of agroecosystems. However, several aspects must be considered when using the closed chamber-based methods, including the following:

(i) **Experimental design** (design of the field experiment, number of replicated plots, plot size, etc.),
(ii) **Chamber construction** (easy to use, easy to transport, but still robust enough to be repeatedly used without being easily damaged. The material used to construct the chamber should be inert and not emitting gases or allow diffusion through the material, but the material can be opaque or transparent),
(iii) **Sampling strategy** (frequency of gas sampling, sample volume, number of samples, soil sampling in addition to gas sampling),
(iv) **Storage technique** (vials for gas sample storage and transportation to laboratory, storage before analysis together with standards stored in the same way),

(v) **Analytical equipment** (gas analysers such as gas chromatograph, CRDS (Cavity Ring-Down Spectroscopy) analyser), and
(vi) **Data analysis and interpretation** (checking for linearity of gas concentration change over time in chambers during measurements. Also, checking for abnormal data points and proper statistical analysis).

Some of the problems with field GHG flux measurements are the large spatial variability of gas fluxes, and the high and often unpredictable temporal changes of fluxes. Other problems associated with field GHG flux measurements include the following:

(i) **Plants**: Sometimes, it is difficult to deal with plants in gas collecting chambers due to their size. So, special chambers need to be designed (such as chambers for a maize field). Plants often consume (and produce) gases (e.g. CO_2, CH_4, and N_2O). Plants also transpire and influence the humidity in the chamber and movement of gases in the soil matrix, e.g. dissolved gas via the transpiration stream.
(ii) **Animals**: It is difficult (if not impossible) to protect the chambers in areas grazed from damage by cattle and other animals. Permanent chambers can be easily damaged and can also cause injuries to animals.
(iii) **Technical and practical challenges**: The large size of a field to be investigated, the frequent long-distance travel to the field, the large number of chambers to be moved around in the field, and a high requirement for manpower for proper GHG sampling.

2.2.2 *Principles and Applications of Chamber-Based Techniques for Gas Flux Measurement*

There is no best technique for GHG flux determination; each technique has its advantages and disadvantages, and no single approach is applicable for all conditions or purposes. An excellent overview of the principles and applications of chamber-based techniques for gas flux measurements was provided by Livingston and Hutchinson (1995). The following text is based on that publication and provides selected information, for practical reasons not every publication mentioned by the authors in their text is cited. More details and references can be found by Livingston and Hutchinson (1995). Some data and experiments taken from Šimek et al. (2014) are also included in this chapter. More recent developments in chamber-based techniques for N_2O flux measurements are discussed and summarised in De Klein and Harvey (2012) and Oertel et al. (2016). The factors causing high temporal and spatial variabilities in GHG emissions are outlined in the following sections.

2.2.3 Gas Exchange Processes

Rates of gas exchange between the soil and the atmosphere are usually extremely variable in time and space, making the determination of gas fluxes very complicated and challenging. Movement of gas molecules is due to either mass flow (advective transport) or molecular diffusion. *Diffusive transport* is described by Fick's law and affected by gas permeability, i.e. the ease with which gases move through soil, varies over several orders of magnitude in relation to the shape, size, orientation of the soil pores and soil water content. *Advective transport* of gas occurs in response to a difference in total pressure between the soil air and the atmosphere, which is described by Darcy's law. *Water* substantially affects the movement of gases in soil. Diffusivity of gases is about 10^4 times smaller in water than in air (although the rates differ for different gases). *Plants* influence the exchange of gases; typically, the presence of plants results in increases in gas fluxes from and to the soil. Plants (mainly vascular plants) function as a direct pathway for the flow of trace gases through (often specialised) plant tissues (aerenchyma system); plants also alter conditions in the rhizosphere and therefore directly or indirectly influence gas formation and transport. Moreover, plants also consume and produce several gases, including CH_4 (Liu et al. 2015) or N_2O (Müller 2003). From a practical point of view, the presence of plants usually makes the gas flux determination even more challenging in comparison with bare soil.

2.2.4 Chamber Types

There are many types of closed/static chambers, typically developed by researchers for specific purposes (De Klein and Harvey 2012; Oertel et al. 2016; Saggar et al. 2007). The chambers may be made from various materials, including metals, plastics, and glass, and can have different designs, sizes, shapes, and volumes; chambers as small as 50 cm^3 and as large as ca 1 m^3 have been used for field flux determination from the soil surface. Obviously, chamber materials should be chemically inert, and thus, neither react with the gases being measured nor emit any contaminants. Recommended materials therefore include stainless steel, aluminium, and glass, while the use of polycarbonate, polyethylene, methyl methacrylate, and polyvinyl chloride should be checked for their suitability before use. The schematic diagram of the most common metal chamber is shown in Plate 2.1. This is similar to the design proposed by De Klein and Harvey (2012).

Metal chambers, as shown in Plate 2.1, represent the best choice for many reasons. Metals are not permeable to gases and are inert if materials such as stainless steel are used and can be manufactured in local workshops. The type of material used is important because galvanised steel or normal steel may alter the soil conditions by releasing zinc and iron ions that have the potential to affect microbial activity; thus, it is recommended to use stainless steel. However, when compared to plastic

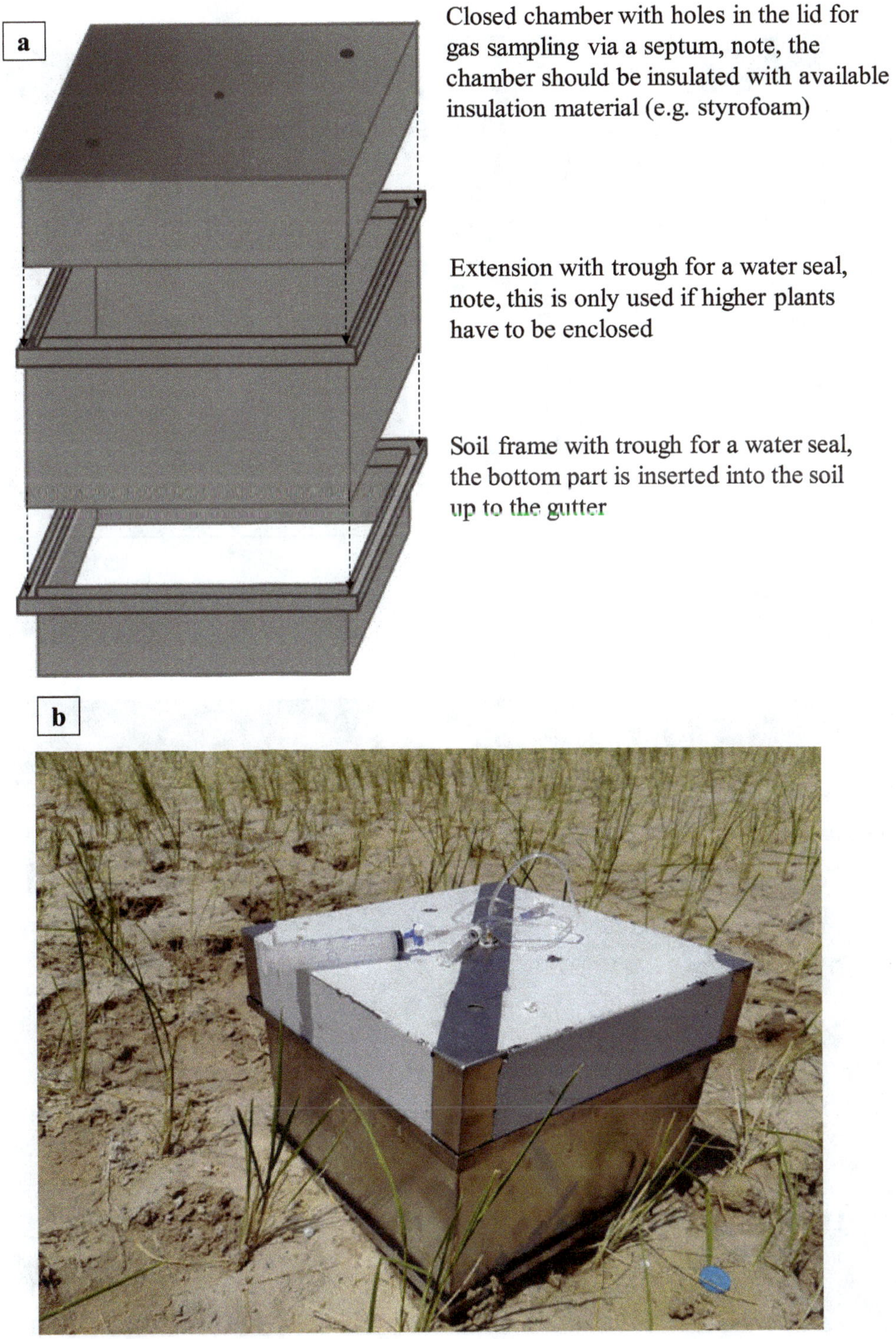

Plate 2.1 A schematic diagram **a** of three parts base frame, extension or enlargement and top lid with GHG sampling ports, a complete metal closed chamber in the paddy field (**b**), and specially designed chamber for maize plants (**c** and **d**)

c

d

Plate 2.1 (continued)

chambers, they can be more expensive, heavier, and less available. Also, insulation is required to minimise temperature fluctuations inside the chamber which in turn would affect fluxes of GHG's, and microbial processes that drive their production. Metal chambers typically consist of two or three parts: the bottom part (also called the base, frame, or soil collar/gutter), the top part (i.e. the chamber), and perhaps a suitable extension (Plate 2.1). The bottom part (frame) should be inserted into the soil at least 2 weeks before the first sampling and permanently installed to minimise soil disturbances effects. Chambers shall be insulated (e.g. using foam or polystyrene with reflective foils) to avoid unnatural heating during chamber closure. This allows repeated gas flux measurements in the same place, e.g. during the whole season. The two/three-part chamber design is strongly recommended, as the disturbances of soils prior to the measurements are eliminated. This, however, means that the frame should ideally be placed into the soil a few weeks before the measurements commence (De Klein and Harvey 2012; Oertel et al. 2016). Precautions should be taken in grazed sites so that neither the chamber nor the animals are endangered.

Plastic chambers have often been used for GHG flux determination (Plate 2.2a, b) (Zaman et al. 2009). The most critical issue for this type of chamber is the nature of the plastic used for making the chamber. Most plastic materials show permeability for gases, as well as the ability to emit some gases or react with them, e.g. hydrocarbons. The advantages of using plastic are (among others) the general availability, easiness to work with, low weight, and able to easily glue different parts together. Plastic chambers consist of two parts (a plastic vessel without a base/collar/gutter) and a lid. The plastic vessel is usually inserted into the soil at least 2–3 days prior to the

Plate 2.2 Plastic chamber made up of two parts: vessel and lid for measuring greenhouse gas emission from **a** pasture soil and **b** arable and vegetable croplands

Plate 2.3 Glass chamber consists of only one part (Šimek, personal communication) for measuring GHG emission from pasture soils

flux measurement. For gas sampling, a lid containing a gas sampling port (rubber septum) connected to a three-way valve is carefully placed on top of the vessel using a gas-tight seal (Plate 2.2a, b). After gas samples are collected, the lid is removed. Plastic chambers can be either transparent or opaque.

Glass chambers have been used less frequently, although the material (glass) is probably the best material to use, considering the inertness and the low gas permeability. However, glass is fragile which makes it very problematic to work with. Therefore, there are more disadvantages than advantages to use glass chambers. Still, one type of glass chambers has been tested for gas flux measurements (Plate 2.3). The major disadvantage of this chamber type is the limited size of the bottles available. Bottle volume is usually between 100 and 2000 cm^3, and surface area covered by the chamber is less than 100 cm^2, which is too small for most uses. Glass chambers consist of a single part (without a base/collar/gutter) and are transparent.

2.2.5 Chamber Design

Critical aspects of the chamber-based methodology include several construction considerations, especially materials, dimensions, gas tightness, and insulation (Table 2.1).

All details for the construction of closed chambers (Table 2.1) shall be taken into consideration in order to maximise the accuracy and precision of the measurements

Table 2.1 Static chamber design requirements and design recommendations (adapted from De Klein and Harvey (2012))

Feature	Requirement/recommendation
Material	Inert to N_2O, CO_2, CH_4, such as stainless steel, aluminium, PVC, acrylic
Area	Recommendation for chamber area: perimeter to be $\geq$10 cm. This equates to a cylindrical chamber of at least 40 cm diameter. As a result, common models have an area smaller than 0.5 m^2, although chambers with an area as large as 2–3 m^2 have been used. The absolute minimum for chamber diameter is ca. 30 cm
Height	Recommended chamber height (cm) to deployment time (h) ratio should be $\geq$40 cm h^{-1}. However, chambers should accommodate crops during measurement, and sometimes chambers higher than 2 m have been used. As a rule of thumb: doubling the chamber height also doubles the cover duration for the same emission rate
Base depth	Ratio of insertion depth: deployment time of $\geq$12 cm h^{-1} to prevent belowground lateral gas transport. Height above soil surface should be as close to the soil surface as practical (<5 cm)
Gas-tight seal	A water trough or rubber/closed-cell foam gasket has mostly been used. Gaskets should have low internal cross-sectional area and be compressible. In general, water seals are effective and often the preferred option
Sampling port	Inert rubber septa or syringe taps inserted through the chamber wall(s)
Venting while placing chamber on base	Opening a vent or sampling port while placing the chamber is recommended
Venting during deployment	If used, vents should be located close to the soil surface, or be designed to avoid mass exchange. Appropriate vent dimensions are dependent on expected wind speed during deployment. Chambers and their vents should be bench-tested to ensure that no Venturi effect occurs
Insulation	Reflective foil, foam, and polystyrene are recommended. Need to consider the effectiveness of insulation, which can be determined by comparing near soil surface temperatures inside and outside the chambers

of soil GHG fluxes. A first factor that shall be considered in terms of dimension is the height of the chamber (De Klein and Harvey 2012). The homogeneity of the air in the headspace can be compromised in higher chambers (e.g. >40 cm). To minimise this effect, small fans connected to rechargeable batteries can be used inside the static chambers or even pumping air from the syringe inside the chamber before gas sampling. However, for the latter care has to be taken to avoid the creation of pressure artifacts in the absence of fans (Christiansen et al. 2011). In addition, higher chambers also lead to poor detection of low gas flux due to the dilution of gas derived from soil with the air of chamber headspace. On the other hand, excessive reduction of the chamber height increases the influence of the chamber deployment on the gas diffusion from soil to chamber causing a bias. In some cases, the use of high chambers or chamber extensions is necessary. For example, when growing plants should be incased, in this case, researchers shall ensure that chambers do not physically injure the plants. If taller chambers (e.g. >60 cm) are used, longer deployment time is necessary to improve the detectability of soil gas fluxes. As a rule of thumb, when the chamber height is doubled, the deployment time should also be doubled.

The chamber base must be inserted into the soil deep enough to prevent gas leakage from chamber headspace. There is not a clear consensus in the literature regarding a minimum depth for the insertion of chamber bases (or frames) into soil. Depths found in the literature range from 5 to 20 cm (Hutchinson and Livingston 2001; Martins et al. 2015; Martins et al. 2017; Zaman et al. 2009). Special care shall be taken regarding the depth of base insertion when GHG emissions are determined from sandy soils because of the higher risk of gas leakage by lateral diffusion. Another very important aspect of the prevention of gas leakage from chamber headspace is base-chamber sealing. The use of a trough soldered on the top of the base and filled with water immediately prior to the base-chamber coupling has been shown to be an efficient method of sealing (De Klein and Harvey 2012). Another option for sealing is the use of gaskets that are compressed by fasteners at the time of base-chamber placement. The advantage of using gaskets is the ability to seal chambers used in areas that are not flat which may exist in natural areas (i.e. forests, hilly pastures) or smallholder cropping systems.

Stainless steel and PVC materials are the most commonly used materials to construct static chambers for field deployment. When soil gas flux is being measured in pasture systems in the presence of animals, cages will need to be used to prevent chambers from being damaged. To avoid unnatural heating in the chamber during gas sample collection, both insulating materials such as foam or polystyrene and reflective foils should cover them. The material for insulation is usually non-expensive and can be easily found in the market. The chamber insulation minimises the changes in air temperature in the chamber headspace, reducing biases due to temperature effects on gas diffusion from soil. The use of small vent tubes is recommended to avoid the effects of pressure difference inside and outside of the chamber on gas diffusion from soil (Xu et al. 2006). Detailed information on how to determine the best diameter and length of vent tubes has been previously presented (Hutchinson and Livingston 2001; Hutchinson and Mosier 1981; Parkin and Venterea 2010). For further reading the paper of the Global Research Alliance by Clough et al. (2020) on chamber design considerations is recommended.

2.2.6 *Chamber Operation, Accessories, Evacuation of Exetainers, and Gas Flux Measurement*

Any chamber, plastic or metal, must be rigid enough to be repeatedly used in the field. The procedure of using static chambers for gas flux measurements from pasture, crop, and vegetable lands is similar. This includes (i) chamber base/collar insertion into the soil, and deployment of the top part if the chamber consists of the two or three parts as shown in Plates 2.1 and 2.2a, b, (ii) closing the chamber (placing either lid/upper part with stopper and septum), and (iii) repeated collection of GHG samples from each chamber using a syringe at specific timings such as 0, 30, and 60 min. Time intervals of gas sampling always depend on factors such as specific conditions, the purpose of the study, and on the gas(es) to be determined.

In addition to chamber design and chamber deployment into soil, having proper gas sampling skills is important to achieve the best quality data for GHG emission (Table 2.2).

Chamber bases shall be installed long enough before measurements commence to allow for conditions to approximate the ambient (Plate 2.4). This might take as little as one hour on coarse-textured soils, while a few days may be needed on clayey soils, provided unvegetated area is investigated. In some cases, even weeks may be required to allow root regrowth. This will avoid any potential impacts of root death, which disrupts C and N cycling with potential effects on CO_2 and N_2O production and consumption in the soil profile. Among annual crops, chamber bases should be installed shortly after seeding, to allow roots to grow into the inner area. Soil water content can impact chamber performance in several ways. Researchers walking around the chambers, especially in very wet conditions, can compact the soil. Chamber bases may also affect lateral surface water flow, and they should be relocated when soil water content inside the chamber differs from the surrounding area. Finally, under very dry conditions, clayey soils may shrink away from the edge of the chamber base. In such circumstances, researchers should carefully loosen and tamp down the soil at the outer edge of the chamber base prior to measurement, to fill the gap and improve the seal between the soil and the chamber base (De Klein and Harvey 2012).

To collect gas samples from each chamber, researchers shall have accessories including syringes (60 ml), three-way taps (Luer-Lock), 12 ml pre-evacuated exetainers, and needle of 0.45 mm × 13 mm (Plate 2.5).

The exetainers are usually pre-evacuated; however, if reused, they must be re-evacuated. However, if an evacuation manifold is not available, also a method is available to use unevacuated vials (for more details, see below).

Various glass vials or vessels have been used for temporary gas sample storage and transportation from field to laboratory before gas analysis. For short-term storage of a large number of samples, inexpensive polypropylene syringes have been used; their use is, however, rather limited because of the possibility of gas sample loss during prolonged storage, for example, when the storage time exceeds 2 days (Rochette

Table 2.2 Recommendations for static chamber deployment (adopted from De Klein and Harvey (2012))

Feature	Requirement/recommendation
Site disturbance	Avoid disturbance of the soil around the chambers. Chamber bases should be inserted at least 24 h prior to the first gas sampling–preferably earlier, if logistics allow a few weeks before the start of measurements
Chamber deployment	Ensure that chambers are sampled by block, rather than by treatment, to ensure each block is sampled in the shortest possible period. For chambers with a maximum height of 20 cm, use a deployment period of ≤30–40 min. For growing rice, a period of 20–30 min, and for fallow a period of 60 min, deployment time is required, at least 3 samples at times 0, 15, and 30 min for the rice-growing period and at times 0, 30, and 60 min for the fallow period. Note, cover duration also depends on the accuracy of the measuring instrument
Number of samples per flux measurement	At least three headspace samples per flux measurement should be collected, especially at times when high emission rates are expected
First air sample (T_0)	Take T_0 sample immediately after chamber top placement on the base
Next air sample (T_1, T_2)	Take T_1 (and subsequently other) sample. The sampling period can be, for example, in the following time: 0, 30, and 60 min, after chamber closure
Time of day	Many studies suggest that between 10 am and 12 noon reflects daily average time of gas sampling. However, it is strongly recommended to determine the diurnal pattern of gas emission to assess time of day that best represents the average daily flux for a given site. In any case, measurements should always be carried out at the same time of day
Placement of chambers	Assess if spatial gradient in fluxes exist, divide farm area into relatively homogenous sections and stratify sampling accordingly. In the absence of spatial structure, place chambers randomly
Treatment replication	An absolute minimum of three replicate plots is needed, preferably more (e.g. 4–5 would be better)
Duration of experiment	GHG emission factor measurement for inventory should ideally cover the full year (12 months)
Frequency of sampling	When emission peak fluxes are expected, sample at least twice per week, or, ideally, daily. During periods of low flux, sample at least once per week. When fluxes have returned to background levels, the sampling interval can be further increased. During agricultural management period (irrigation, fertilisation, etc.), daily measurements are required and the rest of the season or year, at least once a week is recommended, perhaps less frequently during prolonged dry periods
Ancillary measurements	Measure soil texture, bulk density, pH, organic C, and total N content at least once at the start of the experiment. Measure average soil and air temperature and total rainfall hourly or daily. Measure soil water and mineral N content on each sampling day. For the flux calculation, the actual temperature and pressure inside the chamber are required

and Bertrand 2003). If these types of syringes are used, the gas transfer to suitable vials should be done immediately after the gas sampling. Septum-sealed containers of different materials, volumes, and overall qualities have been used for gas storage. The best solution is arguably a glass vial of several ml in volume (8–20 ml), evacuated and sealed with a special gas-tight stopper. In this context, two possible sources of errors exist. First, the quality of the evacuation, even if the vials were bought as evacuated, an additional evacuation prior to use is required. If the vial contains N_2, argon (Ar), or another inert gas such as helium (He) or has been purged with such a gas, it does

Plate 2.4 Metal chamber **a** frame, **b** body and lid, **c** complete chamber for gas flux measurement in the field (material: stainless steel)

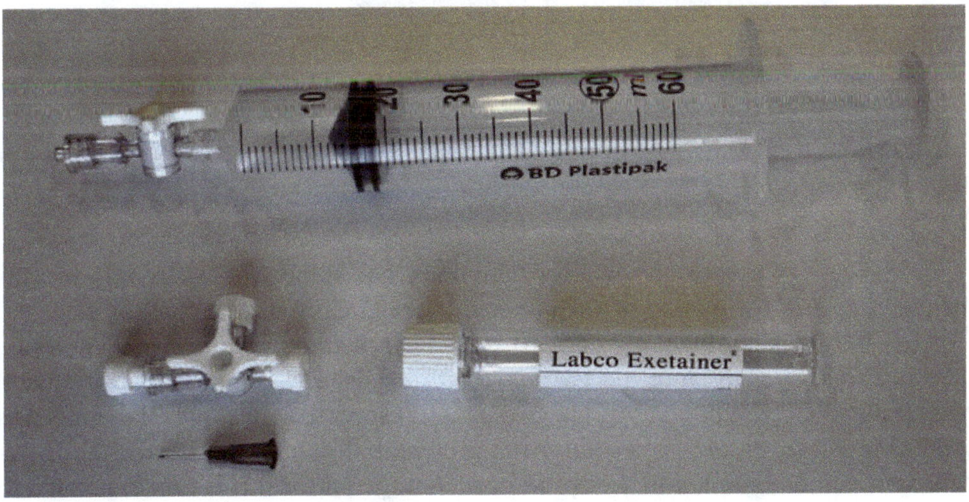

Plate 2.5 Accessories for gas sampling

not create a problem for the analysis of GHGs. However, the vial shall not contain any traces of the gas which is to be determined in the gas sample. There is another related problem: if the amount of inert gas in the vial is too high, the gas sample added to the vial for storage purpose is substantially "diluted", and this affects the gas concentration in the subsample which is taken later for gas analysis. To overcome this potential source of error, vials shall be evacuated before gas sampling using a high vacuum pump (Plate 2.6). This process of evacuation takes about 3–5 min per sampling batch. After ca 5–10 times of repeated use of the vials, the septum should be replaced with a new one. Usually, septa can be separately purchased from the supplier.

It is also a good practice to fill the gas vials first using inert gas (e.g. Ar, He, N_2, depending on the purpose), and then to evacuate them–and to do so (filling, evacuation) repeatedly (3–5 times), or to flush the vials with the inert gas and then use the evacuation system below (Plate 2.6).

Plate 2.6 A high-efficiency vacuum pump connected to a system for simultaneous evacuation of multiple gas vials

In any case, when using vials for temporary gas storage, it is strongly recommended to check the quality of vacuum and overall quality (see below) of vials and stoppers before using them regularly. If gas analyses are not performed right away, then researchers shall store the gas vials with headspace samples in an insulated box (to avoid large temperature changes), and transport the vials to the laboratory as soon as possible, or pack and send them, preferentially in an insulated box for gas concentration determination. For prolonged storage periods, it is recommended to store standard (calibration) gas mixture in the same way as the gas samples. Comparison of direct standard analysis and analysis of several samples containing standard gas mixture, instead of unknown samples, yields the correction factor necessary for sample dilution and gas leakage calculations. For longer storage, it is advisable to store samples with an overpressure in the vial which is often done anyway, e.g. if a sample loop of the analytical instrument has to be filled which is often the case if sample analyses occur via an autosampler.

The second major risk in using glass vials is that the seals or septa are made from plastic material. The stopper must ensure "gas-tightness" in two aspects: it must be nonpermeable to the gas to be analysed (no gas is diffusing to and from the vial) and it must be inert (no gas is generated by the material of the stopper during gas storage). For example, common silicone stoppers are very gas permeable (!) whilst other materials sometimes create a large amount of gases, typically light hydrocarbons including methane (CH_4). Inconsistencies related to gas sampling and storage and possible errors are often ignored which may lead to large errors.

Glass vials (e.g. Exetainer®, Labco Limited, High Wycombe, UK) are now commonly used as air sample containers, and procedures have been developed for their use (Plate 2.7). While different sizes are available, 6- and 12-ml septum-capped glass vials are most commonly deployed with gas sample volumes as small as 1 ml being removed for analysis. Such glass vials have screw-on plastic caps with rubber septa. Experience shows that gas tightness is achieved when the cap is screwed on "finger tight", followed by another quarter-turn. Different septa are available (De Klein and Harvey 2012); as the materials differ in their composition and properties, proper septa must be selected with respect to the gas(es) to be stored in vials and then analysed.

Chamber deployment duration should be long enough to allow flux calculation. This is governed by the accuracy of the analytical instrument used for determining the concentration increment over time (i.e. to determine a flux, the change in gas concentration needs to be higher than the standard deviation of typically 10 repeatedly sampled ambient air standards). However, problems may arise that are associated with changes to the chamber physical environment, and the risk of leaks, which increases with deployment time. Therefore, short chamber deployment periods are recommended (De Klein and Harvey 2012), with each deployment not to exceed 2 h in general. Sampling is often carried out at 0, 30, and 60 min (Zaman et al. 2009). The chamber deployment duration also depends on practical considerations including the number of headspace samples to be taken during the enclosure period, the number of simultaneously deployed chambers, and the number of field operators (De Klein and Harvey 2012).

Sequencing and grouping of chamber measurements vary depending on deployment duration, experimental design, and availability of human resources. The number of chambers that can be handled by one operator increases with deployment duration but decreases with the number of headspace samples to be collected and the distance between chamber installations. Chamber size and height, or stacking requirement (tall crops), may also have an impact on the number of chambers an operator can handle. The time interval between sampling two chambers varies, depending on their location, but it is usually $\geq$60 s (De Klein and Harvey 2012). In the case of a measurement design with repeated treatments, groups of chambers handled together should represent an entire repetition of treatments. Because GHG flux measurement often takes a long time, it is important to sample different treatments within a replicate block in as short a period as possible when there are many chambers to be sampled.

Amount of headspace air to be sampled: The greater the headspace volume to be taken, the better the characterisation of accumulation of trace gases and the less biased each individual flux estimate will be. Generally, 3–4 or more gas samples

Plate 2.7 Gas collecting exetainer for gas storage

are recommended to be taken during deployment, to adequately assess the quality of the calculated flux (detection of outliers and technical problems during handling and analysis of samples), and to account for the increase in non-linear rates of gas concentration with deployment time (De Klein and Harvey 2012; Rochette 2011). Less intensive chamber headspace sampling may be acceptable for certain situations. Any consideration around reducing headspace sampling intensity should be based on minimising the overall uncertainty of the flux estimate. For example, when the spatial variability of fluxes is exceptionally high, it may be preferable to deploy a greater number of less-intensively sampled chambers (two or three samples) to improve plot-level flux estimates, even if this comes at the cost of increased uncertainty in individual chamber estimates. However, if the priority is to generate a representative flux–through the sampling of multiple chambers per plot and assumption of a linear increase in headspace gas concentration, rather than multiple sampling from the headspace of fewer chambers–it is essential to qualify any potential bias introduced by only taking two or three headspace samples per deployment. To reduce the number of samples but still cover the spatial variability, gas pooling techniques are available (Arias-Navarro 2013). Each dataset of four or more headspace samples should be statistically analysed to see if there is non-linearity. At the end of the experimental period, researchers shall summarise this information, provide a percentage of cases when linearity was observed, and then cite this alongside their calculated fluxes. This will provide an indication of the bias–hence confidence–in the results that may have been introduced by assuming linearity in the flux calculation (De Klein and Harvey 2012).

Headspace air sampling usually begins as soon as the chamber is deployed (at time 0), and then in selected intervals: as outlined above (Plate 2.8).

In the case of a measurement design with repeated treatments, groups of chambers handled together should represent the entire repetition of treatments. This avoids temperature-induced biases and enables valid statistical comparisons of fluxes. However, the sampling sequence shall vary between sampling dates, to avoid any potential bias from always sampling in a particular order (De Klein and Harvey 2012). Modern CRDS (Cavity Ring-Down Spectroscopy) techniques for GHG concentration measurements overcome this problem by analysing the gas concentration every few seconds if the analyser is directly connected to the chamber (for more information on the CRDS technique see below, Sect. 2.9).

Daily GHG emissions are often estimated from a single measurement made at the time of the day when the flux is believed to equal its daily mean. For example, in the absence of transient fluxes following a disturbance of soil N_2O producing processes (N application, tillage, and rainfall), diurnal dynamics of fluxes are mostly governed by soil temperature where the main production of the gas occurs (De Klein and Harvey 2012). Research has generally indicated that sampling fluxes when temperature in the plough layer is close to its daily mean are often indicative of the average daily flux. However, data by Šimek et al. (2014) show that diurnal variation in flux rates can be very high and difficult to predict. Periodic measurements of the diurnal pattern in soil N_2O emission during an experiment are the best way to determine when a single sampling time is representative of mean daily fluxes. However, such

Plate 2.8 Gas sampling through a syringe using a plastic/metal chamber

measurements require resources that few projects can afford, and temperature in the plough layer remains the most frequently used index for determining the best single time of flux measurement in a day. Moreover, most experimental designs and measurement protocols assume that diurnal emission patterns are the same in all treatments and throughout the year. However, this may not always be the case. For example, if treatments affect the amount of crop residue retention at the soil surface, the time of daily minimum and maximum soil temperature at a given depth will likely differ among treatments. Similarly, placing N fertilisers at different depths can also produce different temporal patterns in surface N_2O fluxes (De Klein and Harvey 2012).

Seasonal/annual variations of gas fluxes should also be taken into account. As discussed in detail by De Klein and Harvey (2012), the major problem is related to a short period of flux determination (ca. one hour) done with relatively long intervals (from 1 to 14 days), and the need for integrating the data over a much longer period (season or year). Consequently, it is crucial to select an adequate number and time of sampling events when linear interpolation is used between sampling points for temporal integration of emissions (De Klein and Harvey 2012). Theoretically, the GHG flux measurements shall be done every day but for practical reasons, much longer intervals are often selected. If the GHG peaks of the fluxes can roughly be predicted in advance, then sampling at least twice per week, or ideally daily, is recommended. However, after heavy rainfall events or with other rapidly changing conditions when high emission rates are expected (e.g. freezing–thawing or wetting–drying cycles) measurements should be performed immediately after the event and closely followed for the next day or two because peaks might appear once the diffusional constraint due to surface water has subsided. During periods when fluxes are low (e.g. prolonged drought), measurements should be performed at least once a week. When fluxes have returned to background levels, the sampling interval can be increased further.

Spatial integration of fluxes is extremely important due to the enormous spatial heterogeneity of the soils. Together with the temporal variability, the spatial heterogeneity of fluxes represents one of the most difficult features related to integrated gas flux determination from plots and ecosystems. As suggested by De Klein and Harvey (2012), in experiments that determine emissions from a particular practice, selecting small and uniform areas consistent with the measurements being made will help to minimise interference from spatial heterogeneity in background emissions. The location of these relatively homogeneous areas within a landscape–such as a grazed paddock or cropland–can be determined before the experiment, using preliminary flux sampling. However, while this approach usually helps to reduce uncertainty in estimates of the influence of management effects, it does not account for interactions with other soil factors influencing gas dynamics across a given landscape. The number of replicate measurements can often be reduced if preliminary observations have identified the homogeneity of the experimental site.

To deal with large spatial heterogeneity, 2–3 replicated chambers per plot (treatment) have often been used. However, this depends on the available human and financial resources to collect gas samples and their analysis. For further reading the papers of the Global Research Alliance by De Klein et al. (2020a) on Health and safety consideration, by Harvey et al. (2020) on sample collection, storage and analysis and by Charteris et al. (2020) on deployment and source variability are recommended.

2.2.7 *Gas Pooling to Address the Spatial Variability of Soil GHG Fluxes*

Soil–atmosphere exchange of GHGs is notoriously variable at spatial scales. Overcoming this variability is a major issue if fertiliser treatments or other management options need to be compared. The spatial variability of soil GHG fluxes is due to small-scale variability of soil properties, soil environmental conditions, and processes of N and C ecosystem turnover driven by microbes and plants (e.g. Butterbach-Bahl et al. 2002). Spatial variability can be addressed best by increasing the number of replicates and by using larger chambers. But as increasing the number of chambers is not always feasible, one may also use the gas pooling technique as outlined in Plate 2.9.

High-tech equipment such as Cavity Ring-down Spectroscopy (CRDS), Gas Chromatograph (GC), and mass spectrometers and ^{15}N labelled fertiliser are used for measuring GHGs, and their isotopic signatures are expensive and require special

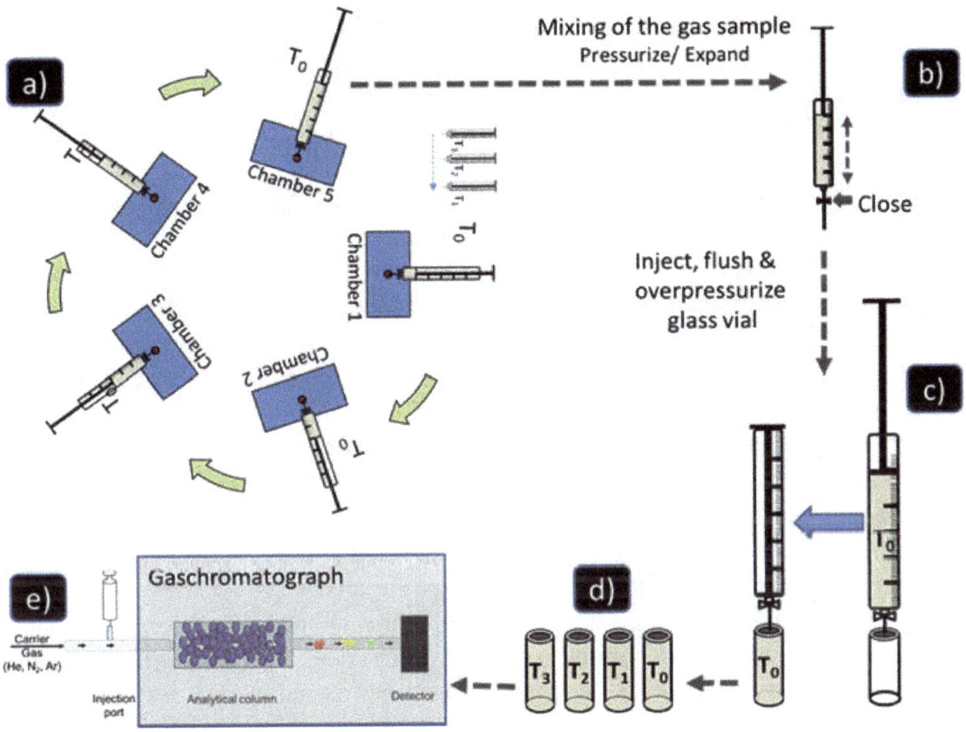

Plate 2.9 Schematic of the gas pooling technique as described in Arias-Navarro et al. (2013). **a** taking of gas samples from five different chambers and mixing of samples within one syringe (**b**). Injecting the mixed gas samples in a vial (**c**) for different sampling times (**d**). Finally, analysis of the gas sample by gas chromatography

technical skills to operate them; therefore, limited field studies have been carried out to quantify GHGs emissions from agriculture worldwide, especially in developing countries. Therefore, it is necessary to identify appropriate methodology and provide suitable guidelines and protocols to help researchers to measure GHGs with greater accuracy and precision.

2.2.8 GHG Measurements in Paddy Rice System

Unlike other field crops, rice is usually grown in flooded conditions. Paddy rice is a large anthropogenic source of CH_4. In recent years, it has become evident that there has been a major increase in the use of N fertilisers in rice agriculture, making rice fields a significant source of N_2O as well.

The closed chamber method, as described earlier, is commonly used to measure GHGs from rice paddy. In comparison to micrometeorological methods, closed chamber techniques are virtually the only available option because of its ease of implementation, low cost, and high logistical feasibility. The United Nations Framework Convention on Climate Change's (UNFCCC) clean development mechanisms (CDM) also recommend carrying out GHG measurements using the closed chamber method (UNFCCC 2008). However, the design of closed chambers for measuring GHG under rice paddy is different from those used for grassland and cropping systems. The chamber should be equipped with a small fan (battery-operated fan to homogenise the air inside the chamber headspace), a thermometer inside the chamber (to monitor temperature changes during the gas sampling period), a vent stopper, a gas sampling port (preferably a tube connected to a valve) (Plate 2.10), and an air buffer bag (1 l Tedlar bag). This air buffer bag compensates for both higher air pressure caused by increased gas production and lower gas pressure caused by gas sampling (Minamikawa et al. 2015). A rectangular chamber (transparent or opaque), with double deck and adjustable height, covering multiple plants of the area occupied by a single rice hill or two hills, is recommended (Plate 2.11). Chamber height should be higher than the rice plant. For the double-deck chamber, a water seal or a suitable gasket between the base and the chamber is required to ensure the gas-tight connection. The belowground depth of the base should be 10–15 cm.

As discussed above (2.2.4), a double-deck chamber should have three components, i.e. chamber base (made of stainless steel and 15 cm deep for rice) that has a trough shouldered in the top of the base filled with water immediately prior to the base-chamber coupling; chamber top facilitating with gas sampling point and a fan; and

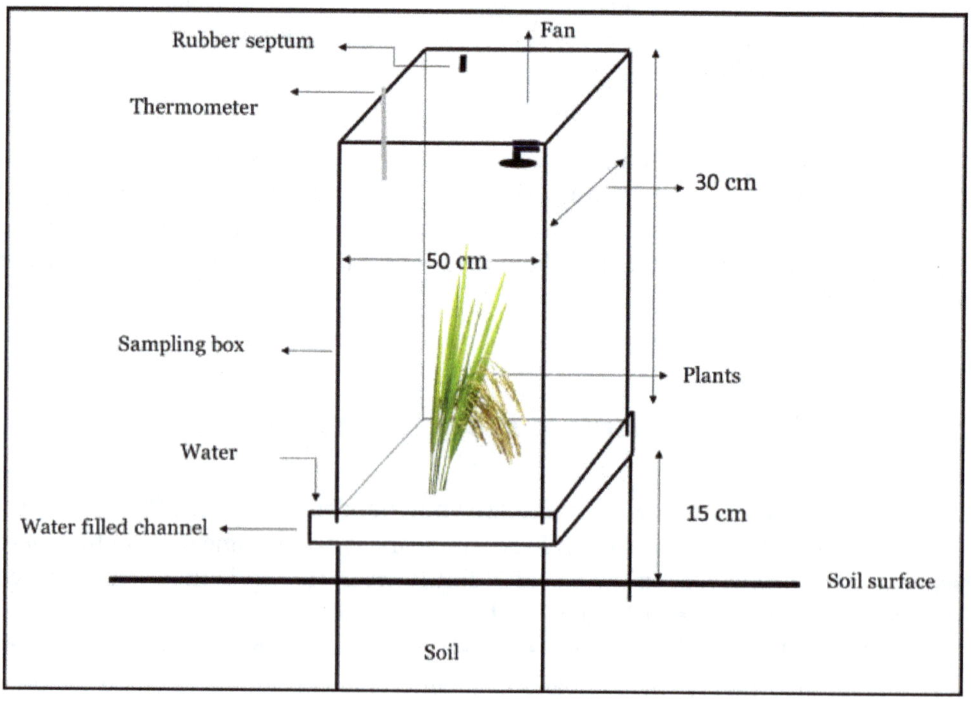

Plate 2.10 Closed chamber used for collection of GHG samples from rice field adopted from Pathak et al. (2013)

Plate 2.11 A metal chamber of 40 × 40 cm is required to cover at least one hill

an extension that connects the chamber lid and the base. The extension and the lid can be made of polycarbonate or stainless steel (Plate 2.10). Handling of polycarbonate is easier than the stainless-steel chambers. Wrapping the polycarbonate chambers with insulating papers may reduce heat increment inside the chambers. Further, when the rice plants are smaller, only the base and the lid can be used without connecting the extension part. It is critical for rice fields, to insert the base to a depth of about 15 cm to restrict lateral flow of nutrients, particularly N, from outside the chamber and vice versa. After chamber installation, the protocol for collection of GHG samples, and sample storage is similar as described above.

2.2.9 *Analysis of GHG Samples on a Gas Chromatograph (GC)*

To avoid changes in concentration during storage, the GHG samples collected from field/lab trials and stored in vials are transported to the laboratory and analysed for trace gas concentrations. An over-pressurisation of the sample gas in the vial ensures that no gas from outside can dilute the sample gas. Crucial recommendations for gas sample collection, storage, and analysis are listed in Table 2.3 (Kelliher et al. 2012). Gas chromatography (GC) is mostly used for analysis of trace gases, including N_2O, CH_4, and CO_2 (Plate 2.12).

Table 2.3 Basic recommendations for GHG sample collection, storage, and analysis (adapted from Kelliher et al. 2012)

Feature	Requirement/recommendation
Sample collection and storage	Use only clean, non-reactive material that can be sealed, the ideal material is glass; container evacuation is recommended
Sample analysis by GC	Commercial GC system; flow control and automated sample injection is recommended
Reference gases, calibration	Confidence in the concentration of all standards. Similar ranges of standards and samples, and many "ambient checks" are strongly recommended
Processing GC data	Determine repeatability (standard deviation) of ambient standard
Sample analysis and gas fluxes	Determine repeatability of standard deviations for the gas samples and associated gas flux

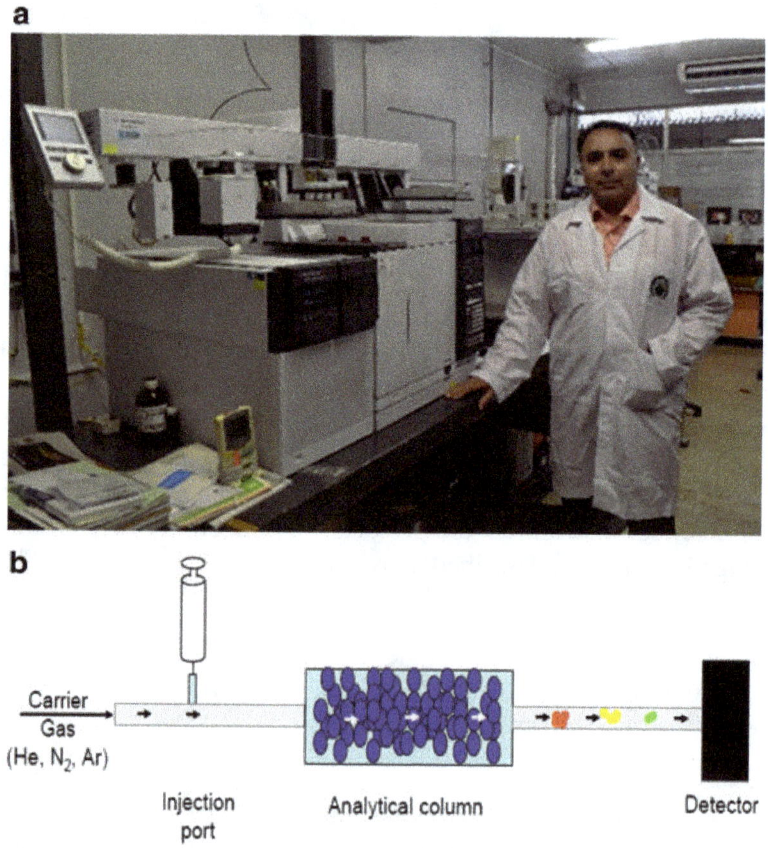

Plate 2.12 A gas chromatograph (**a**) and schematic diagram system for GHG analysis (**b**)

A GC with a sample loop allows the analysis of gas mixtures, and the configuration ensures that the same gas volumes are always analysed under the same condition (pressure and temperature). Besides, gas samples and standard gases are treated always in the same way. Separation of the gas mixture into single gases (CO_2, CH_4, and N_2O) is achieved by passing the sample gas via a carrier gas through a packed column (e.g. a 1/8″ analytical column packed with Haysep Q and/or Molsieve). A carrier gas, usually N_2, He, or Ar, is used, which passes continuously through the system at a constant flow rate. Standard gas chromatographic procedures allow the quantification of CH_4, CO_2, and N_2O in the same sample. To ensure the same conditions for all samples (samples and standards), gas samples are usually injected into a sample loop at constant temperature and pressure (the loop typically has a volume of 0.5–5 ml). After the separation, the gases are analysed with different detectors.

Methane, like all other hydrocarbons, can be burnt, and this feature is used in a specific detector: a flame ionisation detector (FID). After the gas sample enters the FID, it is burnt creating a proportional number of free electrons that generate a current at the collector electrode, which is passed on as an electric signal to the integration unit. When a GC with FID has an attached system with Ni catalysts for conversion of CO_2 to CH_4, it can also be used for the analysis of CO_2 concentration. Otherwise, a GC equipped with a thermal conductivity detector (TCD) is often used to measure CO_2 concentrations. Concentrations of N_2O are analysed with a ^{63}Ni-electron capture detector (ECD) operating at column, injector, and detector temperature of 65, 100, and 280 °C, respectively. An anode is inserted into a small, well-isolated, foil-lined box. The carrier gas (recommended Ar + 5% CH_4) with the gas sample can pass through the detector. The radioactive ^{63}Ni-foil (ß emitter) delivers electrons in the anode interior. The electrons are drawn by the anode in the middle and are "caught"; the number of caught electrons is determined by the electric pulse frequency at the anode. If an electrophile and electron-catching substance (such as N_2O) streams through the space around the anode, it takes up electrons according to its concentration and "electrophilicity". To collect the same number of electrons as before, the electric pulse frequency of the anode must be raised, and this change in pulse frequency is a measure of the amount of the electrophilic substance.

Since the different gases pass through the analytical column at different speeds (e.g. in the order: CH_4, CO_2, and N_2O), it is possible to analyse all three gases in one sample. First, the elution of the column is passed through the FID, and CH_4 is successfully captured by the FID. A switching valve (usually a pneumatical switch) will switch the gas stream from the column from the FID to the ECD detector. Depending on the flow rate of the carrier gas as well as the oven temperature of the GC where the analytical column is located, the analysis time of one sample is typically 3–6 min. In addition, a pre-column is often installed in line with the analytical column to capture all slower moving substances. Once all gases of interest have entered the analytical column, usually slow-moving substances still remain in the pre-column. These substances will then be cleaned from the pre-column via a back-flushing mode with the carrier gas. If that is not done, there is a danger that these substances would appear at some later stage and disrupt the analysis of later samples. To perform the switching, usually a second pneumatically operated 10-port valve is used. For further reading the paper of the Global Research Alliance by Harvey et al. (2020) on gas analysis is recommended.

2.3 Methods to Quantify GHG Production in the Soil Profile

So far we have presented methods to quantify GHG fluxes at the soil–atmosphere interface. However, the various gases are produced in the soil profile and there in sites which are suitable for the activity of microorganisms. Thus, when we are talking about gaseous emissions, we are dealing with two processes that go hand-in-hand: (1) the production of GHGs in suitable soil microsites, and (2) the transport of GHGs from the production site to the soil surface. The transport of GHGs is a diffusion process which is governed by a range of variables such as temperature, soil moisture, soil texture, and the properties of the gas in question. With the help of gas diffusion, based on Fick's law, it is possible to calculate the movement from the production site to the soil surface. The production site is often assumed to be close to the soil surface where most of the management takes place but the main production site can also be deeper in the soil profile (Müller et al. 2004). The gas dynamics in the soil profile can be determined by soil air samplers. Various soil air sampling devices have been developed over the years including (a) stainless steel tubes which are blocked at the end but have close to the tip a radial arrangement of holes for soil air intake (Dörr and Münnich 1987), (b) flexible plastic tubes that allow gas diffusion but are impermeable for water. These can be inserted horizontally at a certain depth (Jacinthe and Dick 1996). The advantage of the second system is that the gas production can be assigned to a specific depth, while gas taken in with the first system could have potentially drawn into the sampler from other depths. For the tube samplers principally two different materials, differing in their diffusive properties, are used: silicone or air permeable, hydrophobic, polypropylene (Accurel®). Both materials can easily be shaped into a coil and inserted at a specific soil depth. However, gas diffusion through silicone is much slower than through Accurel®. Thus, silicone cannot be used for continuous sampling but requires a roughly 24 h equilibration time between samplings (Kammann et al. 2001). This allows discrete gas samplings at a minimum time resolution of approximately one day. The gas diffusion through Accurel® is so quick that continuous sampling is possible (Neftel et al. 2000) which allows for in-field online measurements (Jochheim et al. 2018). The analysis of gas samples is similar to the gas sample analysis from chamber samples.

Plate 2.13 shows an automated setup where Accurel® tubings (Plate 2.13a) are inserted into a soil profile at various depths (Plate 2.13b). The samplers are connected

Plate 2.13 Air sampler setup using Accurel® tubing with **a** soil air sampler with in- and outlet to allow continuous analysis, note the chicken wire around the sampler is there to protect the material from rodent bites, **b** soil profile setup with soil air samplers (right) and soil moisture/temperature sensors (on left) which are connected to a datalogger, **c** manifold system with quick connector gas sampling ports for different depths, **d** discrete sampling with a syringe and an exetainer vial, first the sample will be taken by the syringe and then the three-way-tap will be turned towards the evacuated exetainer and the gas in the syringe will be transferred to the vial, **e** the manifold can also directly be connected to an autoanalyzer arrangement for automated in situ measurements (see also Sect. 3.2.2 for further information)

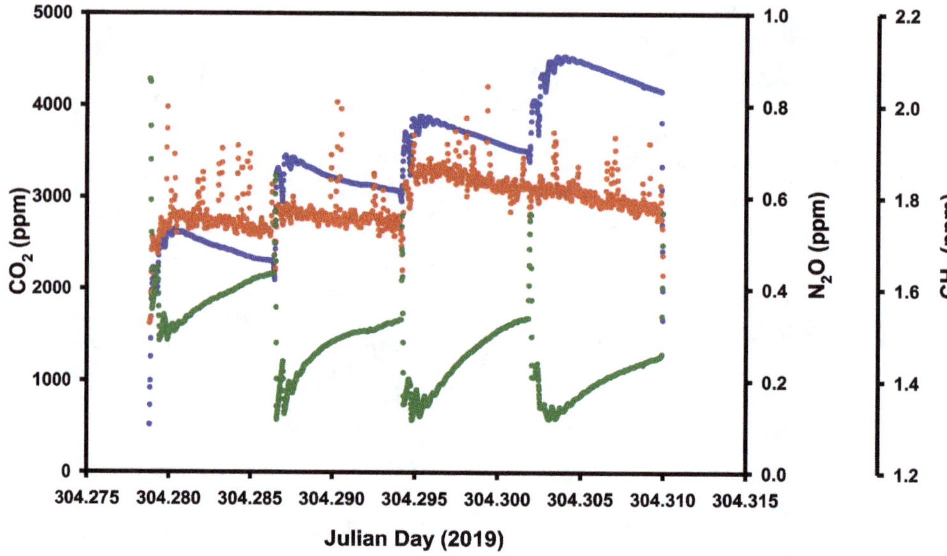

Fig. 2.1 Raw soil gas concentrations of CO_2, CH_4, and N_2O at different depths, analysed with the automated system described in Plate 2.13. (1.11.2019, FACE—Research station, Institute of Plant Ecology, Justus Liebig University Giessen). Highlighted data are used to show the further analysis steps presented in Fig. 2.2

via a teflon tubing to a manifold system at the top of the soil which is fitted with quick-connectors (Plate 2.13c) to allow for discrete sampling using a syringe arrangement (Plate 2.13d) or for connection to an automated arrangement consisting of an LI-COR 8100/8150 multiplexer connected to a CRDS analyzer (Picarro G2508) (Plate 2.13e). For more details on the automated system, see Chap. 3, Sect. 3.2.2.

Figure 2.1 shows a typical output of an in-field measurement campaign. The advantage is that both gas fluxes at the soil surface (if automated chambers are used) together with the soil gas concentrations can be monitored in situ at the same time (see Sect. 3.2.2).

Each air sampler is analysed for a certain period of time (typically 5 min) in a closed-loop system. A decline of the concentration (CO_2, N_2O), or increase under subambient conditions (CH_4) indicates a contamination with ambient air which will be corrected via the following regression analysis. First of all, the results during the time when equilibrium is reached, i.e. between start of sample analysis (t_0) and t_{offset}, are discarded. This period is determined by a moving regression analysis from t_0 till the end of the sample analysis. The dynamics of the intercept of this moving regression indicates the time (t_{offset}) when the adjustment period is completed (see Fig. 2.2b). The most robust measurement is usually CO_2 which will also be used to determine t_{offset} (Fig. 2.2). The concentrations of CO_2, CH_4, and N_2O in the samplers are then determined by a linear regression between t_{offset} and the end of the sample analysis. In the example presented in Fig. 2.2, the resulting concentrations (i.e. the intercept of the regression at t_0) were 2700 ppm for CO_2, 0.571 ppm for N_2O, and 1.494 ppm for CH_4.

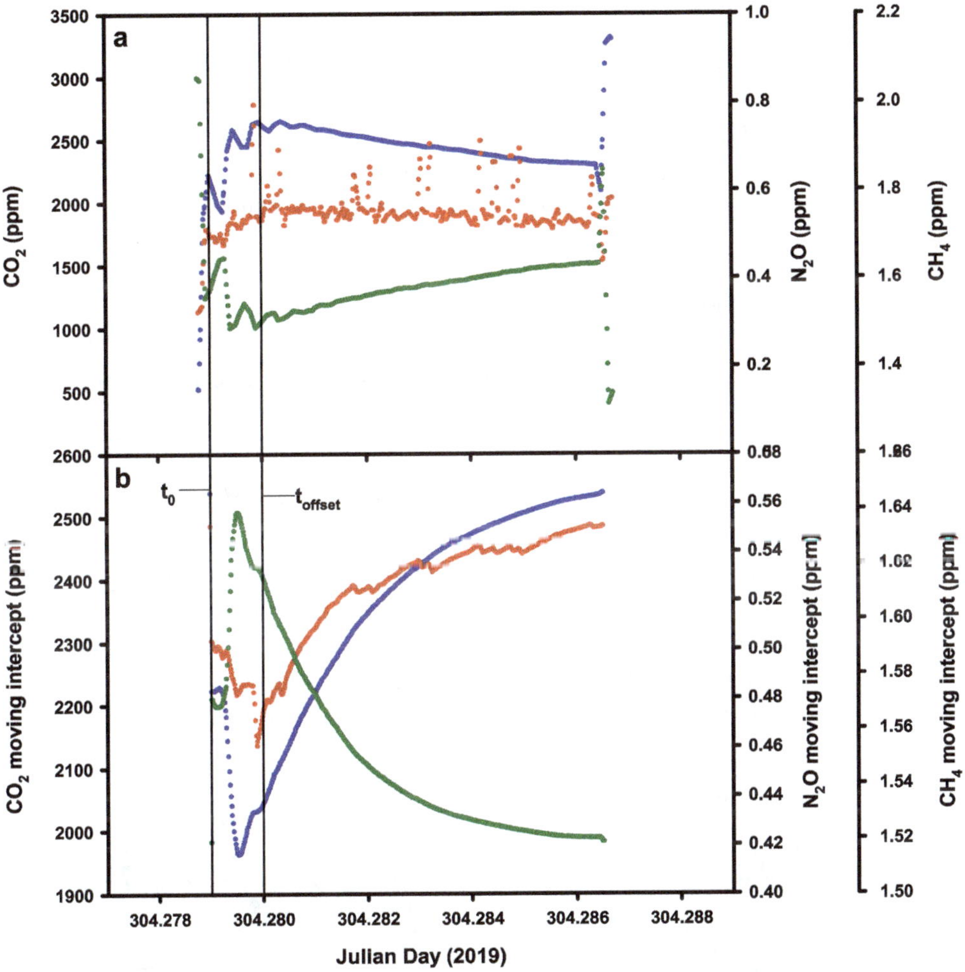

Fig. 2.2 Determination of the soil air sampler concentrations (**a**) is based on a moving regression analysis (**b**) (data were taken from the results indicated by the shaded area in Fig. 2.1)

2.4 Standard Operating Procedure (SOP) for Gas Flux Measurement

2.4.1 Field Gears and Equipment Needed for GHG Sampling

- Closed chamber (would be ideal if the chamber is equipped with a small fan to mix air inside the chamber).
- Wooden block and a hammer.
- Thermometer to record both soil and air temperature during gas sampling.
- Extension chamber if needed to cover tall plants.
- Water supply nearby or water in a container plus a small watering can to add water into the chamber frame to ensure sealing of chamber with its base.

- The accessories for gas sampling include syringes (60 ml), three-way taps (Luer-Lock), 12 ml pre-evacuated exetainers, and needle of 0.45 mm × 13 mm (Plate 2.5).
- Well-labelled pre-evacuated exetainers or gas vials to transfer gas samples collected through syringe(s) from each chamber for storage.
- Timer (at least two) to record sampling time during gas sampling.
- Nitrile gloves.
- Analysis sheet and erasable pen.
- Field clothes and boots.
- First aid kit.
- Sunscreen and insect repellent to protect workers from sunburn and insect bites.

2.4.2 *Step-Wise Procedure (SOP) for GHG Measurements*

- Plan all field and lab activities (designing of the experiment, gas sampling protocol and frequency, etc.) carefully to obtain high-quality data of GHG emission.
- Establish appropriate field plots according to an experimental design (Plate 2.14). Always use at least four blocks (each treatment being replicated at least 4 times or even more) and a control (no treatment and/or standard treatment). Fence in

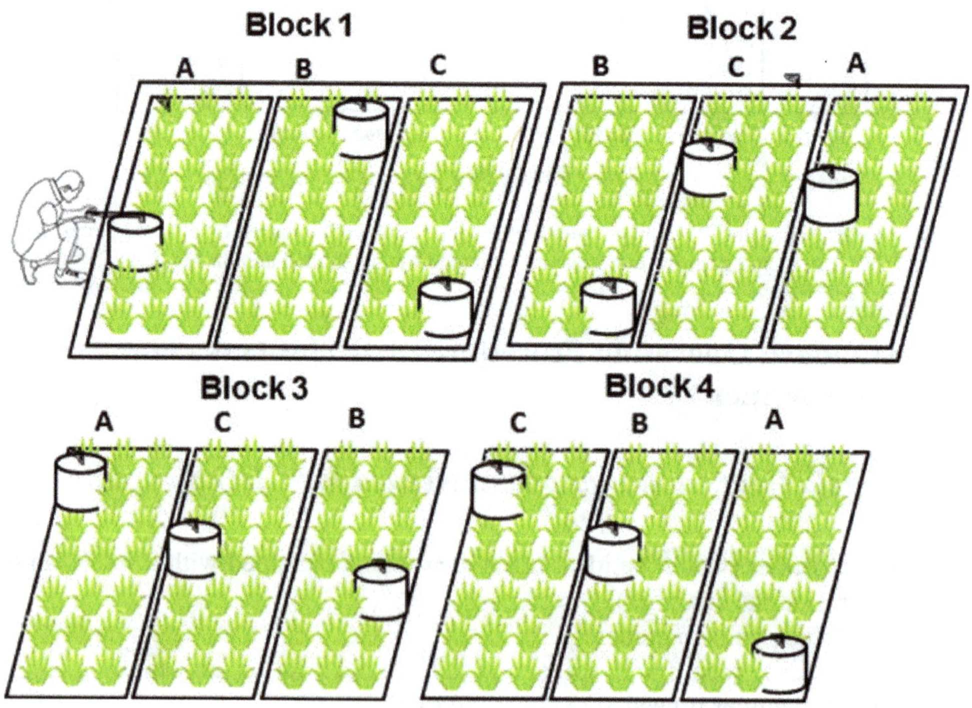

Plate 2.14 A schematic diagram illustrating the collection of gas samples through a large chamber in a pineapple field

the experimental area to protect the field site from animals. In the case of an open grazing system, fence the experimental site 2–3 months prior to treatment application to minimise the effect of animal excreta (urine + dung). If the site is fenced so long in advance, the grazing impact on plants has to be simulated. If possible carry out representative screening of the area including soil analyses, plant analysis, and gas measurements to determine blocks to decide where to place the experimental plots.

- Carefully insert the chamber base/frame using a wooden block and a hammer on the perimeter of each plot. Make sure that the metal trough of the chamber base is not damaged by forceful hammering. After insertion, the chamber base shall be levelled to the soil surface. Ensure that the base is not tilted to any side. This can be checked by pouring water into the trough of the base and observing the water level. If using a plastic chamber, then only the chamber without lid is inserted 2 weeks before measurements.
- In case of rice paddy or wetland (flooded condition), use a wooden boardwalk (above the water level) to reach the gas plots, to avoid soil compaction.
- Chambers should be insulated by wrapping appropriate insulating materials around them.
- Install a weather station at the field to collect data of rainfall, temperature, and moisture at different soil depths (e.g. 5, 10, 15, 20, and 30 cm).
- Prior to fertiliser application, composite soil samples from surface soil (preferably 0–10 cm) from each block shall be collected for basic physical and chemical analyses (texture, bulk density, soil pH, mineral N (NH_4^+ and NO_3^-), total N and total C. In addition, site information regarding latitude, longitude, altitude, soil type, previous, and current farm management practices, shall also be collected.
- Take extreme care by covering the chamber area during fertiliser application to the main field plot. After fertiliser is applied to main field plots, carefully remove the cover and apply the required amount of fertiliser to the area of each chamber.
- Test the linearity of gas fluxes from the given system in advance. Use 2–3 replicate chambers. After chamber deployment, sample headspace air at 0, 20, 40, 60, 80, 100, and 120 min after closing the chamber. Analyse the samples at the laboratory using a GC. In most cases, the gas emissions will be linear for at least 1.5 h. If yes, select the following times: 0, +30, and +60 min, or 0, +20, +40, and +60 for sampling from each chamber.
- Always perform the gas collection process at approximately the same time of the day, e.g. start roughly at 10 a.m. and finish at about 12. Record the temperature outside and inside the chamber at the time of gas sampling. After completing the gas sampling, remove chambers and store them in a suitable and safe place (dry, shaded, and cool).
- For each sampling event, ensure to record air and soil (7.5 cm) temperature using a thermometer, date, and amount of any rainfall or irrigation (mm), and soil moisture content (0–7.5 cm) on water-filled pore space basis (or take data from the datalogger if installed).

2.4.3 Gas and Soil Sampling

- Prepare syringes: label them consecutively (e.g. 1–24 depending on the total numbering of treatments) and additional syringes for air samples (labelled 01, 02, etc.).
- Make sure that the three-way tap or valve is properly connected to each syringe. Always hold the syringe by the three-way valve.
- Place all the syringes needed for each chamber next to the chamber.
- Aerate the chamber before placing it on the frame.
- Before placing the chamber on top of the base, fill the base frame with water using a watering can. The water in the enclosed space between the chamber and the base will act like a seal providing a barrier for gas diffusion. Make sure that enough water is in the gutter; be careful NOT to add any water anywhere else.
- Carefully place the chamber on the frame, make sure that it is sitting properly in the water-filled gutter.
- Connect the syringe to the three-way tap on the chamber (should be an air-tight connection).
- Open the three-way valves, pump 3–4 times and take the gas sample, and then close the three-way valve again.
- Immediately start the timer and leave it running for the entire sampling.
- Note down the date and time on the sampling sheet.
- Note down the air temperature in the chamber.
- Walk to the next chamber and place and repeat the above steps, work out a suitable time interval beforehand (e.g. 2–3 min), and maintain that same interval for the entire sampling period (Plate 2.15).
- After a pre-defined cover period, take the second sample from the chamber. Get ready for the next sampling shortly before the sample time.

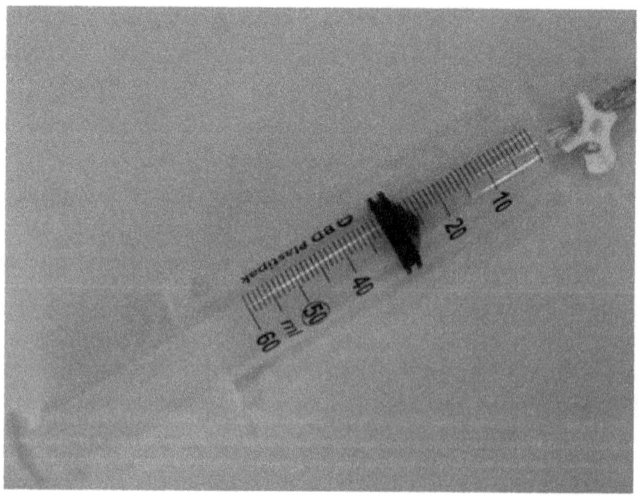

Plate 2.15 Gas collection through a syringe in the field

- In the case of many measuring plots, the second sample may need to be taken before the first round of samples is finished, which would require several people.
- If the sample containers are plastic syringes, gas samples must be analysed within 2 days (if the samples have not been transferred to a pre-evacuated exetainer). For longer storage, always store (and then analyse) calibration gases (gas standard) alongside the samples. Samples should always be pressurised (see above) with sample air (i.e. at least 20 ml in a 12 ml exetainer, ensuring that the overpressure can fill the sample loop if used). For further reading the paper of the Global Research Alliance by De Klein et al. (2020b) on safety measures is recommended.
- Perform gas flux measurements before treatment application to establish the baseline of each plot. Then take gas sampling immediately after fertilisation, other treatments, or extreme events (such as heavy rainfall). After fertiliser/manure/farm effluent application, measure every day for a week, then less frequently (3–5 days) at least once per week until the gas flux from fertiliser plots come to the background (control plot) level.
- To relate N_2O flux to N dynamics, collect soil samples in the surface layer (0–5 cm) to determine mineral N (NH_4^+ and NO_3^- contents) throughout the entire experimental duration (more frequently shortly after the N application).

2.4.4 Safety Measures for GHG Sampling

- Nitrile gloves shall be worn during fieldwork.
- Extreme care should be taken while evacuating exetainer or transferring gas into exetainers to avoid any needle pricks (if not used to keep the needle in the protective cover).
- Tetanus injection record of staff involved in the field collection of gas samples should be up to date.
- Dispose of needles in a special container for needles.
- For GC operation, please refer to the relevant risk assessment and operating manual of the GC.

  For further reading the paper of the Global Research Alliance by de Klien et al. (2020b) on safety measures is recommended.

2.5 Calculation of GHG Fluxes

2.5.1 Overview

1. Analyse reference gases (i.e. gases that are available from commercial companies with a known concentration) using a GC to make a calibration curve (Fig. 2.5).

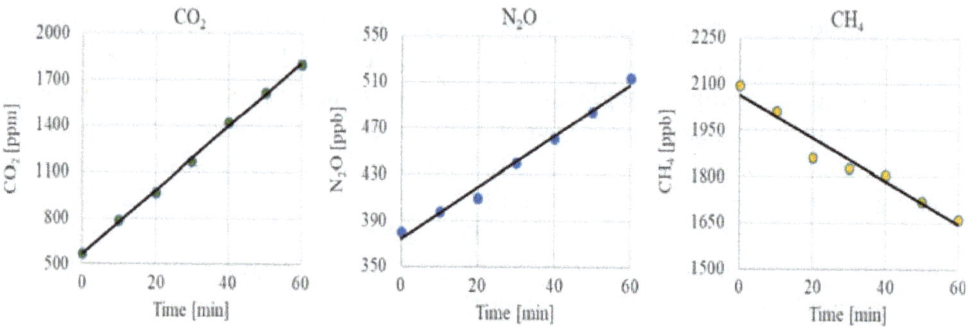

Fig. 2.3 Concentration of CO_2, N_2O and CH_4 in the headspace during the incubation

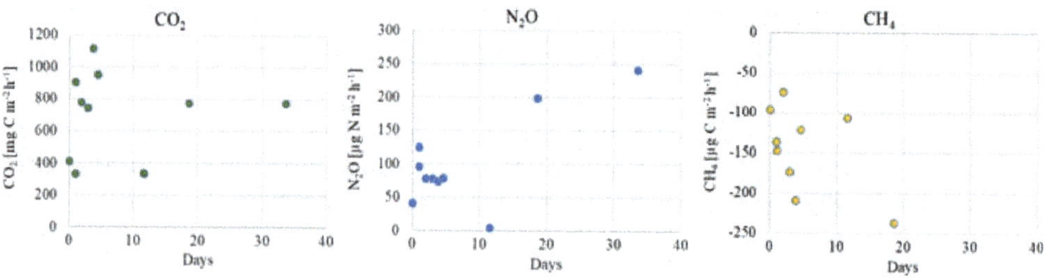

Fig. 2.4 Calculated emissions of CO_2, N_2O and CH_4 during a 35 day measuring campaign

2. With the slope (a) of the regression line, calculate the gas concentration (y) of your samples (Eqs. 2.1 or 2.2). Gas concentration is usually given in ppm (10^{-6}) or ppb (10^{-9}).
3. Based on the concentration changes over time (Fig. 2.3), calculate gas fluxes according to Sect. 2.5.3
4. In the last step of the calculation, convert gas concentration (ppm, ppb) to mass (mole or mg of gas, see Sect. 2.5.4). For each measurement you will get a separate flux (Fig. 2.4), the unit of the gas flux is usually ppm h^{-1} or mg m^{-2} h^{-1}.

2.5.2 Calibration

A **calibration** is a procedure to convert the GC output into a concentration unit, typically parts per million (ppm) or parts per billion (ppb). To develop a calibration curve (Fig. 2.5), normally three to four gas standards of known concentration are injected into a GC and analysed. Standard gas containing gas mixtures at increasing concentrations, contained in gas cylinders, can be ordered from commercial gas companies. With increasing gas concentrations, the GC output also increases. Either a linear increase (e.g. for CH_4) or a non-linear (CO_2 and N_2O) increase is observed which can be described by the following equations (Eqs. 2.1 and 2.2). Note, if the increase is linear, the term "a" in Eq. 2.2 is zero and the entire equation is reduced

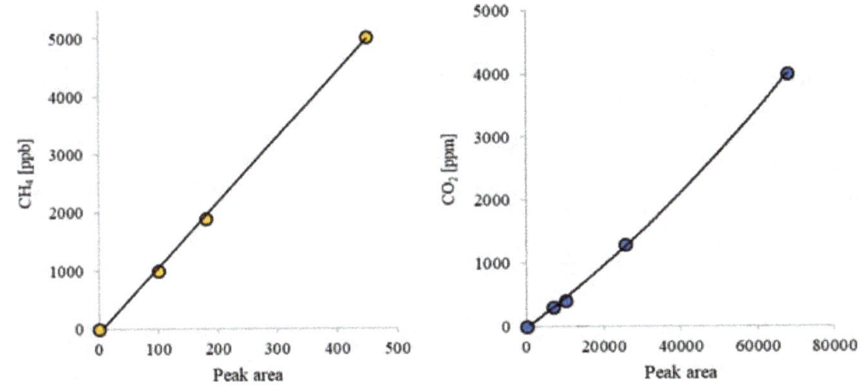

Fig. 2.5 Calibration with a linear (left) or quadratic (right) regression line

to a linear regression.

$$\text{Linear regression: } y = ax + b \tag{2.1}$$

$$\text{Quadratic regression: } y = ax^2 + bx + c \tag{2.2}$$

where
x = area values (area of the standards, output from the GC),
y = % or ppm values (from the standards), and
a, b, c = regression parameters.
Steps of the regression analysis:

- The regression parameters for the appropriate equation suitable for the gas shall be copied into an excel spreadsheet beneath the calibration data.
- The process is carried out for all gases separately.

2.5.3 *Calculation of the Gas Concentration and Fluxes*

After all regression parameters are identified, the calculation of the concentration is, depending on the gas, carried out according to Eqs. 2.1 or 2.2.

From concentration to flux

The gas fluxes under a closed chamber are calculated for the duration of the gas sample collection. To do this, the concentrations are determined at several points in time (Fig. 2.6).

Based on the changes in concentration over time, the slope of the regression line at time = 0 is calculated (this corresponds to parameter "*a*" in Eq. 2.1 and parameter "*b*" in Eq. 2.2). Therefore, the slope of the regression line provides the flux rate as concentration/time. The unit of the flux rate is ppm h^{-1}.

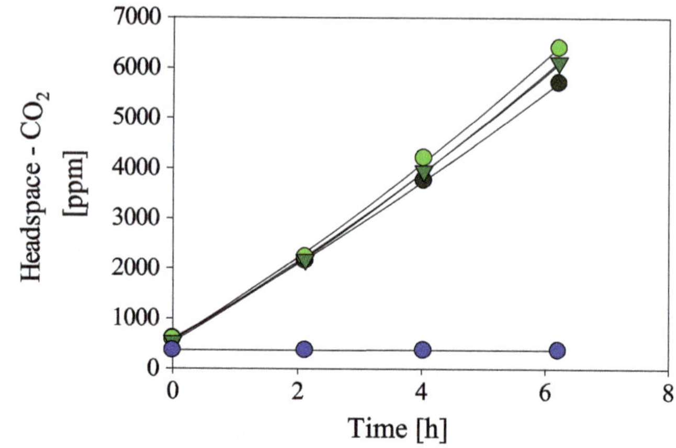

Fig. 2.6 Headspace-CO_2 concentration during the incubation and regression line for each plot. Green dots are flasks containing soil, blue is the control (containing water in the flask)

Note, typically for N_2O and CO_2, we observe a positive slope, i.e. emission from soil to the atmosphere and for CH_4 under aerobic conditions the slope is negative, i.e. uptake of CH_4 by soil.

2.5.4 Conversion from Concentration to Mole

Transformation of concentration (ppm) in mole using the ideal gas law (Eq. 2.3):

$$n = \frac{P \cdot V}{R \cdot T} \tag{2.3}$$

where
n = Number of moles of the examined gas
P = Atmospheric pressure (Pa) [~100,000 = 1000 hPa] (to be measured)
R = Gas constant (J mol^{-1} K^{-1}) [8.314]
T = Temperature (K) [273.15 + t °C] (t is the temperature to be measured)
V = Volume of gas (i.e. N_2O, CO_2 or CH_4) in the chamber (m^3). This is calculated by the multiplication of gas concentration with total chamber volume (V_{tot}) (ppm $\times 10^{-6} \times V_{tot}$).

Why do we convert gas concentration to its mass?

Gas concentration does not provide information about the total amount of gas measured or emitted. The smaller the chamber volume, the higher the concentration increases. Imagine a chamber volume of 1 m^3 where gas concentration increases at 100 ppm h^{-1}. If the chamber volume is only 0.5 m^3, this concentration increase would double up to 200 ppm h^{-1}.

Think: what would be the concentration change if the chamber volume would be 2 m^3?

The answer is 50 ppm h^{-1}

Hence, to know the exact amount of a gas, in addition to the concentration (ppm or ppb), the volume and area of the chamber (V_{tot}, A), the atmospheric pressure (P, the higher the pressure the more gas molecules in the chamber), and the temperature (T, higher temperatures decrease the number of gas molecules per volume) must all be considered (see Eq. 2.3).

Example:
Temperature (T): 20 °C
Temperature: unit transformation °C to K: 20 + 273.15 = **293.15** K
Air pressure (P): 100,000 Pa
Chamber volume (V_{tot}): 0.02 m^3
Chamber area (A): 0.1 m^2
Concentration increase of the gas (CO_2) at $t = 0$ (ΔC): 1000 ppm CO_2 h^{-1}

Molecular weight of CO_2: 44.009 g mol^{-1}

Note, in the ideal gas law the Volume, V, refers to the gas we are interested in, i.e. CO_2, CH_4, N_2O. So, first of all the volume of this gas within the chamber volume, V_{tot}, is calculated:

$V = V_{tot} * \Delta C * 10^{-6} = 0.02 * 1000 * 10^{-6} = 0.00002\ m^3\ h^{-1}$

$$\frac{\Delta n}{\Delta t} = \frac{100,000x0.00002}{8.314x293.15} = 0.00082\,\text{mol}\,CO_2\,h^{-1} \tag{2.4}$$

Multiplied with molar mass of CO_2 (44.009 g mol^{-1}), this corresponds to 0.03607 g CO_2 h^{-1} or 0.00986 g CO_2–C (if only the active C component is applied with a molwt of 12.011 g mol^{-1}). This is now the emission rate from the plot the chamber has covered. To standardise the emission rate, we express it per m^{-2}:

$$F = \frac{0.00986}{0.1} = 0.0986\,\text{g}\ CO_2-C\ m^{-2}\,h^{-1} \tag{2.5}$$

The following information is required for flux calculation:

- Chamber volume (V_{tot}), which can be obtained by multiplying chamber length (L), width (W), and height (H) if it is a square-shaped chamber, for other shapes use the appropriate mathematical equation (or simply fill the chamber with water and measure the volume or weight of water…).
- Mole weight of the gas (as a rule, one converts to N_2O-N and CH_4/CO_2-C).
- Parameters of the gas law [temperature, atmospheric pressure, gas constant, covering time, and time of sampling during the cover period (t_0, t_1, t_2, etc.)].

Units and factors:

ppm: 10^{-6} or 1 μl L^{-1}
ppb: 10^{-9}
ppt: 10^{-12}

Molar masses:

CO_2: 44.009 g mol^{-1}
C: 12.011 g mol^{-1}
N_2O: 44.013 g mol^{-1}
N: 14.007 g mol^{-1}
CH_4: 16.043 g mol^{-1}.

For further reading the paper of the Global Research Alliance by Venterea et al. (2020) on flux calculation is recommended.

For further reading the papers of the Global Research Alliance by De Klein et al. (2020a) on data reporting and further calculations, by Dorich et al. (2020) on gap-filling techniques and by Giltrap et al. (2020) on modelling approaches are recommended.

2.5.5 *Data Analysis*

Data analysis is extremely important to produce realistic emission data. The most appropriate flux calculation method must be selected, and best interpolation of non-continuous measurements adopted to obtain best estimates of emissions and emission factors (EF) (Venterea et al. 2012) (Table 2.4). It is suggested that both gas analyses and data analysis, including appropriate statistical analysis, are performed in a laboratory equipped with staff familiar with all the required skills.

Table 2.4 Basic recommendations for GHG data analysis (adapted from Venterea et al. 2012)

Feature	Requirement/recommendation
Selection and use of a flux calculation method	Method should be matched to the number of headspace samples taken
Estimation of emissions using non-continuous flux data	Daily fluxes can be integrated, using trapezoidal integration. To improve the accuracy of cumulative emissions estimates, maximise sampling frequencies and spatial replication. Repeat experiments over multiple years and consider using spatial or temporal gap-filling procedures
Assessment of minimum detectable flux (MDF)	Determine random measurement error associated with sampling and analysis of replicate standards of known concentration and use the resulting error rates to determine MDF (Sect. 2.9.3. Eq. 2.7, Christiansen et al. (2015))
Statistical considerations for analysing inherently heterogeneous flux data	If treatments are replicated (at least 3–4 replicates), the variability between replicates can be assessed by calculating means of chambers in each replicate. The variability within the replicate can also be determined by assessing the chamber variability

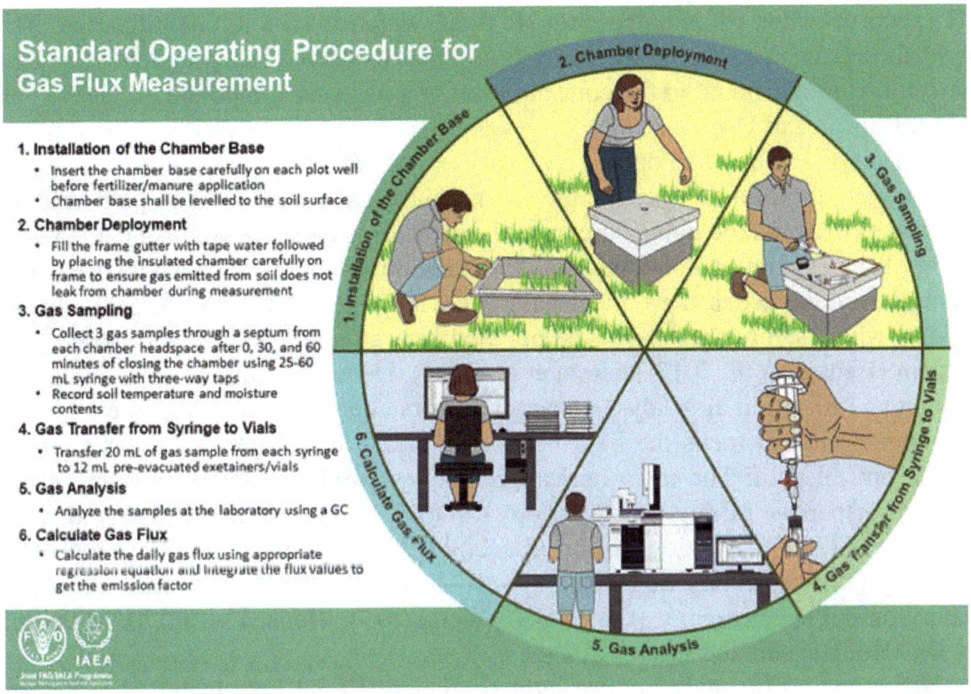

Plate 2.16 A schematic representation of gas sampling, analysis, and interpretation (© FAO/IAEA Mohammad Zaman)

A summary of GHG measurements, analysis, and data interpretation is presented in a schematic diagram below in Plate 2.16.

2.6 Analysis of GHG Samples with Optical Gas Analysers

Gas chromatography is still the most widely used analytical technique for measuring GHGs from chambers (Plate 2.12a, b). It is a well-established, reliable, and robust method; the GC can also be linked to isotope-ratio mass spectrometers (IRMS) for analysis of the abundance of isotopes. In recent years, GCs have become more portable and automated, which makes it possible to run them unmanned in the field in connection with automated chambers. However, the greatest disadvantage of gas chromatography is that one can only measure discrete samples, and it takes several minutes to run a sample, which limits the number of gas samples that can be run (Rapson and Dacres 2014). These disadvantages can be overcome by employing optical gas analysers, which can conduct continuous gas measurements at a high temporal resolution (seconds to Hertz). The measurement principle of optical gas analysers utilises the ability of small molecules (e.g. H_2O, CO_2, CH_4, N_2O, and NH_3) to absorb infrared (IR) and near-infrared (NIR) light at unique wavelengths. Each molecule has a unique combination of atoms and, as a result, produces a unique

IR spectrum when illuminated with IR light, which allows its identification. The light absorption at a specific wavelength, which is measured as light attenuation by a detector, is proportional to the concentration of a gaseous compound (Hensen et al. 2013; Rapson and Dacres 2014).

The general setup of an optical gas analyser consists of a light source from which light travels through the gas sample to an appropriate detector. The path that the light takes between the light source and the detector is called the optical path (Werle 2004). When the optical path lies directly in the outside air, it is called an open path system, whereas when the optical pass is enclosed inside a measurement cell (or cavity) where sample gas has to be pumped into, it is referred to as a closed path system (Hensen et al. 2013; Peltola et al. 2014). Depending on the specific optical technique utilised in an analyser, several mirrors and/or optical filters are added to the optical path to focus the light beam to increase light intensity and the length of the optical path. The main optical techniques employed for quantifying GHG are non-dispersive infrared spectroscopy (NDIR), Fourier-transform infrared spectroscopy (FTIR), photoacoustic spectroscopy (PAS), tunable laser absorption spectroscopy (TLAS), cavity ring-down spectroscopy (CRDS), and off-axis integrated cavity-output spectroscopy (OA-ICOS) (Werle 2004; Hensen et al. 2013; Peltola et al. 2014; Rapson and Dacres 2014).

NDIR and FTIR analysers use light sources that emit broadband IR radiation, e.g. IR lamps or black body light sources. In an NDIR analyser, broadband IR radiation passes unfiltered through the air sample. An optical filter in front of the detector determines which wavelengths are detected and thus which molecules are quantified. NDIR analysers are quite cheap and robust and often are used for quantifying CO_2 and H_2O in the air (Tohjima et al. 2009; Keronen et al. 2014). In FTIR analysers, the source radiation is constantly modulated by a set of mirrors called a Michelson interferometer containing different combinations of frequencies. For each combination, the amount of IR absorbed by the gas sample is measured. A Fourier transform is then applied to the raw data to calculate the absorption at each wavelength for the complete IR spectrum. This method allows the simultaneous quantification of many different gas species in air. Depending on the resolution of the FTIR analyser, it determines only gas concentrations (low-resolution FTIR) or isotopomers simultaneously (high-resolution FTIR) (Griffith et al. 2012; Rapson and Dacres 2014). In PAS, the light source is often a heated black body, but it can also be a laser. In contrast to NDIR and FTIR, PAS takes only indirect measurements of light absorption. The light passes through an optical filter to pre-select a specific wavelength, and a light chopper periodically "switches" the modulated light on and off before it is directed into the cavity via a mirror. Molecules heat up and expand when they absorb the modulated light, resulting in a pressure rise in the cavity. The light chopping creates pressure variations, which in turn generates acoustic signals that can be detected by microphones. The acoustic signals are proportional to the gas concentration of the target gas species (Iqbal et al. 2013; Rapson and Dacres 2014).

Analysers based on NDIR, FTIR, or PAS can operate with the measurement cell at ambient atmospheric pressure. This fact and the chosen light source reduce material costs and lead to lower prices in comparison to laser-based analysers utilising

TLAS, CRDS, or OA-ICOS. Laser-based analysers do not operate with broadband IR radiation, but instead are tuned to the unique absorption line of a specific trace gas (Hensen et al. 2013). The cavity in such an analyser is kept at sub-ambient pressure, which results in narrower IR absorption lines and thus a higher gas species selectivity (Rapson and Dacres 2014). However, it also requires the analysers to be equipped with vacuum pumps and vacuum-proof tubing, tube fittings, and valves. The great advantage is that laser-based analysers are capable of fast and the most sensitive measurements of trace gas concentrations, as well as stable isotope compositions (including isotopomers) in the air (Hensen et al. 2013; Rapson and Dacres 2014; Brannon et al. 2016).

Nowadays, TLAS is the most common laser-based absorption technique for quantifying GHG concentrations in air. Commercially available analysers often employ either tunable diode laser absorption spectroscopy (TDLAS) or tunable infrared laser differential absorption spectroscopy (TILDAS) with quantum cascade lasers (QCL). Detailed information about these techniques is available in Li et al. (2013, 2014). The main disadvantage of TLAS is its susceptibility to instrument noise because the analysers rely on the measurement of a small change in light intensity against a large background light signal. To drastically improve sensitivity, TDLAS and TILDAS analysers are commonly equipped with multipass cells. In multipass cells, the light beam travels several times between aspherical mirrors before exiting the cavity in the direction of the detector, resulting in optical path lengths of several up to hundreds of metres. However, the mirrors are extremely sensitive to changes in alignment and, therefore, more sensitive to vibrations. This has led to the development of high-finesse optical cavities, which allow the build-up of large amounts of light energy in the cavity, boosting analyser sensitivity and robustness, but also decreasing cavity and analyser size. High-finesse optical cavities are the basis of CRDS and OA-ICOS (Rapson and Dacres 2014).

In OA-ICOS analysers, the laser beam enters the cavity at a non-zero angle so that the photons make thousands of passes between high-reflectivity mirrors before leaving the cavity again. This increases the effective optical path length to several thousand metres, enhancing the measured light absorption in comparison to multi-pass cells. The optical path length depends only on optical losses in the cavity, and therefore a precise laser beam alignment is not necessary, and the analysers are also less susceptible to slight changes in mirror alignments (Bakhirkin et al. 2004; Peltola et al. 2014; Rapson and Dacres 2014; Lebegue et al. 2016). In contrast, CRDS does not consider the absolute absorption intensity by the target gas species, but rather the rate of light intensity decay in the cavity. When the laser is turned on, the cavity is quickly filled with NIR light. As soon as a light intensity threshold is reached, the laser is abruptly turned off, and the light leaks out of the cavity with an exponential decay rate through the cavity mirrors. Additional light absorption in the cavity by the target gas species accelerates the light intensity decay rate, also known as ring-down time. The switching on and off of the laser takes place within microseconds, and the ring-down times are used to determine the concentrations of the target gas species in the cavity. Since the actual measurements take place when the laser is turned off, CRDS is not susceptible to laser intensity fluctuations or absolute laser power. The disadvantages, though, are that the mirrors have to stay very clean, and high-speed detection electronics are required (Rapson and Dacres 2014; Brannon et al. 2016). A laser-based optical technique, which has just been recently commercialised for GHG measurement, is laser dispersion spectroscopy (LDS). It tries to overcome the main limitation of TLAS, the detection of a small signal change against a large background, by measuring molecular dispersion instead of molecular absorption. The measurement signal is encoded in the light phase making the analyser more resilient to dirt and water vapour in the cavity because these compounds only affect light intensity (Nikodem and Wysocki 2012).

The past two decades have seen a rapid development of optical techniques and optical analysers for GHG measurements, and this trend is projected to continue over the coming years. One optical technique that might become more important is enhanced Raman spectroscopy (see Sects. 2.8.2 and 2.8.3). The range of commercially available optical gas analysers for different applications can be quite confusing; thus, Table 2.5 summarises the key points one should consider before purchasing a specific optical analyser.

For a comparison of laser spectroscopic analyses of N_2O isotopomers, see Chap. 7, Sects. 7.3.4 and 7.3.5.

2.7 Hands-On Approaches Using a CRDS Analyser

2.7.1 Overview of the CRDS Techniques for Determining GHG Concentrations and Soil Fluxes

As outlined in Sect. 2.6 ("Analysis of GHG samples with optical gas analysers"), the advent of laser-based techniques has enabled real-time measurement of soil gas fluxes in the field. In recent years, CRDS has become of particular interest to researchers due to its ability to measure GHG concentrations with a very high precision in field settings. For example, the Picarro G2508 multi-species gas analyser allows CO_2, CH_4, N_2O, NH_3, and H_2O to be measured simultaneously. CRDS analysers have three key advantages for soil gas flux measurements in comparison to traditional GC measurements: (1) it enables the detection and measurement of low levels of GHG fluxes. Areas with low emission rates, while not significant in the short-term, may be of great interest to calculate the global GHG budget when extrapolated over longer timescales and large surface areas; (2) multi-species measurements allow for direct correlations between gas species. The processes and sources associated with the emission (or uptake) of GHGs may require more than one axis for proper characterisation; and (3) the high sampling frequency of real-time measurements leads to a better characterisation of soil flux profiles and shorter closure times for soil flux chambers. In addition, the real-time nature of this system provides instant feedback on site selection and omits sampling and storage effects.

Table 2.5 Key features of optical analysers

Topics	Key points	Remarks
Target gas species	Combination of gas species	Depending on the type of optical analyser, several gas species can be identified with the same instrument, but not all combinations of gas species are possible because of overlapping absorption lines
	Number of instruments	Measuring as many different gases as possible with one analyser reduces analysis and maintenance costs as well as data analysis time. However, having a separate analyser for each gas species ensures continued measurement of at least some gases when one analyser breaks down
	Customised configurations	Most companies sell preconfigured optical analysers, but some also offer the possibility to customise the target gas species for an analyser
	Stable isotopes	Should the analyser measure only gas concentrations, only isotopic compositions, or perform well for both applications?
	Water vapour	Regardless of the target gas species, water vapour should always be simultaneously quantified by an analyser and spectra corrected for any H_2O interference
	Non-greenhouse gases	Optical analysers can also quantify tracers (e.g. SF_6), O_2, and other non-GHG
Flux measurement method	Response time, precision, and accuracy	For eddy covariance (EC), a fast-response (at least 10 Hz) analyser with high precision and accuracy is mandatory, and thus laser-based absorption techniques are necessary. For chamber measurements, requirements regarding sampling rate, precision, and accuracy are less restrictive. An optical analyser for chamber measurements will always have a higher sensitivity in comparison to the analysis of discrete gas samples with a gas chromatograph (GC)

(continued)

Table 2.5 (continued)

Topics	Key points	Remarks
	Operational range	Expected gas concentrations have to be within the analyser's operational range. For chamber measurements, headspace closure times might have to be adjusted to fulfil this requirement
	Discrete gas sampling	Continuous gas sampling is the great strength of optical analysers, but some are additionally able to analyse discrete gas samples. For discrete gas sample analysis, the required sample volume has to be taken into account in the design of flux studies
	Open versus closed path	For EC, the user has to decide between open path and closed path systems. Open path systems have longer downtimes and produce less data because of mirrors' direct exposure to the environment, especially precipitation, but their power consumption is much lower, and fluxes are not attenuated by any tubing in comparison to closed path systems
Site of operation	Instrument size, mobility, and robustness	Optical analysers come in all sizes–from very small portable analysers enclosed in suitcases or backpacks to large, bulky models (>10 kg) for laboratory bench racks. Important hardware properties to consider are temperature stability, vibration resistance, power consumption, reaction to power outages, and weather resistance. For field applications, additional hardware support systems might have to be constructed (e.g. air-conditioned housing)
	Installation space	For prolonged installations at a site, space might not only be required for the analyser, but also for external pumps, tubing, and gas cylinders. It can save a lot of working time when essential parts of the analyser and other equipment can be accessed for maintenance and potential repairs without having to be moved

(continued)

Table 2.5 (continued)

Topics	Key points	Remarks
Costs	Purchase price	Depends greatly on the optical technique. Fast-response analysers are the most expensive ones, whereas analysers for manual chamber measurements can be cheaper than complete GC systems
	Consumables/maintenance	Optical analysers have to be less frequently calibrated and often do not need continuous carrier gas flows compared to GC systems. For continuous gas measurements, additional costs for discrete gas samplings (e.g. glass vials, septa, syringes) are not required. Other (potential) costs to consider are maintenance times (= working hours), replacement of wearing parts, factory recalibrations, and repair costs
	Distance to company or distributor	In case the instrument has to be sent back to the company for repair or factory recalibration, the cost to be considered also includes shipping and shipping insurance. This is especially important for users outside of the USA, because American companies sell many optical analysers. Companies can sometimes perform software troubleshooting remotely if the analyser is connected to the internet
Maintenance	Time required	Maintenance time depends largely on the required precision and accuracy, as well as the working environment. These factors influence how often the instrument has to be calibrated, how stable the instrument readings are over time, and how often the instrument has to be cleaned
	User serviceability	A crucial part of optical analyser performance is the cleanliness of the cavity and mirrors. Depending on the measurement principle, already small dust particles inside the cavity or condensation on mirrors can cause repairs. There are analysers where the cavity (including the mirrors) can be completely taken apart and cleaned by the user, even under field conditions, whereas for other analysers, the user can change only air filters outside the cavity

(continued)

Table 2.5 (continued)

Topics	Key points	Remarks
	Air filtering	Careful selection of air filters (type, number, replacement interval) is a cost-effective way to reduce/prevent cavity contamination by dust particles. Air filters can also prevent water droplets from entering the cavity (hydrophobic air filters), e.g. if an automated chamber system has problems with water condensation in the tubing
	Remote control	Remote control of an optical analyser can greatly reduce maintenance times (e.g. reduction of field trips, earlier identification of analyser problems)
Manageability	Basic technical understanding	Some optical analysers are designed to be easily installed and operated even by untrained personnel (plug and play systems), whereas other analysers require a certain theoretical and practical knowledge of the optical technique used to ensure proper operation and their use should be supervised by trained technicians
	Analyser software	Degree of analyser configurability and availability of diagnostic tools for analyser performance vary widely between optical techniques and companies. The need for trained personnel increases with increased software complexity
	Data storage	For each optical analyser, a data storage protocol should be in place considering measurement variables to be stored, sampling rate, single file size, file names, folder structure, required storage space, schedule for data retrieval from the instrument, and schedule for data backup/data archiving. Data storage requirements increase with increasing sampling frequency, and a number of measurement variables (e.g. gas species, cavity temperature), and are therefore especially large for fast-response analysers
Data analysis	Raw data	For data analysis, it might sometimes be necessary to re-analyse spectral raw data. Raw data storage (complete or partial) is not always implemented in instruments

(continued)

Table 2.5 (continued)

Topics	Key points	Remarks
	Data analysis software	It is possible to purchase optical analysers which already have software included for the calculation of fluxes in chamber or EC methods
Peripherals	Complete measurement systems	Optical analysers can either be added to chamber and EC systems constructed by users themselves, or many companies also offer complete off-the-shelf systems (e.g. automated chamber systems) and/or a large variety of add-ons to the analysers for different GHG flux measurement applications
	Interfacing of analysers	Some companies that sell complete off-the-shelf systems also provide technical notes for interfacing optical analysers from other companies with their measurement systems. This is usually done to include gas species that are not part of a company's portfolio.
Documentation	Manuals	Quality of analyser manuals varies widely between companies. Some companies post their manuals online for free access. These manuals are a good starting point to get a first detailed impression of the functionality of an analyser before making a purchase decision
	Education	Many companies offer free webinars, technical notes, and other educational resources on their websites. These can often be accessed completely independent of any purchase. It is recommended to have a look at them because they often also provide training on the theoretical background of an optical technique and GHG flux measurement methods

The sub-section below describes the fundamentals of CRDS and discusses how a CRDS analyser may be paired with commercial and home-made closed chambers to create a field and/or laboratory deployable flux system.

2.7.2 Theory: Near-Infrared Absorption Spectroscopy Fundamentals

Nearly every small gas-phase molecule (e.g. H_2O, CO_2, CH_4, N_2O, etc.) has a unique near-infrared absorption spectrum, and this is generally also true for different isotopologues ($^{12}CO_2$ vs. $^{13}CO_2$, $^{12}CH_4$ vs. $^{13}CH_4$, etc.) (for details, see Chap. 7). At atmospheric pressure, these spectra are very broad and continuous, but at sub-atmospheric pressure the contributions from individual vibrational modes are well resolved so that each spectrum is made up of distinct absorption lines; each line is centred at a characteristic wavelength and possesses an extremely predictable line shape. Under these conditions, it is generally possible to find one or more narrow wavelength frequency ranges where the target molecule has strong absorbance features, and the absorbance contributions from other molecules are minimal. Therefore, the spectrum for each molecular species can be thought of as a unique and very information-rich "fingerprint".

After choosing a specific wavelength range, the concentration of the target gas can be measured by determining light absorption. In principle, the concentration could be inferred from the light loss at the absorbance maximum of the target gas; however, in practice, it is far more useful to scan over a specific absorption line by using a tunable, narrow-band laser to measure absorption in small steps over the relevant frequency range. The main advantage of determining the full spectrum of an absorption line is that one can measure multiple gases at the same time and that it also allows to measure the background level of light absorbance in one step. A measured absorption spectrum *s* can then be described as the weighted sum of contributions from individual gases and from the background light loss (Fig. 2.7) and (Eq. 2.6).

$$s = b + a_1 c_1 g_1 + a_2 c_n g_n \tag{2.6}$$

where *b* is the background light loss, *g* is the spectrum specific for each gas, *a* is a scaling factor describing the fundamental absorbance strength of the gas, and *c* is the scaling factor the concentration of each individual gas molecule. If the values of *b*, *a*, and *g* are known, then the gas concentrations *c* can be determined from the observed spectrum *s*. Since *a* and *g* are fundamental properties of the respective gas, they can be established in advance. The background light loss *b* is specific for each instrument but is generally highly stable over time.

2.7.3 Operational Principle of Cavity Ring-Down Spectroscopy

Cavity ring-down spectroscopy allows the concentration and isotopic composition of gases to be measured with very high precision. A key feature of CRDS is that

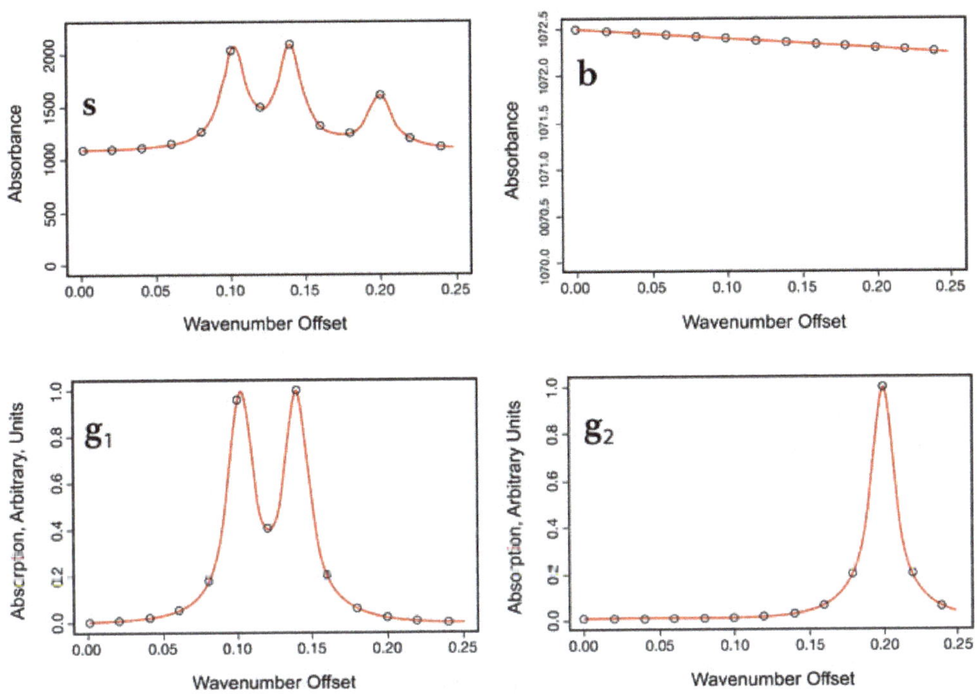

Fig. 2.7 An example spectrum "*s*" (top left) is the sum of background light loss "*b*", and scaled gas-specific absorbance spectra "*g*1" and "*g*2" (bottom). In this fictional example, $s = b + 1000 \times g_1 + 500 \times g_2$

it measures the decay of light in the measurement cell instead of measuring absolute absorption intensities, where the term "ring down" in CRDS is referring to this principle of measuring the light loss in the cavity over time. The cavity itself consists of three high reflectivity mirrors, and the light from a continuous-wave, single-frequency tunable laser diode is reflected between the mirrors (Fig. 2.8). When the laser is turned on, the cavity quickly fills with the circulating laser light. Because the mirrors have slightly less than 100% reflectivity (99.999%), a small amount of light continuously leaks out and is measured with the photodetector. As soon as the light intensity reaches a threshold (within a few tens of microseconds), the laser is abruptly turned off, and the light already within the cavity bounces between the mirrors (about 100,000 times). As the light circulates between the mirrors, light is absorbed by the target gas in the cavity, and the photodetector measures the exponential decay curve, also known as ring downs. In the absence of the absorbing gas, the decay rate is determined solely by the reflectivity of the mirrors.

The ring downs are measured in real time by the photodetector, and the exponential decay curve consists of several thousand individual light measurements. A dedicated signal processor fits this curve to determine the decay rate. To obtain a spectrum, the laser frequency is tuned over a sequence of narrow frequencies, and the ring-down rate is measured for each step (Fig. 2.9). Once a spectrum is obtained, algorithms are used to recover the spectrum contributions from each gas and from background light

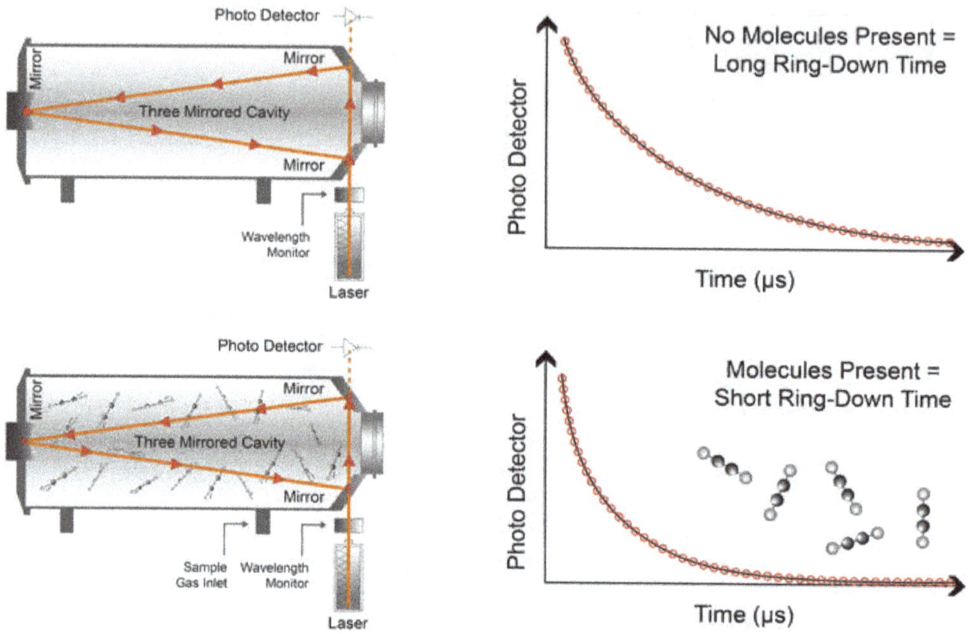

Fig. 2.8 High finesse optical cavity with a three-mirror configuration. Top: no absorbing species in the cavity = long ring-down time. Bottom: absorbing species present in the cavity = short ring-down time

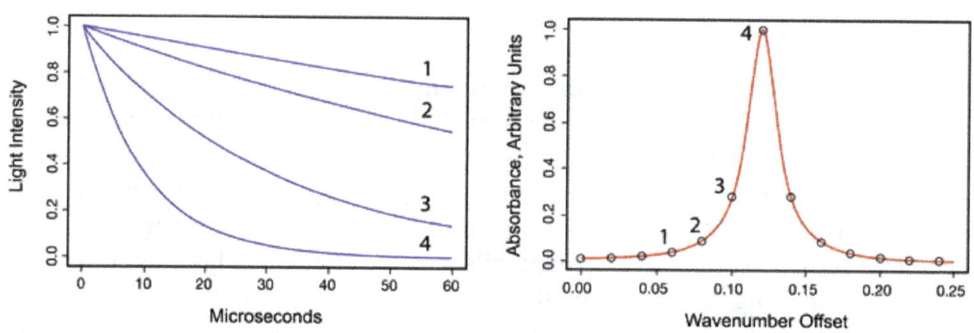

Fig. 2.9 The laser frequency is stepped, and a ring-down measurement is made at each step, generating a series of exponential decay curves (left). The software fits each curve to discover its decay coefficient. Representing these coefficients as absorbance, we obtain an absorbance spectrum (right). Numbers indicate corresponding decay curves and absorbance spectrum data

loss processes, as previously described. Critically, the decay rates are independent of the initial laser light intensity, and as such, the measurement is not affected by laser intensity fluctuations or absolute laser power.

2.7.4 Minimum Detectable Flux (MDF)

CRDS analysers allow GHG concentrations to be measured with a higher frequency and precision than generally possible with GC techniques. These key advantages allow either to determine GHG fluxes with a higher precision or to achieve the same level of precision within a shorter amount of time. The latter is of importance because a reduction in enclosure time minimises negative effects resulting from the use of a closed static chamber.

Christiansen et al. (2015) investigated the relationship between enclosure time and instrument precision and introduced the concept of minimum detectable flux (MDF) to determine the lower limit of flux rates that can be achieved with a given measurement precision (Eq. 2.7):

$$\mathrm{MDF} = \left(\frac{A_{\mathrm{A}}}{t_{\mathrm{c}}}\right)\left(\frac{VP}{ART}\right) \tag{2.7}$$

where A_{A} is the analytical precision in ppm, t_{c} is the closure time of the chamber in hours, V is the chamber volume in L, P is the atmospheric pressure in Pa, A is the chamber area in m^2, R is the gas constant in $\mathrm{L\ K^{-1}\ mol^{-1}}$, and T is the ambient temperature in K. The equation describes that the MDF can be reduced by either increasing the enclosure time or by increasing the analytical precision (reducing the analytical uncertainty).

Nickerson (2016) highlighted the practical importance of the increased measurement frequency of CRDS analysers since an increase in measurement number n will reduce the standard error A_{SE} compared to the raw precision A_{A} of the CRDS analyser (as $A_{\mathrm{SE}} = A_{\mathrm{a}}/\sqrt{n}$). Therefore, he suggested that the effective MDF should consider the measurement/sampling frequency (Eq. 2.8):

$$\mathrm{MDF} = \left(\frac{A_{\mathrm{A}}}{t_{\mathrm{c}}\sqrt{\frac{t_{\mathrm{c}}}{p_{\mathrm{s}}}}}\right)\left(\frac{VP}{ART}\right) \tag{2.8}$$

where p_{s} is the sampling periodicity in hours. For a CRDS analyser, the sampling periodicity is about every 8 s. The GC measurements are based on discrete samples, and the sampling frequency can vary significantly based on the research focus, but typically samples are taken every 5–15 min. Consequently, real-time CRDS analysers achieve a higher precision for a flux estimate than GC measurements, given the same chamber system, deployment time, and assuming a similar raw precision for the two techniques (Fig. 2.10). Or in other words, a prescribed MDF can be achieved within a shorter amount of time with CRDS compared to GC measurements.

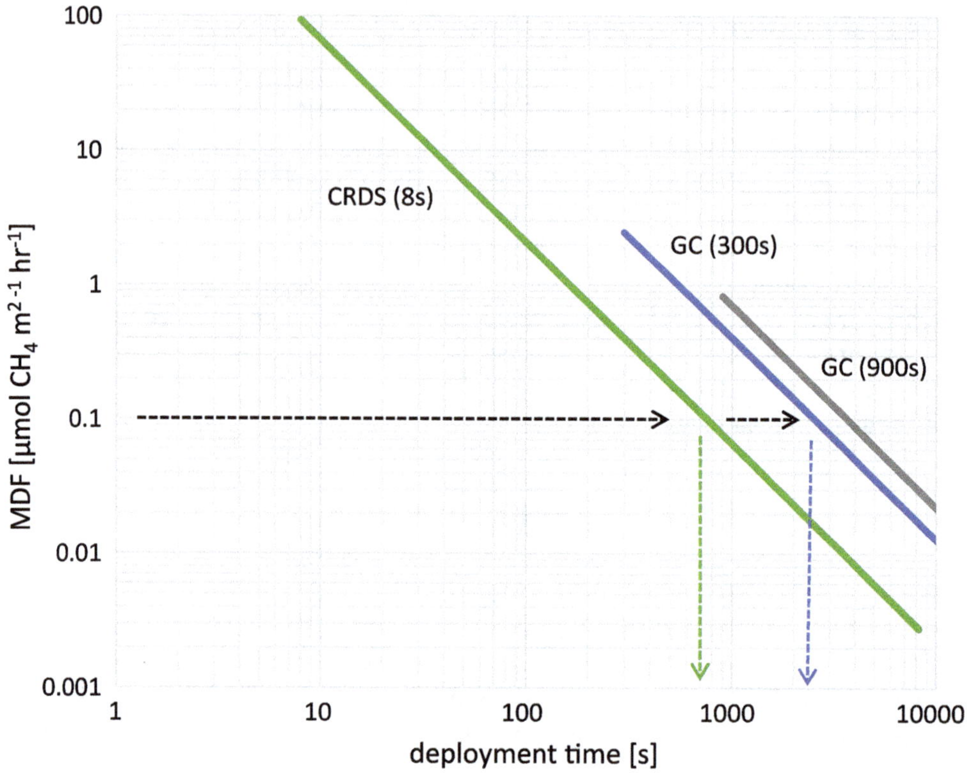

Fig. 2.10 Effect of sampling/measurement frequency on the time needed to reach a certain CH_4 minimum detectable flux (MDF). Real-time CRDS analysers report a concentration measurement about every 8 s, while discrete sampling for GC measurements commonly occurs every 5–15 min (i.e. 300–900 s). Due to the higher integration time, the CRDS analyser achieves a prescribed MDF within a shorter amount of time (an MDF of 0.1 μmol CH_4 m^{-2} h^{-1} is reached within 700 s with CRDS and within about 2300 s for 5 min interval GC sampling). The illustration assumes that the CRDS and GC measurements show the same raw precision. MDF was calculated on the following parameters: $A_a = 0.01$ ppm, $V = 0.5$ m^3, $P = 101325$ Pa, $A = 1$ m^2, $R = 8.31448$ J K^{-1} mol^{-1}, and $T = 295$ K (figure adapted from Nickerson 2016)

2.7.5 *Selecting the Appropriate Flow Path*

CRDS analysers have a continuous flow-through design that pulls gas through the analyser at a fixed flow rate (typical flow rate for a concentration CRDS analyser is ~250 ml min^{-1}). Automated pressure control loops within the analyser can be adjusted to gradual changes in input pressure and flow, but under normal operation, gas enters and leaves the analyser at the same flow rate. This design characteristic makes it suitable for use in recirculation experiments. Depending on the chamber design, the CRDS analyser can act as the primary source of flow or as a secondary sampling source, referred to as in-line or parallel, respectively.

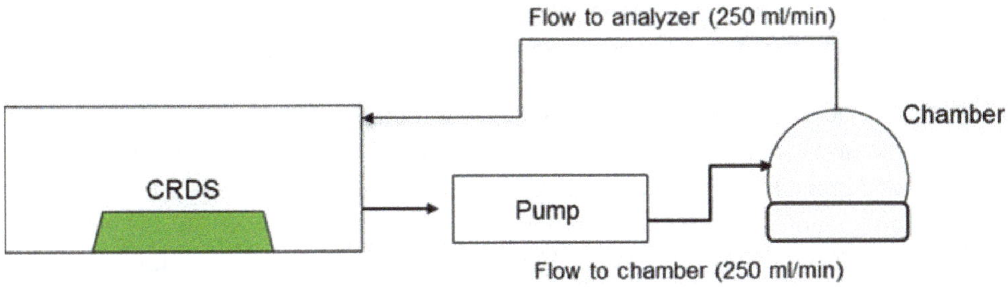

Fig. 2.11 Flow path of an in-line arrangement with a CRDS analyser in combination with a recirculation pump and a soil chamber. Additional communication or control cables are not shown

2.7.6 In-Line Flow Path

In this setup, the CRDS analyser is typically the primary source and control of flow in the recirculation setup (Fig. 2.11). It is commonly used with custom-made chambers or commercial chambers that do not employ their own flow control. When designing such a setup, users need to be mindful of any restrictions or rapid changes of flow that may occur during measurement. Inlet tubing length should not exceed 15 m and should have appropriate filtration in place to protect the analyser from dust and liquid water (see the next section). The CRDS analyser acts as the primary data repository for concentration and external sensor data (chamber temperature and pressure) and can control and coordinate chamber operation (if available).

2.7.7 Parallel Flow Path

The parallel flow arrangement is typically used when the analyser acts as a complementary or additional instrument in an existing recirculation setup. The user may have an existing flow setup that is controlled by a primary pump, or they may be using an existing chamber-analyser system (e.g. LI-COR LI-8100A, for details, see Chap. 3, Sect. 3.2.2). Since the CRDS analyser is no longer the primary source of flow control, it is essential that the flow rate of the main recirculation setup is higher than that of the CRDS analyser. The greater the difference between the two, the smaller the influence of the CRDS analyser will be.

2.7.8 *Multiple Chambers*

It should be noted that many commercial and homemade chambers often come with multiplexers–the ability to deploy and connect multiple chambers to a single analyser. A CRDS analyser can be used with a multiplexer in both the in-line or parallel arrangements.

The connection location of the CRDS analyser should follow the same approach as shown in Figs. 2.12 and 2.13. In either case, the multiplexer is located in the place of the chamber.

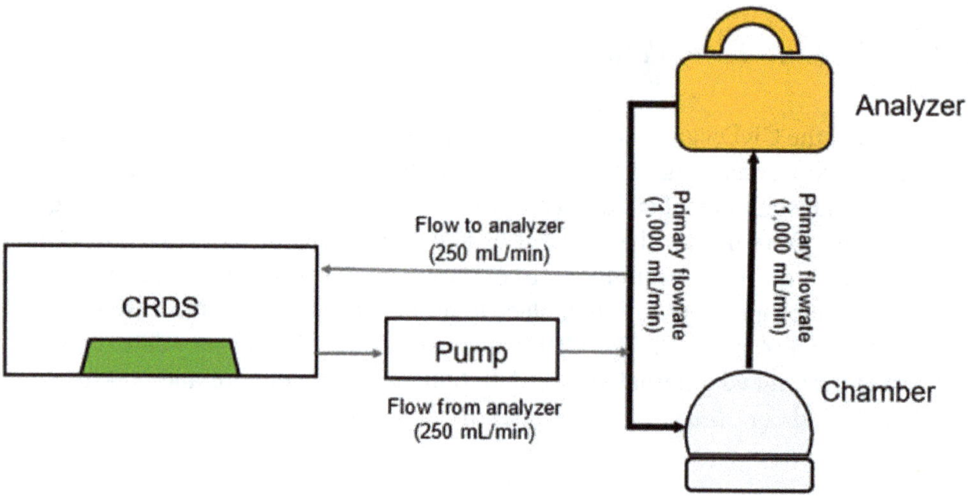

Fig. 2.12 Flow path of a parallel arrangement with a CRDS analyser, a recirculation pump, and a mobile soil gas survey system (LI-COR LI-8100A). Additional communication and control cables are not shown. Note the higher flow rate of the main recirculation setup at 1000 ml min^{-1}, and the parallel connection of the CRDS inlet and return on the return flow of the LI-COR analyser

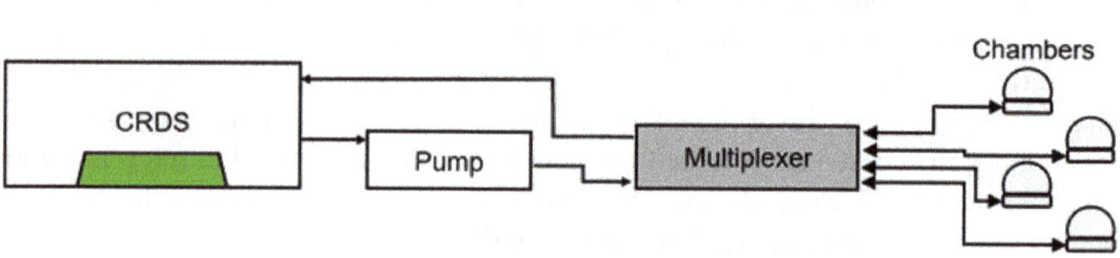

Fig. 2.13 Flow path for a multiple chamber setup. Deployments with multiple chambers are popular when a large surface area is to be monitored over a long period of time

2.7.9 Calibration

The concentration readings of a CRDS analyser behave very stable over time, and therefore, it is not necessary to calibrate the concentrations daily or weekly. However, it is advisable to check the accuracy and performance of a CRDS analyser on a monthly to yearly basis. This can be done by measuring standards from compressed gas cylinders. The concentrations measured by the CRDS analyser are then compared to the accepted value of the gas cylinder. The quality/accuracy of the final analyser calibration reflects (1) the number of standards, (2) the compositional range of the standards, (3) the method of standard measurement, and (4) the accuracy and precision of the standard itself:

1. The minimum number of different standards to calibrate the slope and offset of the analyser is two, and because CRDS instruments are highly linear, this is often enough. However, it is recommended to measure at least three or more standards, to confirm both the linearity of the instrument and the quality of the standards.
2. The range of the standards should generally encompass or exceed the expected concentration range of the sample air and should fall within the guaranteed specification range of the analyser.
3. The gas cylinders can either be connected directly to the CRDS analyser or the gas can be introduced via a gas bag (e.g. Tedlar bag). Direct connections to a tank are preferred as they minimise the chance of leaks and eliminate any dilution of the standard from remnant gas in the bag. Tedlar bags are an acceptable option, but they may lead to a lower degree of confidence in calibration. For applications where 1–10 ppm levels of accuracy are required, a direct connection to a tank is recommended.
4. It is commonly accepted that primary standards, when compared to secondary or even tertiary standards, have higher levels of accuracy. Having confidence in the value of the standards leads to a higher degree of confidence in the calibration of the instrument. It is advisable to assess whether the accuracy and precision of the standards are high enough for the research needs.

2.7.10 Advanced Application Considerations: Filtration of Gas Samples

The precision of an absorption measurement within a CRDS analyser can degrade under the presence of foreign particles that scatter or absorb light. To overcome this potential issue, the inlet of a CRDS analyser is generally equipped with two 4 nm particulate filters. One is user-replaceable, and the other can only be replaced in a cleanroom environment. To prevent long-term damage to the sample handling of the analyser and maintain a healthy cavity, external filters are recommended.

Of primary concern are particles that can potentially be pulled into the sample handling of the analyser via the inlet port. In order to select and build a proper filtration system, the user must first assess which, if any, particles are potentially present in the field. Arid environments with high wind speeds or other mechanical means of increasing particle load (walking, driving, animals, etc.) pose a much greater danger to the analyser than vegetated surfaces. In most cases, a cheap, user-replaceable Acrodisc®, 0.2–1 μm is enough to remove most particles. The frequency of replacement should be dictated by the particle load of the gas sample and the flow rate through the filter.

2.7.11 Liquid Water and Water Vapour

One of the biggest risks to any field-based system is exposure to liquid water. In this section, we focus on the introduction of liquid water through the inlet of the analyser as a result of (1) condensation or (2) actual sampling of liquid water (precipitation, ground saturation, etc.).

Long-term deployments of CRDS analysers can subject them to environments with temperature variations that exceed the operating range of the system (10–35 °C). Custom or commercially available enclosures are used to provide a measure of temperature control and/or protection from the elements (rain, snow, wind, etc.). The temperature difference between the exterior and interior of the enclosure can lead to condensation in sampling lines. This problem is particularly evident in scenarios where hot and humid gas is brought into a cold enclosure. To reduce the likelihood of condensation, desiccants such as Drierite or magnesium perchlorate can be used to reduce the water vapour level in the sample gas. Such desiccants are also practical if water vapors need to be reduced below 4 vol%, the operational limit for a CRDS analyser.

If condensation, or the presence of liquid water through direct sampling, cannot be avoided, then a water trap must be used before the inlet of the analyser. The volume of a water trap should be appropriate to handle any input of liquid water. It is also possible to install water traps with automatic draining capability so that the system may operate in a user-free fashion. It is important to note that a water trap will add additional volume to the recirculation system, something that needs to be considered when flux calculations are performed.

2.7.12 CRDS-Specific Considerations

A CRDS analyser is a laser absorption spectrometer. Under specific conditions, the composition of the sample gas may have unintended effects on the spectra. If levels of ethane, ethylene, acetylene, or hydrogen sulphide exceed atmospheric levels, the user should consult manufacturer documentation to determine the effects of these

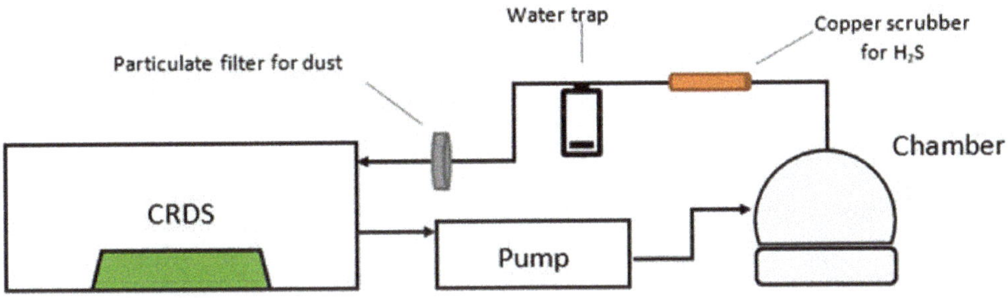

Fig. 2.14 Example setup of a CRDS analyser for field deployment in a forest environment. The copper scrubber is in place as trace amounts of H_2S are expected, the water trap is used to minimise the change of liquid water entering the system due to ground saturation during long rainfall events, and the particulate filter is there to prevent particles that may enter the air stream. Not shown on this image is an enclosure for the system and a stable supply of power

species on their measurements of greenhouse gases. In many cases, the installation of an additional scrubber may be used to remove an interfering species. One such example is the removal of hydrogen sulphide using laboratory-grade copper filings (Malowany et al. 2015). Figure 2.14 illustrates how a recirculation setup can be adjusted for specific field conditions by adding a particulate filter, a water trap, and a copper scrubber.

The composition of the carrier gas, specifically any significant changes to the N_2–O_2 ratio of the carrier gas, will have pressure broadening effects on the spectra (Nara et al. 2012). This can result in a degradation of accuracy. The installation of an additional oxygen sensor inside a CRDS analyser, or any external oxygen sensor, can help the user detect changes in the N_2–O_2 ratio and apply a correction. The manufacturer also offers a pre-configured mode for pure N_2 carrier gases. It should be noted that major changes to Argon (Ar) or Helium (He) will have the same effects. More importantly, a pure He carrier can permanently damage a CRDS analyser.

2.7.13 Datalogging and Flux Processing

All concentration data obtained using a CRDS analyser is automatically stored on the hard drive of the analyser. This raw data can then be processed using the CRDS software or other, third-party compatible software that is usually supplied by the chamber manufacturer to obtain flux values. In order to convert concentrations into fluxes, the software will ask the user to enter the chamber and tubing volume, chamber surface area, chamber and soil temperature, and chamber pressure. Depending on the software package used, it may be possible to monitor the accumulation of the gases in real time or apply post-processing using three or more fitting algorithms: linear (Eq. 2.9), quadratic (Eq. 2.10) (Wagner et al. 1997), or Hutchinson and Mosier (1981) (Eq. 2.11).

Linear:

$$y = \text{slope} \times X + \text{background} \tag{2.9}$$

Quadratic:

$$y = \text{a} \times X^2 + \text{slope} \times X + \text{background} \tag{2.10}$$

Hutchinson and Mosier:

$$y = \text{background} + \text{step}\left(1 - \text{e}^{-\text{slope} \times X/\text{step}}\right) \tag{2.11}$$

Custom fitting options are available in certain software packages (Fig. 2.15).

CRDS analysers are compatible with the streaming of data to external datalogger using RS232, Ethernet, or analogue communication protocols. In addition, it is also possible to incorporate external sensor data such as pressure and temperature probes within the soil flux chamber. Detailed support and integration documentation are available from the manufacturer.

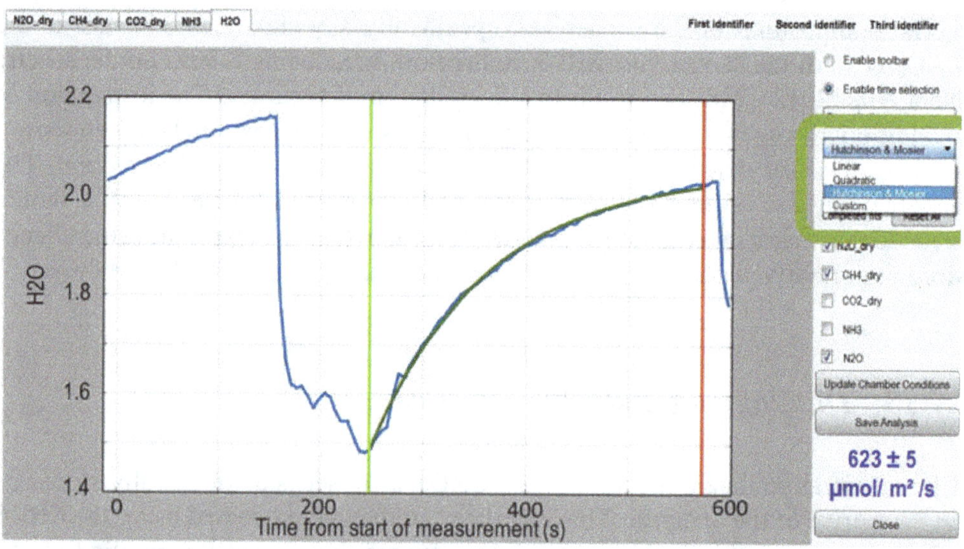

Fig. 2.15 Example of a Hutchinson and Mosier fitting algorithm applied to a measurement of H_2O on a CRDS analyser

2.8 Enhanced Raman Spectroscopy of Greenhouse Gases

Enhanced Raman spectroscopy emerged in recent years as a powerful analytical tool for highly selective identification and simultaneous quantification of multiple components in complex gas mixtures (Hanf et al. 2014; Frosch et al. 2013a; Jochum et al. 2015a; Hippler 2015; Kiefer 2015; Knebl et al. 2017). Conventional Raman spectroscopy was already discovered in the 1920th (Long 2002; Smekal 1923; Raman and Krishnan 1928; Landsberg and Mandelstam 1928; Kramers and Heisenberg 1925), but its application in real-world applications was very limited until recently, due to the inherently weak signal intensities and technical limitations.

2.8.1 Raman Spectroscopy of Gases

Raman spectroscopy is based on inelastic scattering of light. When monochromatic laser light is guided through a gas sample, most photons pass the analyte volume without any interaction and thus provide no gas specific chemical information. The scattering cross section σ is only around 10^{-27} cm^2 for typical gases. Thus, only a small fraction of the photons is scattered by the gas molecules which also have a very low number density. The major part of the scattered radiation results from elastic scattering (Rayleigh scattering) and has the same frequency as the excitation light. Inelastic scattering occurs with approximately three orders of magnitude smaller intensity. The frequency shift of the inelastically scattered light is caused by the energy transfer in the rotational, vibrational, or rotational–vibrational states of the molecules and is thus the basis of the unique chemical selectivity of Raman spectroscopy. Raman scattering and the more established IR absorption are complementary effects that probe molecular vibrations and rotations. Raman scattering depends on the changes in the polarisability of the molecule during the rotation/vibration, while IR absorption relies on changes in the (permanent) dipole moment. Thus, IR absorption spectroscopy is not sensitive to diatomic homonuclear molecules such as oxygen (O_2), hydrogen (H_2), or nitrogen (N_2) which provide no permanent dipole moment for light interaction. In contrast, inelastic Raman scattering does not require a permanent dipole moment of the molecules, and the specific molecular transitions must not be matched with the excitation wavelength. Thus, Raman spectroscopy is a powerful method for the analysis of all but noble gases with only one monochromatic laser source. A whole suite of biogenic gases, including CO_2, N_2O, O_2, CH_4, H_2, and N_2, can be identified and quantified simultaneously (Fig. 2.16).

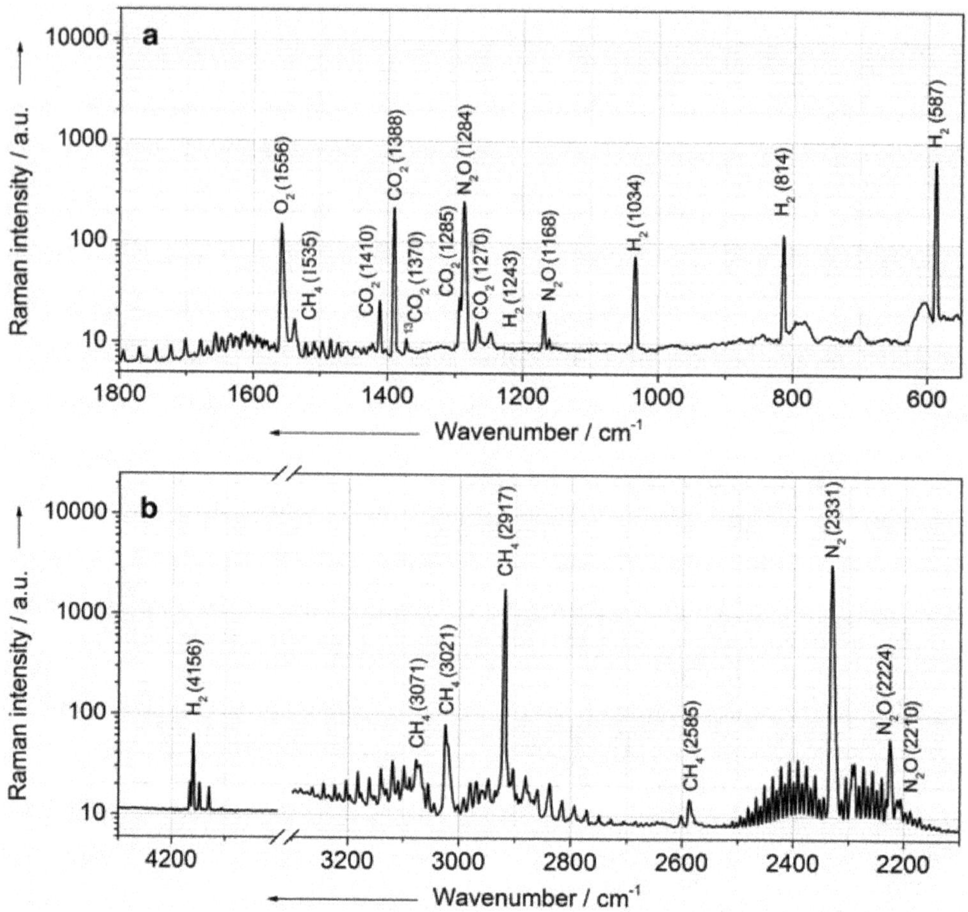

Fig. 2.16 Raman spectrum of the rotational and rovibrational bands of a complex mixture of biogenic gases, including CO_2, N_2O, O_2, CH_4, H_2, and N_2. Reprinted with permission from Hanf et al. (2015a).

2.8.2 *Enhanced Raman Gas Spectroscopy*

Conventional Raman spectroscopy is an extremely weak process. Recent advances in laser as well as detector technology, new highly efficient optical filters, and optical components develop Raman spectroscopy into a comprehensive technique for multigas analysis. The Raman Stokes intensity I_R depends on the laser intensity I_0, the angular frequencies of the laser ω_L and the scattered light ω_R as well as the polarisability α of the molecule, and the number N of molecules of the measured gas (Eq. 2.12).

$$I_R = \text{const} * N * I_0 * (\omega_L - \omega_R)^4 * |\alpha|^2 \tag{2.12}$$

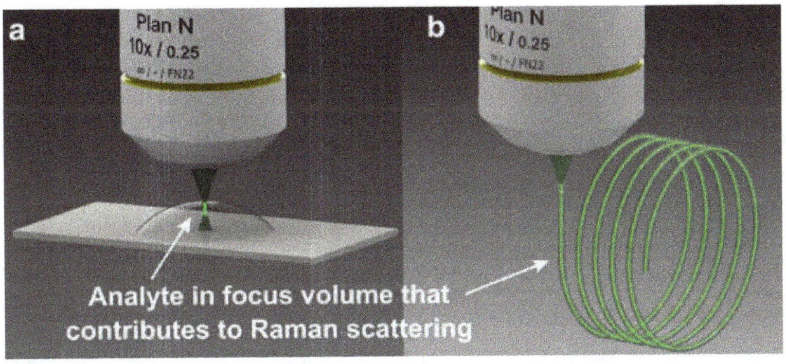

Fig. 2.17 Comparison of conventional Raman spectroscopy (**a**) and fibre-enhanced Raman spectroscopy (**b**). Conventional Raman setups suffer from the limited interaction volume of excitation light and analyte gas (**a**). In FERS, the excitation light is guided over an extended range within the hollow core of the optical fibre, which also functions as minimised analyte container (**b**). An excellent light–analyte interaction is achieved in fibre-enhanced Raman spectroscopy and a high number of molecules contribute to the Raman signal (Frosch et al. 2013b). Reprinted with permission from Frosch et al. (2013b).

The originally small inelastic scattering signals of the gas molecules must be increased with the development of enhancement techniques based on the above-given parameters in order to achieve high sensitivities. Raman gas sensors are nowadays based on specific multi-pass cavities, high-pressure cells, or the application of high-power lasers (Li et al. 2008; Kiefer et al. 2008; Schiel and Richter 1987). Two very promising techniques, which were recently developed, are fibre-enhanced Raman spectroscopy (FERS) (Hanf et al.2014; Knebl et al. 2018) and cavity-enhanced Raman spectroscopy (CERS) (Frosch et al 2013a).

In FERS, the number N of gas molecules that contribute to the Raman signal (Eq. 2.10) is extremely increased. In conventional Raman spectroscopy, the signal is excited and collected from a small scattering volume (Fig. 2.17a) (Frosch et al. 2013b). This limitation can be overcome with the help of hollow-core optical sensor fibres. In elaborated micro-structured hollow-core optical fibres, the light can be guided with extremely low attenuation within a certain spectral range (Russell 2003; Knight 2003; Hartung et al. 2015; Yan et al. 2017a, b). By filling the gas in the same inner hollow core of the fibre, light and analyte are both guided in the same volume and thus the analyte is efficiently excited over an extended path length (Fig. 2.17b) and the number of scattering molecules is strongly increased (Frosch et al. 2013b). In other words, the Raman fibre sensor can be seen as optimised cuvette and provides strongly enhanced analytical sensitivity (Hanf et al. 2014; Frosch et al. 2013b; Knebl et al. 2019; Jochum et al. 2016; Hanf et al. 2015a; Boegoezi et al. 2015; Sieburg et al. 2019). A highly efficient FERS setup was developed for gas analysis (Figs. 2.17 and 2.18).

In CERS, the weak inelastic scattering signals of the gas molecules can be increased by up to six orders of magnitude with the help of a high finesse optical cavity. One of the developed gas sensors consists of a miniaturised laser diode with $\lambda_{exc.} = 650$ nm (Frosch et al. 2013a; Keiner et al. 2013, 2014) which is passively frequency locked and feedback coupled to a high finesse cavity, enabling a power

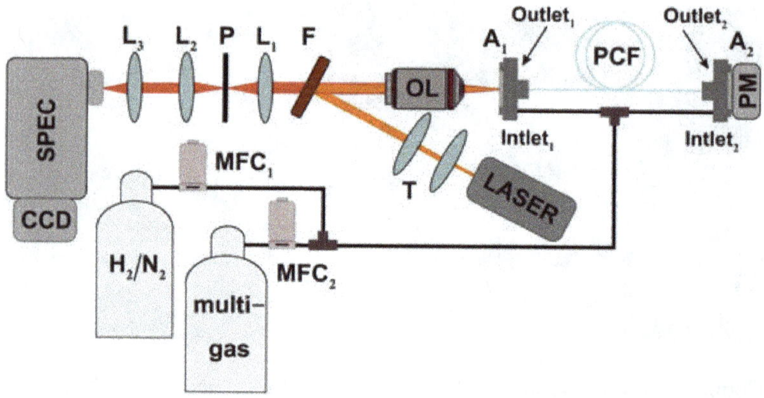

Fig. 2.18 Design of the optical setup for fibre-enhanced Raman spectroscopy (FERS) (Hanf et al. 2014) consisting of Laser, telescope (T), long-pass dichroic beam splitter (DC), objective lens (OL), fibre adapter assembly (A1 and A2), hollow-core photonic crystal fibre (PCF), power metre (PM), pinhole (P), edge filter (E), aspheric lens (L), spectrometer (SPEC), and CCD detector. Reprinted with permission from Hanf et al. (2014).

build-up to 100 W (Frosch et al. 2013a). The cavity components are aligned for spatial mode matching of the input beam and the Gaussian beam supported by the power build-up cavity (PBC), while the facet of the laser diode helps in stabilising mode matching by spatial filtering (Frosch et al. 2013a). Such arrangement of the PCB is extremely stable to mechanical vibrations (Frosch et al. 2013a), and it has been shown that concentration fluctuations of about 50–100 ppm can be monitored within measurement times of one second (Frosch et al. 2013a).

Main advantages of enhanced Raman spectroscopic gas sensing include the following:

- All gases, except noble gases, can be analysed, including diatomic homonuclear molecules such as oxygen (O_2), hydrogen (H_2), and nitrogen (N_2).
- A whole suite of gases can be identified simultaneously (Fig. 2.16) with no cross sensitivities and the use of only one excitation laser.
- A priori unexpected gases can easily be identified in the Raman spectra of complex multigas compositions, in contrast to simple gas sensors that must be designed for all expected gases beforehand.
- As direct spectroscopic method, which is based on the intrinsic molecular vibrations, Raman gas sensing does not need any labels or transducers and does not suffer from saturation, bleeding, poisoning, etc. This optical technique does not need direct physical contact with the analyte gas. Measurements can be performed through an optical window and remote.
- The technique is non-consumptive. The gas concentrations are not disturbed due to the measurement process, and small gas exchange processes can be monitored continuously online in closed cycle experiments.

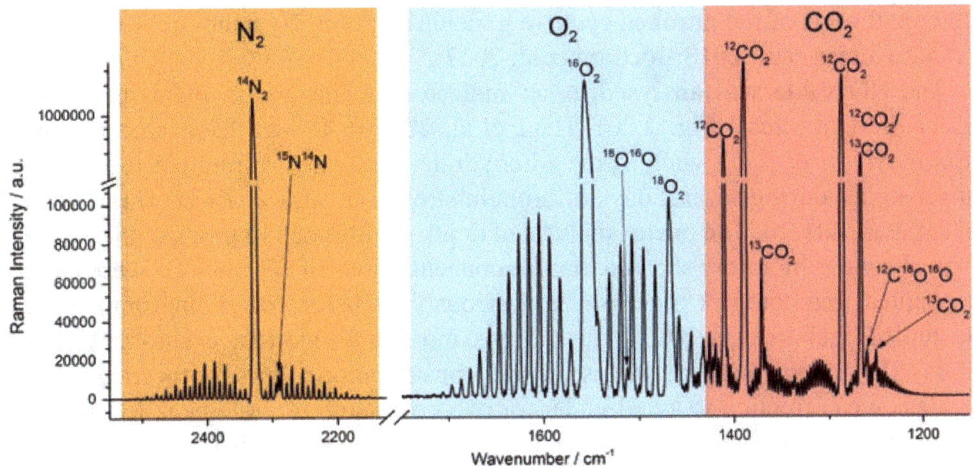

Fig. 2.19 Combined Raman spectra containing the rovibrational bands of $^{14}N_2$, $^{14}N^{15}N$, $^{16}O_2$, $^{18}O^{16}O$, $^{18}O_2$, $^{16}O_2$, $^{12}C^{16}O_2$, $^{13}C^{16}O_2$, and $^{12}C^{18}O^{16}O$. Adapted with permission from Knebl et al. (2019).

- The Raman scattering intensity offers perfect linearity with the analyte concentration (Eq. 2.10) and allows for robust instrument calibration over a broad concentration range from ppm to pure compounds. Trace gases can be quantified on the background of a higher concentrated gas matrix.
- The fast Raman spectroscopic measurement enables online monitoring of temporal changes in gas concentrations during process.
- Stable gas isotopes can be distinguished (due to changes in the reduced mass and thus spectral position, see Fig. 2.19) and be used as tracers to follow specific pathways.
- Raman devices can be highly miniaturised for field deployment.

2.8.3 *Enhanced Raman Spectroscopic Analysis of Greenhouse Gases*

Research regarding enhanced Raman spectroscopy of biogenic gases can be divided into three areas: (i) The capability for unambiguous identification and quantification of various gas mixture components to follow the formation, spread, and exchange as well as storage of gases (Frosch et al. 2013a; Jochum et al. 2015a, b; Keiner et al. 2013, 2014, 2015a, b; Sieburg et al. 2017, 2019; Hanf et al. 2015b). One focus is the analysis of the respiratory quotient (RQ, CO_2 released per O_2 consumed during respiration) as an indicator of changes in substrate use and metabolism (Jochum et al. 2015a; Keiner et al. 2013; Sieburg et al. 2017; Hanf et al. 2015b); (ii) the monitoring of stable isotopes ($^{12/13}CO_2$, $^{14/15}N_2$, $^{14/15}N_2O$, and $^{16/18}O_2$) alongside unlabeled gases as tracers for specific pathways (Knebl et al. 2019; Keiner et al. 2014, 2015a, b); and (iii) the unique ability for direct quantification of nitrogen at natural background to study pathways of the nitrogen cycle (e.g. denitrification, N_2 fixation) (Keiner et al. 2015a; Kumar et al. 2018; Jochum et al. 2017).

The RQ value was analysed as an indicator of changes in plant metabolism under drought stress (Fig. 2.20) (Hanf et al. 2015b). It was discovered that pine (*Pinus sylvestris*) can switch from carbohydrate-dominated respiration to a mixture of substrates during several days of drought stress, but spruce (*Picea abies*) cannot (Hanf et al. 2015b). The onsite analysis of depth profiles of soil gases in the Hainich critical zone exploratory showed that the concentrations of O_2 and CO_2 were largely decoupled, and complex processes in previously uncharacterised environments can be studied (Sieburg et al. 2017). The ability to monitor the inert tracer sulphur hexafluoride (SF_6) alongside biogenic gases under consideration allows for thorough online gas leakage correction to avoid under- or overestimation of biological activity such as respiration or photosynthesis (Fig. 2.21) (Jochum et al. 2015b).

The discriminatory power of Raman spectroscopy was used to monitor several stable gas isotopes simultaneously to investigate the labelling of young poplar trees

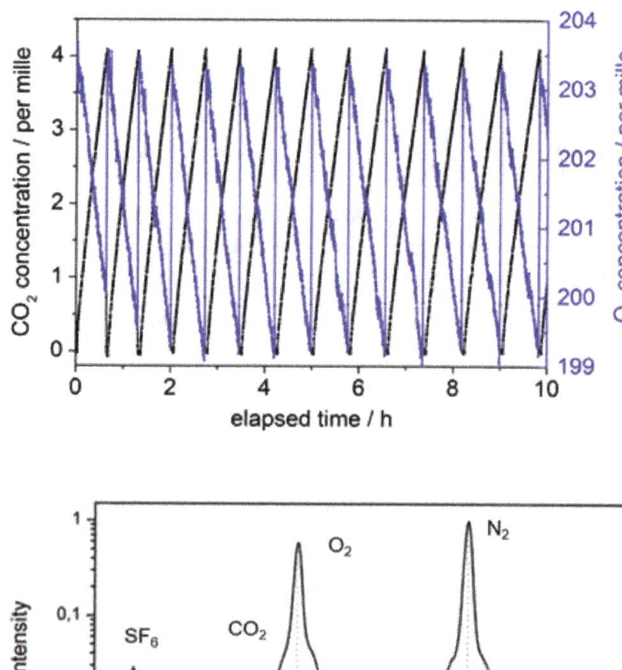

Fig. 2.20 Example for monitoring dark respiration rates of a branch of untreated pine. The individual respiration rates (both O_2 and CO_2) are used to calculate subsequent RQ values (Hanf et al. 2015b). Adapted with permission from (Hanf et al. 2015b).

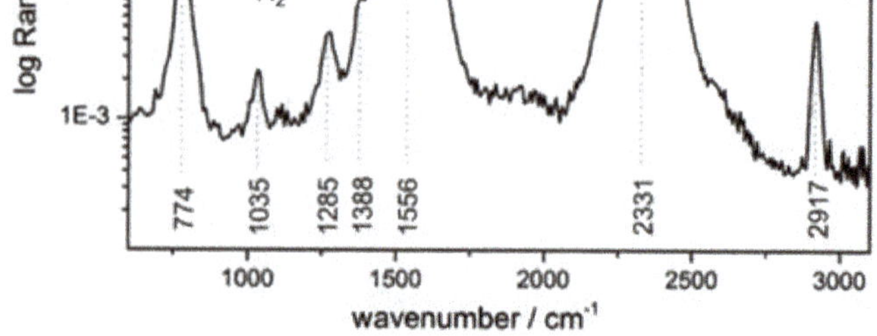

Fig. 2.21 Example of an experimentally acquired multigas Raman spectrum, consisting of the biogenic gases O_2, CO_2, N_2, H_2, CH_4, and the tracer gas SF_6. Adapted with permission from Jochum et al. (2015b).

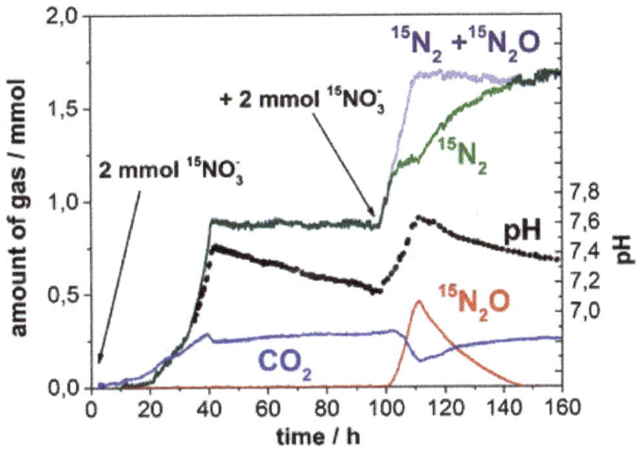

Fig. 2.22 Raman spectroscopic multigas monitoring of the denitrification of ^{15}N-nitrate by *Pseudomonas stutzeri* (Keiner et al. 2015a). The concentration courses of $^{15}N_2$, CO_2, $^{15}N_2O$, and the calculated pH value are shown as well as the sum of the nitrogen gases $^{15}N_2 + {}^{15}N_2O$. Adapted with permission from Keiner et al. (2015a).

under aphid infestation with $^{13}CO_2$ to analyse the possible incorporation of ^{13}C in defense compounds (Keiner et al. 2014, 2015b). A combination of ^{13}C-labelling and RQ analysis was applied to investigate the microbial degradation of ^{13}C-labelled benzene in soil against the background of the heterotrophic soil respiration (Jochum et al. 2015a). By combining $^{13}CO_2$ and $^{12}CO_2$ as well as $^{18}O_2$ and $^{16}O_2$ measurements in one setup, it was proposed to use carbon dioxide and oxygen isotopologues to track and disentangle different overlaying processes and to help elucidating the contributions of photosynthesis, photorespiration, and respiration to the net gas exchange of plants (Knebl et al. 2019).

The nitrogen evolution was continuously monitored over the stepwise enzymatic denitrification of labelled and unlabeled nitrate by *Pseudomonas stutzeri* (Fig. 2.22) (Keiner et al. 2015a). The simultaneous quantification of the whole gas phase also enabled the contactless and sterile online acquisition of the pH changes in the *P. stutzeri* culture by the stoichiometry of the redox reactions during denitrification and the CO_2-bicarbonate equilibrium. Continuous pH-monitoring–without the need to insert an electrode into a sterile solution–elucidated an increase in the slope of the pH value coinciding with an accumulation of nitrite, which in turn led to a temporary accumulation of N_2O, due to an inhibition of N_2O reductase (Keiner et al. 2015a). The gas quantification was complemented with the analysis of nitrate and nitrite concentrations for the online monitoring of the total nitrogen element budget (Fig. 2.23) (Keiner et al. 2015a). In an investigation of the thiosulfate- and hydrogen-driven autotrophic denitrification by a microbial consortium enriched from groundwater of an oligotrophic limestone aquifer, the turnover reactions of electron donors (thiosulfate and H_2) were traced, as well as electron acceptor (nitrate), gaseous intermediates, and end products ($^{15}N_2$, $^{15}N_2O$, CO_2, H_2, $^{14}N_2$, and O_2) in the headspace, using Raman gas spectroscopy (Kumar et al. 2018). N_2 production and H_2 consumption rates were calculated under denitrifying conditions and followed the electron donor usage of the bacterial consortium. Recently, the biological nitrogen fixation of a *Medicago sativa–Rhizobium* consortium was, for the first time, directly investigated at natural background and without a proxy or isotopic labelling, by continuously

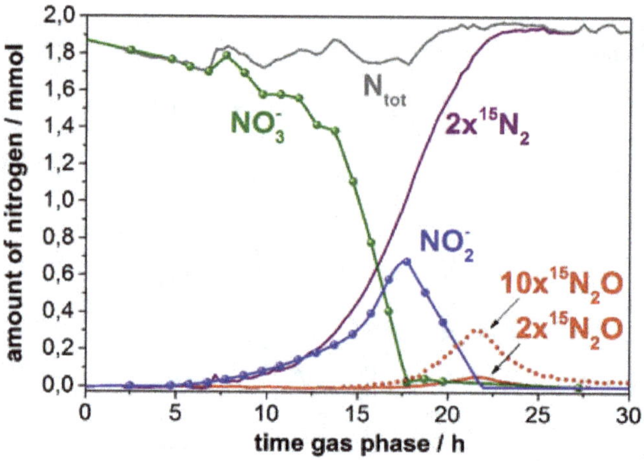

Fig. 2.23 Concentration courses of nitrate, nitrite, nitrous oxide, and dinitrogen during the successive reduction $NO_3^- \rightarrow NO_2^- \rightarrow NO \rightarrow N_2O \rightarrow N_2$ (Keiner et al. 2015a). Continuous Raman gas concentrations of $^{15}N_2$, $^{15}N_2O$, and CO_2 are represented by solid lines. The total nitrogen balance from all nitrogen components was calculated and is given as grey line (N_{tot}) (Keiner et al. 2015a). Adapted with permission from Keiner et al. (2015a).

analysing the amount of atmospheric N_2 in static environmental chambers (Jochum et al. 2017).

Enhanced Raman gas spectroscopy combines the unmatched analytical prowess of Raman spectroscopy with the enhancement of small signals through the sophisticated use of optical cavities (CERS) and hollow-core optical fibres (FERS). Enhanced Raman spectroscopy is a powerful technique for simultaneous multigas analysis, including N_2, N_2O, O_2, H_2, CH_4, CO_2, and stable isotopes (^{13}C, ^{15}N, and ^{18}O). This high selectivity enables the study of complex gas exchange processes, including pathways of the nitrogen cycle.

2.9 GHG Fluxes from Aquatic Systems

Following the first set of IPCC guidelines being produced (IPCC 1995), further developments were needed to improve national inventory calculations. These included a quantifiable way to determine N_2O production after the leaching and runoff of N from agricultural soils, in order to improve the methodology for calculating annual N_2O emissions from agricultural soils at a national level (Mosier et al. 1998). The second phase approach assumes that all N cycles within 1 year and makes no allowance for potential sequestration within the soil, which may be subsequently released on longer timescales (Mosier et al. 1998).

The approach taken in the second phase of inventory development for determining N_2O emissions from agricultural soils was to include direct emissions from agricultural fields, direct emissions from animal production systems, and indirect emissions. Indirect emissions result from volatilisation and subsequent deposition of NH_3 and NO_x, leaching and runoff of N, and human consumption of crops followed by sewage treatment. Nitrogen leaching and runoff from agricultural systems (NLEACH) include N applied as fertiliser (NFERT; kg N year^{-1}), and N derived from animal excretion (NEX; kg N year^{-1}) so that manure produced during grazing is captured. The fraction of fertiliser and excreta N that moves through the leaching and runoff pathways (FRACLEACH) was estimated by Mosier et al. (1998) to range from 0.1 to 0.8. Some countries have developed country-specific values for FRACLEACH (Thomas et al. 2005). Thus, the flux of N leached (NLEACH, kg N year^{-1}) (Eq. 2.13) is

$$\text{NLEACH} = [\text{NFERT} + \text{NEX}] \times \text{FRACLEACH} \tag{2.13}$$

Production of N_2O occurs when the leached N moves into the groundwater and surface drainage, rivers, and estuaries with emission factors for these three zones designated as $EF_{5\text{-}g}$, $EF_{5\text{-}r}$, and $EF_{5\text{-}e}$, respectively (Mosier et al. 1998). The sum of these three components provides the N_2O emission factor (EF_5) for N_2O originating from NLEACH, where N_2O (l) is the N_2O emissions associated with agricultural N lost via leaching and runoff (kg N year^{-1}) (Eq. 2.14):

$$N_2O(\text{l}) = \text{NLEACH} \times EF_5 \tag{2.14}$$

Supersaturation of N_2O in ground and surface waters occurs due to either N_2O leaching from the soil or because microbial processes utilise NLEACH to form N_2O in the ground and surface waters.

The approach taken to determine EF_5 seeks to compare the emission based on the N loading, which is comparable with calculating direct emission factors from agricultural soils. For soils, a direct emission factor is determined by applying a quantity of N to plots and measuring the resulting N_2O emissions relative to control plots. A different approach has been taken to determine EF_5. Briefly, the N_2O concentration in a water sample is measured, and the degree of N_2O saturation is determined. The EF_5 value is then determined based on the ratio of N_2O-N: NO_3^--N or on the basis of assumed N_2O emissions using the degree of saturation to predict a flux. To use the IPCC methodology to determine EF_5 values requires that the concentration of dissolved N_2O is first established so that the flux of N_2O can be determined.

2.9.1 Determining Dissolved N_2O Concentrations

There are several methods in the literature that are available to measure dissolved gas depending on the source of the water sample. Water samples may be taken from groundwater, drains, and open water bodies such as rivers, lakes, and estuaries. When sampling water bodies for dissolved gases, it is important to also record the water temperature in situ, the atmospheric pressure, and to take some gas samples of the atmosphere at the sampling site. Atmospheric pressure will vary with altitude and weather conditions. A further requirement is to know the salinity of the water being sampled since this also affects the solubility of gases.

Nitrous oxide produced at depth within a soil can be readily dissolved in soil pore water or water draining from the soil in question. When water samples are collected using automated samplers, there remains the need to prevent degassing of the sample. In an ideal situation, drainage water samples would be collected as drainage occurred. However, this is not always feasible. Roper et al. (2013) developed a technique to facilitate dissolved N_2O sampling from automated water sampling bottles. A 10 ml volumetric pipette is placed, inverted inside the sample bottle, within the autosampler. The inversion of the pipette allows for simple modifications for length and enables faster filling while simultaneously minimising turbulence during filling (Roper et al. 2013). The inverted pipette means that the water contained in the pipette has a reduced surface area compared to the larger sample bottle. A water sample for dissolved gas determination is taken manually by withdrawing 5 ml of the 9 ml of water from inside each pipette using a syringe equipped with a 20-gauge 30.5 cm needle (Popper®, Fisher Scientific) and placing the sample in a 12 ml Exetainer (Labco International, UK). The Exetainers were previously flushed, using ultrapure helium, and brought to atmospheric pressure with helium in advance of water sampling. Exetainers also had a microbial inhibitor, placed inside prior to evacuation, to prevent microbial activity altering dissolved N_2O concentrations. Generally, the microbial inhibitor of choice is mercuric chloride ($HgCl_2$). The inhibitor must not physically or chemically alter the water sample in a manner that could alter the N_2O concentration. Once sampling is complete, the dissolved N_2O embodied in the water equilibrates with the gaseous phase in the headspace of the Exetainer. It is thus important to record the laboratory temperature when this equilibrium is attained prior to sampling the gas headspace.

A headspace sample equilibration technique can also be used to sample dissolved N_2O in groundwater. Clough et al. (2007a) used this technique to study dissolved N_2O in water samples obtained from piezometers. Using a Masterflex L/S portable peristaltic pump (Cole Parmer, Vernon Hills, Il, USA) with the pumping rate adjusted to minimum flow, to avoid out-gassing, water was pumped from the piezometers (Plate 2.17). The water was pumped to waste, until a volume of water equal to several internal piezometer volumes had been pumped, and then diverted into a 60 ml plastic syringe, by way of a two-way stopcock, until 20 ml had been collected. The resulting 20 ml sample was then injected into a pre-evacuated (−0.93 atm) 160 ml serum bottle fitted with a rubber suba seal. Then using a helium-flushed syringe, fitted with a stopcock and hypodermic needle, connected to a helium-filled Tedlar bag, helium was injected into the serum bottle until atmospheric pressure was reached. This was visually determined by inverting the serum bottle so that the water sample covered the helium needle inlet: when bubbling ceased, the serum bottle headspace was at atmospheric pressure. The sample is thus ready for equilibration (Plate 2.18).

A third example of where water samples are commonly taken for headspace equilibration is that of open waterways. For example, Beaulieu et al. (2012) measured dissolved N_2O in river water by taking 140 ml polypropylene syringes equipped with

Plate 2.17 Groundwater being sampled from a piezometer tube using a peristaltic pump. Note the syringe is kept under water to prevent any contamination by air while it is being filled

stopcocks, and slowly filling these with water at 5 cm depth. The water sampling was performed slowly to avoid degassing the sample water. These samples were stored under water during transit to the laboratory whereupon a 20 ml headspace was created by transferring 20 ml of high purity helium to the sample syringe.

Plate 2.18 A water sample has been previously injected into the serum bottle. Here, the He is being injected, with the bubbles indicating He gas is still entering the serum bottle, once bubbling ceases the headspace will be at atmospheric pressure, the needle will be removed, and equilibration of the water and gas phases can occur

In all these previous examples, headspace equilibration was achieved by gently shaking the sample for >5 min so that the N_2O dissolved in the water sample equilibrates with the gas phase in the headspace. The addition of a microbial preservative prevents N_2O consumption and/or production of N_2O. This equilibration, as we will see below, is temperature-dependent. So, it is important to also measure the laboratory temperature at which the equilibration is performed. This means that the water in the sample needs to be at the same temperature as the laboratory, so if the sample has been in a fridge it must come to room temperature.

After shaking and equilibration, the headspace gas can then be sampled and analysed on a gas chromatograph (GC) to determine the N_2O concentration. Typically, this involves using a syringe equipped with a stopcock and a hypodermic needle to transfer a suitable volume of gas to a GC sample vial, for example, an Exetainer. In the first example above, the water was equilibrated in an Exetainer, and further gas transfer was not required. Even with a small volume of water, the GC analyses were able to be performed. Often, however, a gas sample is taken from the serum bottle or syringe headspace to transfer into a GC sample vial.

The actual GC analysis of a gas sample is discussed elsewhere in this publication. The output from the analysis will provide the concentration of N_2O in the equilibration vessel's headspace (C_g), typically with units of $\mu l\ l^{-1}$.

The next step is to convert this value to one with units of $\mu mol\ l^{-1}$. This is performed using the ideal gas equation (Eq. 2.15):

$$PV = nRT \tag{2.15}$$

where P is the pressure (atm.), V is the volume (l), n is moles, R is a constant (0.0821 l. atm. mol^{-1} K^{-1}), and T is the temperature (K). Rearranging equation (2.15) as follows gives moles: if the volume of gas is input as μl, then the output is μmol (Eq. 2.16):

$$n = \frac{PV}{RT} \tag{2.16}$$

What we wish to determine is how "saturated" in N_2O the water sample is, with respect to river water in equilibrium with the atmosphere. So first we must calculate the N_2O concentration in the water that would be in equilibrium with the atmosphere (C_{eq}; $\mu mol\ l^{-1}$): we must know the partial pressure of the N_2O in the atmosphere at the sampling site, the water temperature at the sampling site, and the atmospheric pressure at the sampling site. Then, C_{eq} is determined using (Eq. 2.17):

$$C_{eq} = \beta \times P_{gas} \times P_{barometric} \tag{2.17}$$

where P_{gas} is the partial pressure of the gas (atm atm^{-1}) and β is the Bunsen coefficient (mol l^{-1} atm^{-1}). Values for β are found in the publication of Weiss and Price (1980), who measured the solubility of N_2O in pure water and seawater over a temperature range of 0–40 °C and a salinity range of 0–40‰. $P_{barometric}$ is the atmospheric pressure during field sampling (atm).

For example, assume a river temperature of 12 °C and a salinity of zero, then β equals 3.741 $\times$ 10^{-2} mol l^{-1} atm^{-1}, atmospheric partial N_2O pressure of 0.32 $\times$ 10^{-6} atm atm^{-1}, and a barometric atmospheric pressure of 1 atm, and then C_{eq} equals 0.012 μmol l^{-1}.

The next step is to determine how saturated in N_2O the water sample (C_{water}) is with respect to the atmospheric equilibrium concentration (C_{eq}) just established. To do this, we must know the N_2O concentration in the headspace of the equilibration vessel (C_g). This is provided from the GC results, with units of μl l^{-1}. We also need to know the water temperature during equilibration so that the correct value of β can be used. The unit of β in Weiss and Price (1980) is moles l^{-1} atm^{-1}, so the value of β is first converted to units of moles l^{-1} atm^{-1} in order to be compatible with the units of C_g. Finally, we need to know the atmospheric pressure inside the equilibration vessel. This would be set at 1 atmosphere if equilibrating the headspace with atmospheric pressure (e.g. the Tedlar bag method discussed above), but if the atmospheric pressure when measured differs, then the measured value should be used. Then the following calculation is used (Eq. 2.18):

$$C_{water} = \beta \times C_g \times P_{barometric} \tag{2.18}$$

For example, if we have a laboratory equilibration temperature of 20 °C, and salinity remains at zero, then β equals 2.875 $\times$ 10^{-2} mol l^{-1} atm^{-1} (Weiss and Price 1980) which, using the ideal gas law equals 0.644 l l^{-1} atm^{-1}. If the concentration of N_2O in the equilibration headspace is 0.35 $\times$ 10^{-6} μmol l^{-1}, and the barometric atmospheric pressure is 1 atm, and then C_{water} equals 0.2255 μl l^{-1}, which, using the Ideal Gas Law, equates to 0.0101 μmol l^{-1}.

Then following a mass balance equation (Hamilton and Ostrom 2007; Beaulieu et al. 2012), the total amount of N_2O present in the equilibration vessel is calculated as follows (Eq. 2.19):

$$\left(C^{o}_{water}\right)(V_{water}) = (C_{water})(V_{water}) + \left(C_{gas}\right)\left(V_{gas}\right) \tag{2.19}$$

where V_{water} and V_{gas} are the volumes of water (l) and gas in the equilibration vessel, respectively, and C_{water} and C_{gas} are the respective concentrations (μmol l^{-1}) in the water and gas phases following equilibration. Then, since all the N_2O came from the original water sample, the concentration of N_2O in this water sample $\left(C^{o}_{water}\right)$, with units of μmol l^{-1}, is calculated as (Eq. 2.20)

$$\left(C^{o}_{water}\right) = \frac{(C_{water})(V_{water}) + \left(C_{gas}\right)\left(V_{gas}\right)}{(V_{water})} \tag{2.20}$$

Thus, using the value for C_{water} determined above, with equilibration volumes for water and gas of 0.12 l and 0.02 l, respectively, the value of C^{o}_{water} equals 6.93 $\times$ 10^{-2} μmol l^{-1}. The degree of saturation, or saturation ratio, in this example, is thus expressed as a ratio of C^{o}_{water} to C_{eq}. For the example above, this equates to 5.70.

2.9.2 Determining N_2O Fluxes from a Water Body

Gas fluxes from a water body may be calculated using the following equation (Beaulieu et al. 2012), where k is the gas transfer velocity with units of distance per unit time, e.g. cm h^{-1} (Eq. 2.21):

$$F = k\left(C^{o}_{\text{water}} - C_{\text{eq}}\right) \tag{2.21}$$

If the value of C_{eq} has not been determined, but atmospheric N_2O concentration and water temperature in situ are known, then the flux of N_2O off the water surface may also be calculated as follows (Reay et al. 2003) (Eq. 2.22):

$$F = k\left(C^{o}_{\text{water}} - \frac{C_{\text{a}}}{K'_{\text{h}}}\right) \tag{2.22}$$

where K'_{h} is Henry's law constant (Weiss and Price 1980) for N_2O, which is calculated as the equilibrium concentration in the gas phase (moles per unit volume air) divided by the equilibrium concentration in the water phase (moles per unit volume water), and C_{a} is the concentration of N_2O in air (moles per unit volume air). Theoretical background and commonly used quantities and units for K'_{h} can also be found in Sander (2015). To derive sensible flux units, the units in Eqs. (2.21) and (2.22) must be consistent with those used for the value of k. So, if k has units of cm h^{-1}, the units of concentration must be in mol cm^{-3} with the resulting flux given in mol cm^2 h^{-1}. Similarly, if k and concentrations are in units of m d^{-1} and mol m^{-3}, respectively, then the flux of N_2O has units of mol m^2 d^{-1}.

2.9.3 Determining Gas Transfer Velocity (K)

The value of k varies due to the turbulent mixing of the water body at the water–atmosphere interface. Turbulence within the water body may result from shear stresses on the bed of streams or rivers, or tidal currents and is thus dependent on the depth of the water body and the speed at which it flows. It may also result from the wind. Thus, it can be expected that small, shallow, and sheltered water bodies, such as agricultural drains, will have values of k more strongly influenced by water depth and/or speed (current), while deeper large open expanses of water such as estuaries, large rivers, lakes, and oceans will be more strongly influenced by wind, although tidal currents in estuaries and winds that oppose the direction of the water body current can all influence turbulence.

Methods to derive values for k include (i) measuring the N_2O flux using floating chambers (or fixed chamber) and back calculating k, (ii) using published relationships between k and wind speed to derive the value of k, (iii) tracer gas methods, and (iv) modelling of water body and micrometeorological parameters, or a combination of these approaches.

Measuring the N_2O flux with chambers that float on the water body surface is commonly performed. Besides the requirements of chamber design that optimise sensitivity of the measurement, discussed elsewhere for land-based gas measurement (see Sect. 2.2), the chamber must obviously float, but with minimal intrusion into the water body when floating, have a vent to allow equilibration of the headspace pressure during the placement of the chamber on the water surface, be insulated to prevent temperature changes inside the chamber, and have a low profile to reduce the influence of wind moving the chamber across the water surface at directions opposed to the current. Such an effect creates additional turbulence and can lead to an overestimation of N_2O fluxes (Clough et al. 2007b). Similarly, tethering the chamber so that it cannot drift in the current can also increase turbulence inside the chamber and lead to erroneous fluxes being derived (Hartman and Hammond 1984). If the water body is large, the chamber can be tethered to a boat, with a slackline, while both drift in the current (Beaulieu et al. 2012). Procedures for gas sampling, analysis, and calculation of fluxes are consistent with those for land-based chambers (Sect. 2.2): best practice involves taking several chamber headspace samples over time to enable the fitting of models to determine if the change in headspace N_2O concentration fits a linear or non-linear trend, with fluxes calculated accordingly (Beaulieu et al. 2012; Hutchinson and Mosier 1981) (Plate 2.19).

For large water bodies, the value of k can be estimated using its relationship to wind speed (de Wilde and de Bie 2000; Barnes and Owens 1998) using equations that derive k from wind speed (Liss and Merlivat 1986; Wanninkhof 1992; Clark et al. 1994; Wanninkhof 2014). This methodology is best suited to deeper large open expanses of water, such as estuaries, lakes, and oceans. However, caution is required, as water currents can still affect turbulence in large open water bodies as demonstrated by Beaulieu et al. (2012): the value of k in the Ohio River was attributed to not just wind but both current and wind speeds, with 46% of k at low wind speeds (0.5 m s^{-1}) resulting from water currents which were reduced to 11% at higher wind speed ($>2.0 \text{ m s}^{-1}$).

The principle of the **tracer gas method** is to release a soluble gas at a given point in the water body under study and to follow the dissipation of the injected gas over distance and time. Trace gas studies have been performed in lakes (Cole and Caraco 1998), oceans (Wanninkhof 1992), estuaries (Clark et al. 1994), and streams and drainage ditches (Harrison and Matson 2003; Premaratne et al. 2017). Trace gases should be absent from the study environment, and sulphur hexafluoride (SF_6), methyl chloride (CH_3Cl), propane (C_3H_8), krypton (Kr), and Freon-12 have been used. Studies by Jin et al. (2012) and Premaratne et al. (2017) provide examples of the technique using CH_3H_8, which is briefly described below. The tracer gas is first bubbled through a large container of water to saturate the water body with C_3H_8. Typically, this is a large carboy with the exact volume-dependent on the size of the water body under study and the ensuing release rate of the gas saturated water into the water body, assumed here to be a stream. Also dissolved in the water is an inert chemical tracer that can be used to estimate stream flow rate and dilution of the

a

b

Plate 2.19 Floating chambers being used to determine the N_2O flux from the LII river in Canterbury, New Zealand. Note the high number of replicates. Floating chambers are drifting freely in the current, but loosely tethered to a person walking alongside on the river bank who can retrieve the chambers for sampling (**a**), and fixed chamber being used to determine the GHGs flux from the river in China that can be raised or lowered with the fluctuation of water level (**b**)

added water. For example, this inert tracer might be bromide (Br^-) or a chloride salt (Cl^-). If the background conductivity of the water body is high, then a dye such as the rhodamine dye may be used. The water is injected at a known constant rate into the stream, using a peristaltic pump. Combining gas and chemical tracers in the one vessel prevents uneven mixing of the tracers in the stream following injection, which might occur if they are added from separate containers. However, for larger water bodies, this will be impractical, and the gas tracer will be delivered by gently releasing gas directly into the water body. For example, in the Iroquois River, Laursen and Seitzinger (2004) introduced tracer gas by staking a perforated garden hose to the stream bed perpendicular to the channel flow direction. Sampling stations are set up at intervals downstream of the tracer injection site, dependent on-stream speed and flow, with the first site several metres downstream to ensure adequate mixing of the injected water and to allow for any ebullition of tracer gas bubbles to be completed. Again, with larger rivers, these sampling stations will be at greater distances (e.g. kilometres) from the tracer injection site. At the sampling stations, water samples are taken to determine trace gas concentrations using the equilibration technique, and floating chambers may also be deployed to measure trace gas fluxes. In addition, water chemistry, micrometeorological parameters (e.g. wind speed, air temperature, and solar irradiance) and hydrological parameters are measured: depth, h (m); width, w (m); and velocity, U (m s^{-1}) enable the volumetric flow (Q, m^3 s^{-1}) to be calculated.

If floating chambers are deployed, an estimate of k can be made using the equilibrium concentrations of the trace gas and the measured flux, thus, for C_3H_8 (Eq. 2.23):

$$k_{C_3H_8} = F_{C_3H_8} / \left(C^o_{\text{water}} - C_{\text{eq}}\right) \tag{2.23}$$

Using this CH_3H_8 transfer velocity, the N_2O transfer velocity $\left(k_{N_2O}\right)$ can be estimated as follows (Eq. 2.24):

$$k_{N_2O} = \left(Sc_{N_2O} / Sc_{C_3H_8}\right)^{-n} \times k_{C_3H_8} \tag{2.24}$$

where Sc is the temperature-dependent Schmidt number for CH_3H_8 and N_2O, and n is the Schmidt exponent (equal to 0.5 for surfaces with waves and 0.67 for surfaces without waves (Jähne et al. 1987). A Schmidt number represents the dimensionless ratio of the kinematic viscosity to the diffusion coefficient of the gas of interest, which is temperature-sensitive. Thus, it is a common practice to normalise the value of k_{N_2O} to a value of $Sc = 600$, which is the Schmidt number for carbon dioxide at 20 °C in freshwater (Wanninkhof 1992) (Eq. 2.25):

$$k_{600-N_2O} = \left(600 / Sc_{N_2O(T)}\right)^{-n} \tag{2.25}$$

where $Sc_{N_2O(T)}$ is the Schmidt number for N_2O at temperature T, and k_{600-N_2O} is the standardised gas transfer velocity for N_2O at $Sc = 600$. The Schmidt number can be calculated from published equations for C_3H_8 and N_2O according to Wanninkhof

(1992), where

$$Sc_{C_3H8} = 1911.1 - 118.11 \times T + 3.4527 \times T2 - 0.04132 \times T3 \quad (2.26)$$

$$Sc_{N_2O} = 2055.6 - 137.11 \times T + 4.3173 \times T2 - 0.05435 \times T3 \quad (2.27)$$

Raymond et al. (2012) revisit these Schmidt numbers following a meta-analysis of 563 experiments.

The dissolved tracer gas (C_3H_8) concentration data for each sampling station can also be used to estimate *k*. As the tracer gas moves downstream some of this gas is lost across the air–water interface and this can be defined in terms of a first-order transfer rate (Clark et al. 1994; Laursen and Seitzinger 2004) as follows (Eq. 2.28):

$$F = hK\left(C_{\mathrm{m}} - C_{\mathrm{eq}}\right) \quad (2.28)$$

where C_m is the mean concentration of the tracer gas in the water, *h* is the mean water depth, and *K* is the gas exchange coefficient (Clark et al. 1994). By combining Eqs. (2.21) and (2.28), gas transfer velocity may be defined as (Eq. 2.29)

$$k = h\frac{\left(C_{\mathrm{m}} - C_{\mathrm{eq}}\right)}{\left(C^{o}_{\mathrm{water}} - C_{\mathrm{eq}}\right)}K \quad (2.29)$$

The slope of the ratio of tracer gas (C_3H_8): chemical tracer (Br^-) ratios versus time allows the gas exchange coefficient (*K*) to be determined (Wilcock 1988; Chapra and Wilcock 2000). Following corrections for background C_3H_8 or Br^-, $K_{C_3H_8}$ can be determined by plotting the ln(C_3H_8 concentration/Br concentration) versus water travel time (Jin et al. 2012). The assumption is made that the ratio of C_3H_8: Br during transit in the stream is altered solely as a result of C_3H_8 loss to the atmosphere. Then, the C_3H_8 exchange coefficient ($K_{C_3H_8}$) is used to derive the N_2O exchange coefficient (K_{N_2O}) as follows (Eq. 2.30):

$$K_{N_2O} = \left(Sc_{N_2O}/Sc_{C_3H_8}\right)^{-n} \times K_{C_3H_8} \quad (2.30)$$

The N_2O exchange coefficient (K_{N_2O}) is then multiplied by *h* to determine the air–water N_2O transfer velocity (K_{N_2O}) that can be used to calculate the N_2O flux (Eq. 2.28), where it is assumed that the water column is well mixed. The mean value of *h* may be determined as follows (Eq. 2.31):

$$h = \frac{Q}{wU} \quad (2.31)$$

2.9.4 Models for Determining N_2O Fluxes from Water Bodies

Numerous models exist in the literature for estimating *k*, and it is beyond the scope of this section to cover all of these. However, N_2O emissions from agricultural drains and first-order streams are recognised hotspots for N_2O emissions (Reay et al. 2003; Beaulieu et al. 2008; Outram and Hiscock 2012; Turner et al. 2015; Davis and David 2018), and these sites readily allow the collection of variables required for modelling. One model which has been used extensively is that of O'Connor and Dobbins (1958) who introduced the model to predict *k* based on water turbulence (Eq. 2.32):

$$k = \left(\frac{DU}{h}\right)^{0.5} \tag{2.32}$$

where U is the stream velocity (m s^{-1}), h is the average river depth (m), and D (m^2 s^{-1}) is the diffusion coefficient of the respective gas in water at the water's given temperature. This model has been widely used. More recently, in a meta-analysis of 563 direct gas tracer release experiments, Raymond et al. (2012) found that the gas transfer velocity scaled with the product of stream slope and velocity, and they reported further models for estimating *k* and Schmidt numbers in streams. Other models (e.g. Schwarzenbach et al. 1993) have also been used to determine *k* by partitioning the individual contributions of water and wind-driven turbulence effects (Beaulieu et al. 2012; Hamma-Aziz et al. 2016).

2.9.5 Other Factors to Consider

Besides measurements of dissolved gases in the water body, consideration should also be given to reporting other parameters or features at the site that will enable interpretation of the results, extrapolation of the results, modelling, and comparison with other studies. Water bodies such as drains and streams are unlikely to have constant chemistry throughout the day, and diel variation is commonly observed in temperature, dissolved oxygen (DO), and pH (Harrison et al. 2005; Clough et al. 2007b). Similarly, seasonal changes also occur. Such changes have implications for N cycling, and thus the amount of N_2O dissolved in the water. A key driver of DO, besides respiration, is the rate of photosynthesis. This is affected by how clear the water is; so measurements of turbidity and sunshine hours should also be reported. Similarly, wind speed and direction are other critical parameters to gather. Dissolved organic C and inorganic-N species, both known to affect N_2O production or consumption transformations, should also be reported.

2.9.6 *Determining EF5*

Once the N_2O flux has been established, the EF_5 value can be established. First, the N_2O flux is converted to an N_2O-N flux. This will have units of mass of N per unit area per unit time. To establish the EF_5 value, the amount of NO_3^- moving through a given area per unit time is then calculated, which then enables the EF_5 ratio to be determined. For example, if a 1-m-wide stream is 1 m deep and has a speed of 1 m s^{-1}, then the stream flows through one square metre every second. If the NO_3^--N concentration in the water is 0.2 mg l^{-1}, then 0.2 g of NO_3^--N flows through underneath the 1 m^2 surface every second. If an N_2O-N flux is 0.5 μg m^{-2} s^{-1}, then the ratio is 0.5×10^{-6}: 0.2, or 2.5×10^{-6}. The EF_5 value is then determined based on the ratio of N_2O-N: NO_3^--N or on the basis of assumed N_2O emissions using the degree of saturation to predict a flux.

Units of N_2O-N flux and NO_3-N can be adjusted for the time period concerned. Obviously, stream chemistry and N_2O fluxes are dynamic, and so it is expected that EF_5 could also vary. Thus, modelling of these parameters at varying scales is useful for predicting EF_5 over different periods. For example, Reay et al. (2003) successfully modelled the N_2O flux along a drain with N_2O emissions calculated on a "per metre of stream" basis, with predicted N_2O losses from the preceding 1 m stretch being subtracted from the dissolved N_2O concentration for each 1 m section of the drain.

2.10 Indirect GHG Emissions–Ammonia Emissions

Ammonia (NH_3) itself has no direct greenhouse effect. It is a gas with relatively low residence time in the atmosphere (2–10 days) compared to some GHG, such as CO_2 (3–4 years), CH_4 (9 years), and N_2O (150 years) (Hobbs 2000). However, after NH_3 enters the atmosphere and reacts with acids forming salts, it returns to the earth's surface and causes N_2O emissions similar to a fertiliser-N application. When the soil is subjected to conditions near to optimal for urease activity (e.g. pH close to neutrality, soil moisture near the field capacity, temperature > 30 °C), the N losses through NH_3 volatilisation from urea-based fertilisers applied on soil surface can be as high as 50% (Rochette et al. 2013; Martins et al. 2017). Therefore, the measurement of NH_3 emission is important to estimate indirect N_2O emissions derived from soil amendments, such as urea-based fertilisers, green manures, animal excreta, or ammonium-based fertilisers in different soils. A default emission factor defined by IPCC, known as EF_4, can be applied for the estimation of indirect N_2O emissions derived from volatilisation of NH_3 and other nitrogen oxides (NO_x) (de Klein et al. 2007). The mean value of EF_4, considering the N volatilisation and consequent re-deposition, is 0.01 kg N_2O-N per kg N volatilised as $NH_3 + NO_x$ with an uncertainty ranging from 0.002 to 0.05 (de Klein et al. 2007). Management options that reduce NH_3 volatilisation from soils are considered mitigating practices because they reduce indirect N_2O emissions (IPCC 2014; Lam et al. 2017).

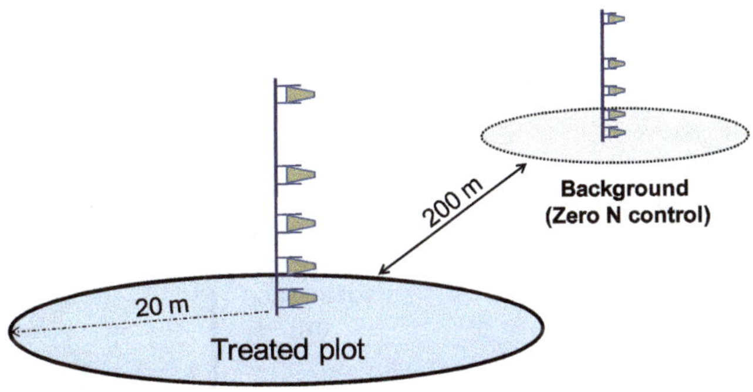

Plate 2.20 Schematic representation of circular plots for measurements of NH_3 volatilisation using mass balance integrated horizontal flux (IHF) method. A mast with NH_3 samplers installed at different heights is placed in the centre the plot. A mast placed in an unfertilised area is used to quantify the background soil NH_3-N emissions (zero N control)

There is a considerable range of non-isotopic methods that can be used to quantify the NH_3 volatilisation from soils. Chamber-based methods are the most common. They are generally inexpensive compared to other more sophisticated methods, such as micrometeorological, and are easier to install and handle. On the other hand, the micrometeorological method has the advantage of being more representative for larger areas and therefore are more representative of the field scale. The classical micrometeorological method to measure NH_3-N emissions is the mass balance integrated horizontal flux (IHF) method (Leuning et al. 1985). This method is based on the assumption that "vertical NH_3-N flux from a fertilised area of limited upwind extent" is equal to "the integrated horizontal NH_3-N flux at known downwind distance" (Wilson et al. 1983). This method uses NH_3 samplers attached to masts placed in the centre of a circular field plot (Plate 2.20).

The NH_3 samplers with mounting pivots and fins are installed in different heights of the mast above the soil surface (Leuning et al. 1985; Misselbrook et al. 2005; MSU 2019). Therefore, the samplers are pointed into the wind quantifying the horizontal NH_3 fluxes at different levels (Plate 2.21), which are used in the final calculation of the vertical NH_3-N flux (VF), in μg N m^{-2} s^{-1}, as shown in the following equation (Eq. 2.33):

$$VF = \frac{1}{r}\left[\int_0^z \left(\frac{MN}{EAS \times t}\right)_{\text{treated}} \times dz - \int_0^z \left(\frac{MN}{EAS \times t}\right)_{\text{background}} \times dz\right] \quad (2.33)$$

where r is the plot radius (fetch length), MN is the mass of N captured in the sampler, EAS is the effective cross-sectional area of the sampler, t is the sampling period, and z is the height of the sampler (MSU 2019).

The use of relatively sophisticated techniques, such as the mass balance IHF method described above, is often not possible due to limitations of financial resources or lack of skilled technicians. Thus, chamber-based methods are often employed instead, because of their viability and ease to work with. Part of the procedures

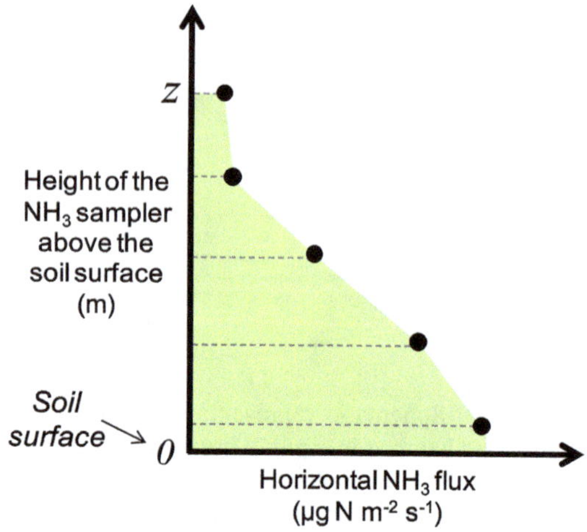

Plate 2.21 Example plot showing the integration of NH_3 fluxes based on the height of the sampler above the soil surface (*y*-axis) versus horizontal NH_3 flux measured at each height (*x*-axis)

described above for the chambers used for measurement of N_2O fluxes is also valid for the chambers used to quantify NH_3 volatisation (Tables 2.6 and 2.7). A difference between these two types of chambers is the way how total N losses over the period of measurements are estimated. The usual continuous NH_3 captured by acid traps during the entire period of measurement provides direct results of time-integrated N losses, i.e. there is no collection of temporally separated samples for a final calculation of fluxes as usually performed for N_2O measurements.

Chamber-based methods have been used since the mid-twentieth century to test the effectiveness of N management practices aiming at a reduction of NH_3-N losses (Volk 1959). An example of a chamber design used to collect volatilised NH_3 is the semi-open collector described by Nômmik (1973). A concise schematic representation of this chamber with the foam discs with an acid solution used for NH_3 trapping is shown in Plate 2.22.

Another common type of chamber is based on the use of a low vacuum system to collect NH_3, which consists of tubes connecting the chamber to a vacuum pump and a small flask containing an acid solution used as NH_3 trap (Kissel et al. 1977).

To warrant the precision and accuracy of measurements, i.e. reducing the uncertainty for extrapolation of NH_3-N losses to larger areas, some procedures of installation and handling the chambers under field conditions should be warranted. For instance, the installation and replacement of the acid traps should be performed in a short period to avoid a time-dependent bias. Methods employing soil collars should be preinstalled at least 24 h before the beginning of measurements to avoid undesirable effects of soil physical disturbance on the NH_3 volatilisation process. Moreover, the efficiency of NH_3 captured by some type of chambers is usually not 100%, demanding the use of correction factors for the final calculation of NH_3-N losses. These correction factors are usually predefined in previous studies in which calibration is performed using reference methods, such as the ^{15}N-balance method (Araujo et al. 2009; Jantalia et al. 2012; Martins et al. 2021b).

The advantages and disadvantages of usual methods for NH_3 volatilisation using non-isotopic techniques are shown in Table 2.6.

Table 2.6 Some methods used to quantify the NH_3 volatilisation from soils

Method	Reference	Representativeness[a]	Area requirement	Replicability	Requirement of high-skilled technicians	Portability	Start-up cost
Micrometeorological	Misselbrook et al. (2005)	High	High	Very low	High	Low	High
Wind tunnels	Lockyer (1984)	Medium	Medium	Low	High	Low	High
Closed chambers	Ernst and Massey (1960)	Low	Low	Medium	Medium	Low	Medium
Semi-open static	Nômmik (1973), Marshall and Debell (1980)	Low	Low	High	Low	High	Low
Plexiglass chamber	Kissel et al. (1977)	Low	Low	Medium	Medium	High	Medium
Open chamber	Alves et al. (2011)	Low	Low	High	Low	High	Low
PVC "Sewer-hatch" chamber	Saggar et al. (2004)	Low	Low	Medium	Medium	High	Medium
"Plastic bottle" open chamber	Araujo et al. (2009)	Low	Low	High	Low	High	Very low

[a]Degree of certainty for extrapolation to larger areas

Table 2.7 Example of a spreadsheet for the calculation of NH_3-N losses from the data of N concentration in the solution extracted from the acid traps (foams) used in the simple design of open chamber ("plastic bottle" chamber)

Time after fertiliser application (days)	N concentration in the extraction solution (mg l^{-1})	Volume of the extraction solution (ml)	NH_3 captured in the acid trap (foam) (mg N)	"Outside" NH_3 (blank chamber) (mg N)	Correction factor[a]	Volatilised NH_3 (mg N/chamber)	Area enclosed by the chamber (m^2)	Volatilised NH_3 (kg N ha^{-1})	N rate (kg N ha^{-1})	NH_3-N volatilisation (fraction of the fertiliser-N)[b]
Control (Zero N)										
0–3	1.8	50	0.09	0.01	1.7	0.14	0.007854	0.17	–	–
3–7	2.0	50	0.10	0.01	1.7	0.15	0.007854	0.19	–	–
7–12	2.4	50	0.12	0.02	1.7	0.17	0.007854	0.22	–	–
12–17	2.3	50	0.12	0.02	1.7	0.16	0.007854	0.21	–	–
Sum			*0.43*	*0.06*		*0.62*		*0.79*	–	–
N-fertilised treatment (150 kg N ha^{-1})										
0–3	306.5	50	15.33	0.01	1.7	26.04	0.007854	33.15	150	***22.0%***
3–7	166.2	50	8.31	0.01	1.7	14.11	0.007854	17.97	150	***11.8%***
7–12	33.4	50	1.67	0.02	1.7	2.81	0.007854	3.57	150	***2.2%***
12–17	2.4	50	0.12	0.02	1.7	0.17	0.007854	0.22	150	***0.0%***
Sum			*25.43*	*0.06*		*43.12*		*54.90*		***36.0%***

[a] See Araújo et al. (2009) and Martins et al. (2021b); [b]assuming that no significant priming effect occurred, i.e. no significant interaction of fertiliser-N and soil N

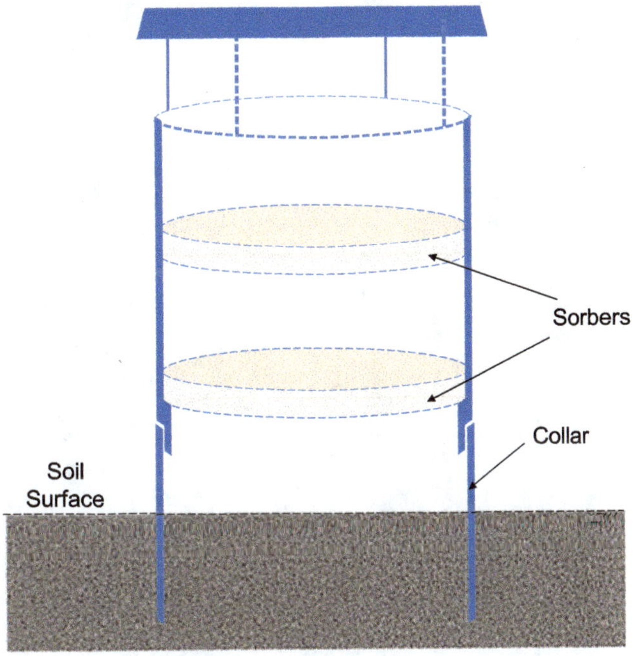

Plate 2.22 Schematic representation of the Nômmik's chamber for NH_3 measurement

2.10.1 A Simple Low-Cost Chamber to Quantify NH_3 Volatilisation

Among all methods for NH_3 volatilisation measurements, a low-cost chamber has been used successfully under field conditions (Araújo et al. 2009; Nichols et al. 2018). This chamber costs less than one US dollar and can be manufactured in any region of the world, including those regions with limited financial resources for scientific research. Studies conducted under field conditions have been showing that this chamber can be used with sufficient accuracy and precision in experiments aiming at testing management practices for NH_3-N loss abatement. Details of the design of this chamber were presented by Araújo et al. (2009), Jantalia et al. (2012), Nichols et al. (2018), and Martins et al. (2021a,b). Briefly, this chamber is made by cutting the bottom of a plastic soda bottle. After cutting, the bottom is attached to the top of the bottle finish, serving as protection against rainwater. A vertically hung foam strip (250 mm long, 25 mm wide, and 3 mm thick) pre-soaked in a 10 ml solution of sulphuric acid (1 mol/L) plus glycerol (2%, v/v) is placed inside the chamber to capture NH_3. A small plastic pot is used inside the chamber to retain the remaining acid solution not absorbed by the foam. A basket made of wire and hung from the bottle finish is used as a support for the plastic pot and the foam inside the chamber. The steps to manufacture and anchor the chamber in the soil are shown in Plates 2.23 and 2.24.

Metallic supports can be used to anchor the chamber (Plate 2.24). The amounts of "outside NH_3", i.e. NH_3 not directly derived from soil, are usually not significant

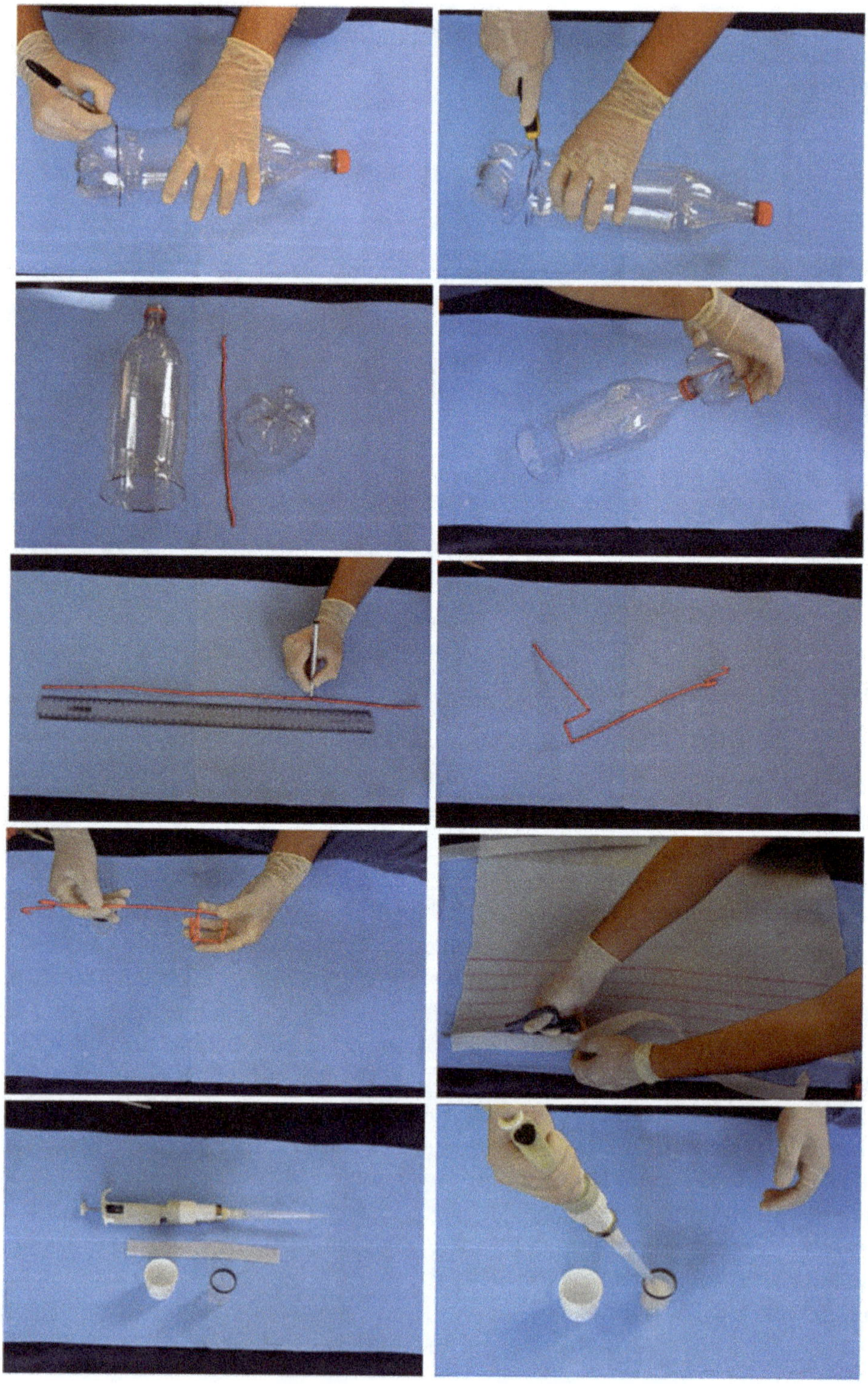

Plate 2.23 Steps to manufacture a low-cost chamber to measure ammonia volatilisation under field conditions (Araujo et al. 2009; Martins et al. 2021b)

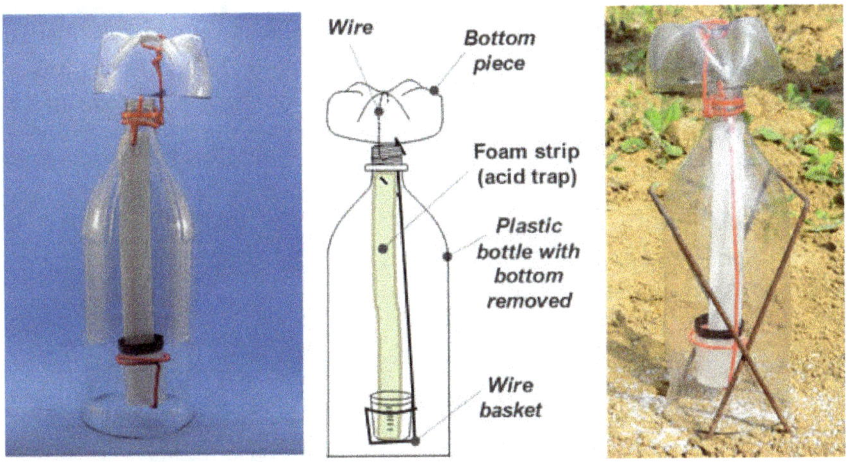

Plate 2.24 A simple design of an open chamber after the assemblage of the parts. A correction factor of 1.7 is used to convert the amount of NH_3 trapped in the total volatilised NH_3 (Araujo et al. 2009; Martins et al. 2021b)

in this type of chamber, even though the non-significance of the amount of "outside NH_3" can be checked by installing some blank chambers in the area of NH_3 measurements (e.g. one "blank" chamber per experimental block). Some eventual NH_3 detected in these "blank" chambers can be subtracted from the total NH_3 trapped in the chambers installed in treated and control plots. This "blank" chamber is the same described above, but the soil surface is covered using an impermeable barrier (e.g. an inert and impermeable plastic sheet).

Considering that the area circumscribed by chambers (<0.1 m^2) is usually much smaller than the final area used for extrapolation of the NH_3-N losses (e.g. hectare), the minimisation of the inherent uncertainty depends on the use of some strategies. For instance, one way of reducing uncertainty is the employment of multiple chambers per field plot. This strategy can increase the capacity of the detection of significant NH_3-N loss abatement by adequate N management practices. In a field plot experiment, Martins et al. (2021a) estimated that, by using five chambers per plot (31 m^2), the margin of error was less than 15% of the real NH_3-N loss. To optimise the fieldwork and the laboratory analyses, the use of multiple chambers per plot can be employed only in the critical periods of emissions, i.e. focusing sampling in the period of a high probability of occurrence of significant NH_3 fluxes and associated errors. For example, after easy hydrolysable N sources are deposited on the soil surface (e.g. fertilisers, excreta), the bulk of NH_3-N losses usually occurs in the first 2 weeks after N deposition. Therefore, using multiple chambers in this period has a high probability of significantly reducing errors associated with the use of chambers. It is worth noting that the conventional use of multiple plots per treatment (replicates) in field experiments intrinsically improves the precision of NH_3-N loss estimates, even when using a single chamber per plot.

In addition, a significant part of the errors in measurements of NH_3 volatilisation can be reduced if it is ensured that the amounts of fertiliser applied in the relatively small area circumscribed by the chamber are equivalent to the N rate tested in the area of the plot, in kg ha^{-1}. The use of microplots inside the main field plots reduces this kind of uncertainty. An example calculation to define the amount of fertiliser applied in a microplot with a chamber is presented below.

Example calculation:

- N rate: 150 kg N ha^{-1}
- Fertiliser type: Urea
- Application method: Broadcast
- N content: 46%
- Size of a field plot: 5 m × 5 m (25 m^2)
- Size of the microplot: 1.0 m × 0.9 m (0.9 m^2)

The amount of urea-N required for the entire plot is (Eq. 2.34)

$$\begin{aligned}
&\frac{10{,}000\,\text{m}^2}{25\,\text{m}^2} = \frac{150{,}000\,\text{g N}}{\text{Amount of N in the plot}}\\
&\text{Amount of N in the plot} = 15\,\text{g N} \times 25\\
&\text{Amount of N in the plot} = 375.0\,\text{g N}\\
&\frac{46\,\text{g N}}{100\,\text{g urea}} = \frac{375.0\,\text{g N}}{\text{Amount of urea in the plot}}\\
&\text{Amount of urea in the plot} = \frac{375.0 \times 100\,\text{g urea}}{46}\\
&\text{Amount of urea in the plot} = 815.2\,\text{g urea}
\end{aligned} \tag{2.34}$$

The amount of urea-N required for the microplot is

$$\begin{aligned}
&\frac{10{,}000\,\text{m}^2}{0.9\,\text{m}^2} = \frac{150{,}000\,\text{g N}}{\text{Amount of N in the microplot}}\\
&\text{Amount of N in the microplot} = 15\,\text{g N} \times 0.9\\
&\text{Amount of N in the microplot} = 13.50\,\text{g N}\\
&\frac{46\,\text{g N}}{100\,\text{g urea}} = \frac{13.50\,\text{g N}}{\text{Amount of urea in the microplot}}\\
&\text{Amount of urea in the microplot} = \frac{13.50 \times 100\,\text{g urea}}{46}\\
&\text{Amount of urea in the microplot} = 29.35\,\text{g urea}\\
&\text{Amount of urea in the remaining area of the plot} = 815.20 - 29.35\\
&\text{Amount of urea in the remaining area of the plot} = 785.85\,\text{g urea}
\end{aligned}$$

Another way to avoid a bias due to a slight unevenness in the fertiliser distribution in the plot is the application of a precise amount of fertiliser in the area circumscribed by the chamber. This procedure ensures a proper extrapolation of the NH_3–N loss captured by the chamber to the total NH_3–N loss per hectare. An example calculation of the amount of urea using the data of the calculations above, considering the "plastic bottle" chamber method, is presented as follows (Eq. 2.35):

- Diameter of the chamber: 100 mm

$$
\begin{aligned}
&\text{Area circumscribed by the chamber} = \pi \times \frac{\text{diameter of the chamber}^2}{4} \\
&\text{Area circumscribed by the chamber} = 3.14159 \times \frac{\left(100 \times 10^{-3}\right)^2}{4} \\
&\text{Area circumscribed by the chamber} = 0.007854\,\text{m}^2 \\
&\frac{10,000\,m^2}{0.007854\,\text{m}^2} = \frac{150,000\,g\,N}{\text{Amount of N in the area of the chamber}} \\
&\text{Amount of N applied in the area of the chamber} = 15\,\text{g N} \times 0.007854 \\
&\text{Amount of N applied in the area of the chamber} = 0.11781\,\text{g N} = 117.81\,\text{mg N} \\
&\frac{46\,\text{mg N}}{100\,\text{mg urea}} = \frac{117.81\,\text{mg N}}{\text{Amount of urea in the area of the chamber}} \\
&\text{Amount of urea in the area of the chamber} = \frac{117.81 \times 100\,\text{mg urea}}{46} \\
&\text{Amount of urea in the area of the chamber} = 256.11\,\text{mg urea}
\end{aligned}
\tag{2.35}
$$

The foams capturing NH_3 in the "plastic bottle" chamber should be periodically replaced in the course of the measurement period. The frequency of replacement depends on the purpose of the NH_3-N loss quantification. When the measurements are performed to understand the temporal dynamics of NH_3-N losses, more frequent replacements of the foams should be adopted (e.g. every 2 days or on a daily basis). By contrast, when the purpose is to assess only the total amount of NH_3-N loss, the intervals of replacements of the foams can be increased, but not exceeding 5 days (Araújo et al. 2009). Considering the amount of sulphuric acid in the solution, each foam has the NH_3 trapping capacity equivalent of more than 300 kg N ha^{-1}. Considering also that the foams are replaced several times in the course of the measurement period, the saturation or reduction of trapping efficiency is very unlikely for the usual NH_3 fluxes measured in soils.

When an easily hydrolysable N source is applied, such as urea-based fertilisers in agricultural soils or urine in pasture areas, the significant fluxes of NH_3 are usually very intense in the first days and usually return to the background NH_3 emissions in the first two weeks (Martins et al. 2021a). On the other hand, the application of more recalcitrant N sources causes prolonged NH_3 fluxes above the background emissions, demanding measurements for longer periods. For instance, the measurements of NH_3-N losses derived from plant residues with high C/N ratio usually should be performed for several weeks or months after the amendment. If promptly performed after collection in the field, the analysis of N in the foams can indicate the need for further NH_3-N measurements.

Plate 2.25 Orbital shaker with the Erlenmeyer flasks used for the extraction of NH_4^+ from the foams coming from the chambers

The plastic pots with the collected foam strips should be transported to the laboratory for the extraction of N trapped in the foams (NH_4^+ form). To extract the NH_4^+ from the foams, the content of the plastic pots (foam + solution) is transferred to an Erlenmeyer flask, and then deionised water (40 ml) is added for the extraction of the NH_4^+. Alternatively, KCl solutions can also be used (Jantalia et al. 2012), but some tests indicate that pure distilled water is able to extract the total quantity of N trapped in the foams. After that, the Erlenmeyer flasks are shaken on an orbital shaker, usually for less than 30 min (Plate 2.25).

After the NH_4^+ is extracted from the foams, the N concentration of the extraction solution can be analysed using one of the many methods available (Bremner 1965; Keeney and Nelson 1982). Some examples of usual methods for the determination of NH_4^+-N in solution are continuous flow analysis, steam distillation, and colorimetric techniques. Continuous flow system, such as flow injection analysis (FIA), is quite adequate because they are usually rapid and very sensitive, which are significant advantages when working with a high number of chambers and with many replacements of foams during the period of measurement of NH_3 volatilisation. Standard solutions of ammonium salt (e.g. sulphate) should be analysed along with the extraction solutions. An example spreadsheet for illustrating the calculation for the conversion of the N concentration in the extraction solutions to the total NH_3-N loss per hectare is detailed in Table 2.7. In the example shown below, considering a total NH_3-N loss of 54.90 kg N ha^{-1} in the N-fertilised treatment and subtracting 0.79 kg N ha^{-1} (Zero N treatment, i.e. "*background*" emission), we estimated that 36% of the applied N was lost as volatilised NH_3.

2.11 Gas Production Processes in Terrestrial Ecosystems

The total production of a certain gas from the soil is often the sum of several soil fluxes that can occur in diverse microsites of the soil (e.g. aerobic and anaerobic microsites), for instance, soil microbial processes for N_2O production, including nitrification, denitrification, co-denitrification, and organic N oxidation to mineral N (Zaman et al. 2012; Butterbach-Bahl et al. 2013; Müller et al. 2014; Smith 2017). The microbial processes of N_2O and non-greenhouse gas N_2 production in soils, sediments, and groundwater across the landscape may occur simultaneously and depend on the physical (moisture contents or O_2 level and permeability) and chemical conditions [N form (i.e. organic N, NH_4^+-N, NO_2^--N, and NO_3^--N), pH and organic C contents] (Zaman et al. 2012). Such complexity of gas production processes in soil makes the precise measurements of GHG flux, especially across different agroecosystems, very difficult and challenging. The identification of process and pathway-specific GHG fluxes can be tackled by suitable stable isotope-tracing techniques which are described in detail in Chap. 7.

References

Alves AC, Oliveira PP, Herling VR, Trivelin PC, Luz PH, Alves TC, Rochetti RC, Barioni Jr W (2011) New methods to quantify NH_3 volatilization from fertilized surface soil with urea. Braz J Soil Sci 35:133–140

Araujo ES, Marsola T, Miyazawa M, Soares LHB, Urquiaga S, Boddey RM, Alves BJR (2009) Calibration of a semi-opened static chamber for the quantification of volatilized ammonia from soil (in Portuguese, with English abstract). Pesquisa Agropecuaria Brasileira 44:769–776

Arias-Navarro C, Díaz-Pinés E, Kiese R, Rosenstock TS, Rufino MC, Stern D, Neufeldt H, Verchot LV, Butterbach-Bahl K (2013) Gas pooling: a sampling technique to overcome spatial heterogeneity of soil carbon dioxide and nitrous oxide fluxes. Soil Biol Biochem 67:20–23

Bakhirkin YA, Kosterev AA, Roller C, Curl RF, Tittel FK (2004) Mid-infrared quantum cascade laser based off-axis integrated cavity output spectroscopy for biogenic nitric oxide detection. Appl Opt 43:2257–2266

Barnes J, Owens NJP (1998) Denitrification and nitrous oxide concentrations in the Humber Estary, UK and Adjacent Coastal Zones

Beaulieu JJ, Arango CP, Hamilton SK, Tank JL (2008) The production and emission of nitrous oxide from headwater streams in the Midwestern United States. Global Change Biol 14:878–894

Beaulieu JJ, Shuster WD, Rebholz JA (2012) Controls on gas transfer velocities in a large river. J Geophys Res 117:G02007

Boegoezi T, Popp J, Frosch T (2015) Fiber-enhanced Raman multi-gas spectroscopy—what is the potential of its application to breath analysis? Bioanalysis 7(3):281–284

Brannon EQ, Moseman-Valtierra SM, Rella CW, Martin RM, Chen X, Tang J (2016) Evaluation of laser-based spectrometers for greenhouse gas flux measurements in coastal marshes. Limnol Oceanogr Methods 14:466–476

Bremner JM (1965) Total nitrogen. In: Black CA, Evans DD, White JL, Ensminger LE, Clark FE (eds) Methods of soil analysis. Part 2. Chemical and microbiological properties, ASA, Madison, Wisconsin, pp 1149–1176

Butterbach-Bahl K, Gasche R, Willibald G, Papen H (2002) Exchange of N-gases at the Höglwald forest—a summary. Plant Soil 240:117–123

Butterbach-Bahl K, Baggs E, Dannenmann M, Kiese R, Zechmeister-Boltenstern S (2013) Nitrous oxide emissions from soils: how well do we understand the processes and their controls? Phil T R Soc B Biol Sci 368:1–13

Chapra SC, Wilcock RJ (2000) Transient storage and gas transfer in lowland stream. J Environ Eng 126:706–712

Charteris AF, Chadwick DR, Thorman RE, Vallejo A, De Klein CAM, Rochette P, Cárdenas LM, (2020) Global Research Alliance N_2O chamber methodology guidelines: Recommendations for deployment and accounting for sources of variability. J Environ Qual 49:1092–1109

Chirinda N, Plauborg F, Heckrath G, Elsgaard L, Thomsen IK, Olesen JE (2014) Carbon dioxide in arable soil profiles: a comparison of automated and manual measuring systems. Commun Soil Sci Plant Anal 45(9):1278–1291

Christiansen JR, Outhwaite J (2015) Comparison of CO_2, CH_4 and N_2O soil-atmosphere exchange measured in static chambers with cavity ring-down spectroscopy and gas chromatography. Agric Forest Meteorol 211–212:48–57

Christiansen JR, Korhonen JFJ, Juszczak R et al. (2011) Assessing the effects of chamber placement, manual sampling and headspace mixing on CH_4 fluxes in a laboratory experiment. Plant Soil 343, 171–185

Clark JF, Wanninkhof R, Schlosser P, Simpson HJ (1994) Gas exchange rates in the tidal Hudson river using a dual tracer technique. Tellus 46B:274–285

Clough TJ, Addy K, Kellogg DQ, Nowicki BL, Gold AJ, Groffman PM (2007a) Dynamics of nitrous oxide in groundwater at the aquatic-terrestrial interface. Glob Change Biol 13:1528–1537

Clough TJ, Buckthought LE, Kelliher FM, Sherlock RR (2007b) Diurnal fluctuations of dissolved nitrous oxide (N_2O) concentrations and estimates of N_2O emissions from a spring-fed river: implications for IPCC methodology. Glob Change Biol 13:1016–1027

Clough TJ, Rochette P, Thomas SM, Pihlatie M, Christiansen JR, Thorman RE (2020) Global Research Alliance N_2O chamber methodology guidelines: Design considerations. J Environ Qual 49:1081–1091

Cole JJ, Caraco NF (1998) Atmospheric exchange of carbon dioxide in a low-wind oligotrophic lake measured by the addition of SF_6. Limnol Oceanogr 43:647–656

Davis MP, David MB (2018) nitrous oxide fluxes from agricultural streams in East-Central Illinois. Water Air Soil Pollut 229–354

De Klein CAM, Harvey MJ (2012) Nitrous oxide chamber methodology guidelines. Global Research Alliance on agricultural greenhouse gases. Ministry for Primary Industries, New Zealand, p 146

De Klein CD, Novoa RSA, Ogle S, Smith KA, Rochette P, Wirth TC, McConkey BG, Mosier A, Rypdal K, Walsh M, Williams SA (2007) N_2O emissions from managed soils, and CO_2 emissions from lime and urea application. In: Eggleston S, Buendia L, Miwa K, Ngara T, Tanabe K (eds) 2006 IPCC guidelines for National Greenhouse Gas Inventories. IGES, Hayama, Japan

De Klein CAM, Alfaro MA, Giltrap D, Topp CFE, Simon PL, Noble ADL, Van der Weerden TJ (2020a) Global Research Alliance N_2O chamber methodology guidelines: Statistical considerations, emission factor calculation, and data reporting. J Environ Qual 49:1156–1167

De Klein CAM, Harvey MJ, Clough TJ, Petersen SO, Chadwick DR, Venterea RT (2020b) Global Research Alliance N_2O chamber methodology guidelines: Introduction, with health and safety considerations. J Environ Qual 49:1073–1080

De Wilde HPJ, De Bie MJM (2000) Nitrous oxide in the Schelde estuary: production by nitrification and emission to the atmosphere. Mar Chem 69:203–216

Denmead OT (2008) Approaches to measuring fluxes of methane and nitrous oxide between landscapes and the atmosphere. Plant Soil 309:5–24

Diel J, Vogel HJ, Schlüter S (2019) Impact of wetting and drying cycles on soil structure dynamics. Geoderma 345:63–71

Dorich CD, De Rosa D, Barton L, Grace P, Rowlings D, De Antoni Migliorati M, Wagner-Riddle C, Key C, Wang D, Fehr B, Conant RT (2020) Global Research Alliance N_2O chamber methodology guidelines: Guidelines for gap-filling missing measurements. J Environ Qual 49:1186–1202

Dörr H, Münnich KO (1987) Annual variation in soil respiration in selected areas of the temperate zone. Tellus 39B:114–121

Ernst JW, Massey HF (1960) The effects of several factors on volatilization of ammonia formed from urea in the soil. Soil Sci Soc Am J 24:87–90

Frosch T, Keiner R, Michalzik B, Fischer B, Popp J (2013a) Investigation of gas exchange processes in peat bog ecosystems by means of innovative Raman gas spectroscopy. Anal Chem 85(3):1295–1299

Frosch T, Yan D, Popp J (2013b) Ultrasensitive fiber enhanced UV resonance Raman sensing of drugs. Anal Chem 85(13):6264–6271

Giltrap D, Yeluripati J, Smith P, Fitton N, Smith W, Grant B, Dorich CD, Deng J, Topp CFE, Abdalla M, Liáng LL (2020) Global Research Alliance N_2O chamber methodology guidelines: Summary of modelling approaches. J Environ Qual 49:1168–1185

Griffith DWT, Deutscher NM, Caldow C, Kettlewell G, Riggenbach M, Hammer S (2012) A Fourier transform infrared trace gas and isotope analyser for atmospheric applications. Atmos Meas Tech 5:2481–2498

Hamilton SK, Ostrom NE (2007) Measurement of the stable isotope ratio of dissolved N_2 in ^{15}N tracer experiments. Limnol Oceanogr 43:647–656

Hamma-Aziz Z, Hiscock KM, Cooper RJ (2016) Indirect nitrous oxide emission factors for agricultural field drains and headwater streams. Environ Sci Technol 51(3):01–307

Hanf S, Keiner R, Yan D, Popp J, Frosch T (2014) Fiber-enhanced Raman multigas spectroscopy: a versatile tool for environmental gas sensing and breath analysis. Anal Chem 86(11):5278–5285

Hanf S, Bögözi T, Keiner R, Frosch T, Popp J (2015a) Fast and highly sensitive fiber-enhanced Raman spectroscopic monitoring of molecular H_2 and CH_4 for point-of-care diagnosis of malabsorption disorders in exhaled human breath. Anal Chem 87(2):982–988

Hanf S, Fischer S, Hartmann H, Keiner R, Trumbore S, Popp J, Frosch T (2015b) Online investigation of respiratory quotients in *Pinus sylvestris* and *Picea abies* during drought and shading by means of cavity-enhanced Raman multi-gas spectrometry. Analyst 140(13):4473–4481

Harrison J, Matson P (2003) Patterns and controls of nitrous oxide emissions from waters draining a subtropical agricultural valley—art. no. 1080. Global Biogeochem Cycles 17:1080–1080

Harrison JA, Matson PA, Fendorf SE (2005) Effects of a diel oxygen cycle on nitrogen transformations and greenhouse gas emissions in a eutrophied subtropical stream. Aqua Sci 67:308–315

Hartman B, Hammon DE (1984) Gas exchange rates across the sediment-water and air-water interfaces in South San Francisco Bay. J Geophys Res 89:3593–3603

Hartung A, Kobelke J, Schwuchow A, Bierlich J, Popp J, Schmidt MA, Frosch T (2015) Low-loss single-mode guidance in large-core antiresonant hollow-core fibers. Opt Lett 40(14):3432–3435

Harvey MJ, Sperlich P, Clough TJ, Kelliher FM, McGeough KL, Martin RJ, Moss R (2020) Global Research Alliance N_2O chamber methodology guidelines: Recommendations for air sample collection, storage, and analysis. J Environ Qual 49:1110–1125

Hensen A, Skiba U, Famulari D (2013) Low cost and state of the art methods to measure nitrous oxide emissions. Environ Res Lett 8:025022

Hippler M (2015) Cavity-enhanced Raman spectroscopy of natural gas with optical feedback cw-diode lasers. Anal Chem 87(15):7803–7809

Hobbs P (2000) Introduction to atmospheric chemistry. Cambridge University Press, Cambridge, UK, p 150

Hutchinson GL, Livingston GP (1993) Use of chamber systems to measure trace gas fluxes. In: Agriculture ecosystem effects on trace gases and global climate change. ASA Special publication number 55, ASA and SSSA, Madison, Wisconsin, USA, pp 63–78

Hutchinson GL, Livingston GP (2001) Vents and seals in non-steady-state chambers used for measuring gas exchange between soil and the atmosphere. Europ J Soil Sci 52:675–682

Hutchinson GL, Mosier AR (1981) Improved soil cover method for field measurement of nitrous oxide fluxes. Soil Sci Soc Am J 45:311–316

Hutchinson GL, Rochette P (2003) Non-flow-through steady-state chambers for measuring soil respiration: numerical evaluation of their performance. Soil Sci Soc Am J 67:166–180

IPCC (1995) IPCC guidelines for National Greenhouse Gas Inventories, Prepared by the National Greenhouse Gas Inventories Programme (eds). IGES, Japan

IPCC—Intergovernmental Panel on Climate Change (2014) Climate change 2014: mitigation of climate change. Contribution of Working Group III to the Fifth Assessment Report of the Intergovernmental Panel on Climate Change. Cambridge University Press, Cambridge, UK and New York, NY, USA

Iqbal J, Castellano MJ, Parkin TB (2013) Evaluation of photoacoustic infrared spectroscopy for simultaneous measurement of N_2O and CO_2 gas concentrations and fluxes at the soil surface. Glob Change Biol 19:327–336

Jacinthe P-A, Dick WA (1996) Use of silicone tubing to sample nitrous oxide in the soil atmosphere. Soil Biol Biochem 28:721–726

Jähne B, Heinz G, Dietrich W (1987) Measurement of the diffusion coefficients of sparingly soluble gases in water. J Geophys Res 92(C10):10767–10776

Jantalia CP, Halvorson AD, Follett RF, Alves BJR, Polidoro JC, Urquiaga S (2012) Nitrogen source effects on ammonia volatilization as measured with semi-static chambers. Agron J 104:1595–1603

Jin HS, White DS, Ramsey JB, Kipphut GW (2012) Mixed tracer injection method to measure reaeration coefficients in small streams. Water Air Soil Pollut 8:5297–5306

Jochheim H, Wirth S, von Unold G (2018) A multi-layer, closed-loop system for continuous measurement of soil CO_2 concentration. J Plant Nutr Soil Sci 181:61–68

Jochum T, Michalzik B, Bachmann A, Popp J, Frosch T (2015a) Microbial respiration and natural attenuation of benzene contaminated soils investigated by cavity enhanced Raman multi-gas spectroscopy. Analyst 140(9):3143–3149

Jochum T, von Fischer JC, Trumbore S, Popp J, Frosch T (2015b) Multigas leakage correction in static environmental chambers using sulfur hexafluoride and Raman spectroscopy. Anal Chem 87(21):11137–11142

Jochum T, Rahal L, Suckert RJ, Popp J, Frosch T (2016) All-in-one: A versatile gas sensor based on fiber enhanced Raman spectroscopy for monitoring postharvest fruit conservation and ripening. R Soc Chem, 141: 2023–2029.

Jochum T, Fastnacht A, Trumbore SE, Popp J, Frosch T (2017) Direct Raman spectroscopic measurements of biological nitrogen fixation under natural conditions: an analytical approach for studying nitrogenase activity. Anal Chem 89(2):1117–1122

Kammann C, Grünhage L, Jäger HJ (2001) A new sampling technique to monitor concentrations of CH_4, N_2O and CO_2 in air at well-defined depths in soils with varied water potential. Eur J Soil Sci 52:297–303

Keeney DR, Nelson DW (1982) Nitrogen-inorganic forms. In: Page AL, Miller RH, Keeney DR (eds) Methods of soil analysis, part 2. Chemical and microbiological properties. American Society of Agronomy, Madison, Wisconsin, USA, pp 643–698

Keiner R, Frosch T, Massad T, Trumbore S, Popp J (2014) Enhanced Raman multigas sensing—a novel tool for control and analysis of (13)CO(2) labeling experiments in environmental research. Analyst 139(16):3879–3884

Keiner R, Frosch T, Hanf S, Rusznyak A, Akob DM, Kusel K, Popp J (2013) Raman spectroscopy-an innovative and versatile tool to follow the respirational activity and carbonate biomineralization of important cave bacteria. Anal Chem 85(18):8708–8714

Keiner R, Herrmann M, Kuesel K, Popp J, Frosch T (2015a) Rapid monitoring of intermediate states and mass balance of nitrogen during denitrification by means of innovative cavity enhanced Raman multi-gas sensing. Anal Chim Acta 864:39–47

Keiner R, Gruselle MC, Michalzik B, Popp J, Frosch T (2015b) Raman spectroscopic investigation of $^{13}CO_2$ labeling and leaf dark respiration of *Fagus sylvatica* L. (European Beech). Anal Bioanal Chem 407:1813–1817

Kelliher FM, Sherlock RR, Clough TJ, Premaratne M, Laughlin RJ, McGeough KL, Harvey MJ, McMillan AMS, Reid A, Saggar S (2012) Air sample collection, storage and analysis. In: de

Klein CAM, Harvey MJ (eds) Nitrous oxide chamber methodology guidelines. Global Research Alliance on agricultural greenhouse gases. Ministry for Primary Industries, New Zealand, pp 56–72

Keronen P, Reissell A, Chevallier F, Siivola E, Pohja T, Hiltunen V, Hatakka J, Aalto T, Rivier L, Ciais P, Jordan A, Hari P, Viisanen Y, Vesala T (2014) Accurate measurements of CO_2 mole fraction in the atmospheric surface layer by an affordable instrumentation. Boreal Environ Res 19:35–54

Kiefer J (2015) Recent advances in the characterization of gaseous and liquid fuels by vibrational spectroscopy. Energies 8(4):3165

Kiefer J, Seeger T, Steuer S, Schorsch S, Weikl MC, Leipertz A (2008) Design and characterization of a Raman-scattering-based sensor system for temporally resolved gas analysis and its application in a gas turbine power plant. Meas Sci Technol 19(8):085408–085408

Kissel DE, Brewer HL, Arkin GF (1977) Design and test of a field sampler for ammonia volatilization. Soil Sci Soc Am J 41:1133–1138

Knebl A, Yan D, Popp J, Frosch T (2017) Fiber enhanced Raman gas spectroscopy. TrAC Trac-Trend Anal Chem

Knebl A, Yan D, Popp J, Frosch T (2018) Fiber enhanced Raman gas spectroscopy. Trac-Trend Anal Chem 103:230–238

Knebl A, Domes R, Yan D, Popp J, Trumbore S, Frosch T (2019) Fiber-enhanced raman gas spectroscopy for (18)O-(13)C-labeling experiments. Anal Chem 91(12):7562–7569

Knight JC (2003) Photonic crystal fibres. Nature 424(6950):847

Kramers HA, Heisenberg W (1925) Über die Streuung von Strahlung durch Atome. Z Phys A Hadrons Nuclei 31(1):681–708

Kumar S, Herrmann M, Blohm A, Hilke I, Frosch T, Trumbore SE, Küsel K (2018) Thiosulfate- and hydrogen-driven autotrophic denitrification by a microbial consortium enriched from groundwater of an oligotrophic limestone aquifer. FEMS Microbiol Ecol 94(10):141

Lam SK, Suter H, Mosier AR, Chen D (2017) Using nitrification inhibitors to mitigate agricultural N_2O emission: a double-edged sword? Glob Change Biol 23:485–489

Laursen AE, Seitzinger SP (2004) Diurnal patterns of denitrification, oxygen consumption and nitrous oxide production in rivers measured at the whole reach scale. Freshw Biol 49:1448–1458

Landsberg G, Mandelstam L (1928) Eine neue Erscheinung bei der Lichtzerstreuung in Krystallen. Die Naturwissenschaften 16(28):557–558

Lebegue B, Schmidt M, Ramonet M, Wastine B, Kwok CY, Laurent O, Belviso S, Guemri A, Philippon C, Smith J, Conil S (2016) Comparison of nitrous oxide (N_2O) analyzers for high-precision measurements of atmospheric mole fractions. Atmos Meas Technol 9:1221–1238

Leuning R, Freney JR, Denmead OT, Simpson JR (1985) A sampler for measuring atmospheric ammonia flux. Atmos Environ 19:1117–1124

Li X, Xia Y, Zhan L, Huang J (2008) Near-confocal cavity-enhanced Raman spectroscopy for multitrace-gas detection. Opt Lett 33(18):2143–2145

Li JS, Chen W, Fischer H (2013) Quantum cascade laser spectrometry techniques: a new trend in atmospheric chemistry. Appl Spectrosc Rev 48:523–559

Li JS, Yu B, Zhao W, Chen W (2014) A review of signal enhancement and noise reduction techniques for tunable diode laser absorption spectroscopy. Appl Spectrosc Rev 49:666–691

Liss PS, Merlivat L (1986) Air-sea gas exchange rates: introduction and synthesis. In: Buat-Menard P (ed) The role of air-sea exchange in geochemical cycling. D Reidel, Dordrecht, pp 113–127

Liu J, Chen H, Zhu Q, Shen Y, Wang X, Wang M, Peng C (2015) A novel pathway of direct methane production and emission by eukaryotes including plants, animals and fungi: an overview. Atmos Environ 115:26–35

Livingston GP, Hutchinson GL (1995) Enclosure-based measurement of trace gas exchange: application and sources of error. In: Matson PA, Harriss RC (eds) Biogenic trace gases: measuring emissions from soil and water. Blackwell Science, Oxford, pp 14–51

Lockyer DR (1984) A system for the measurement in the field of losses of ammonia through volatilisation. J Sci Food Agric 35:837–848

Long DA (2002) The Raman effect: a unified treatment of the theory of Raman scattering by molecules. Wiley, New York

Malowany K, Stix J, Van Pelt A, Lucic G (2015) H2S interference on CO2 isotopic measurements using a Picarro G1101–i cavity ring-down spectrometer. Atmos Meas Technol 8(10):4075–4082

Marshall VG, Debell DS (1980) Comparison of four methods of measuring volatilization losses of nitrogen following urea fertilization of forest soils. Can J Soil Sci 60:549–563

Martins MR, Jantalia CP, Polidoro JC, Batista JN, Alves BJR, Boddey RM, Urquiaga S (2015) Nitrous oxide and ammonia emissions from N fertilization of maize crop under no-till in a Cerrado soil. Soil Tillage Res 151:75–81

Martins MR, Sant'Anna SAC, Zaman M, Santos RC, Monteiro RC, Alves BJR, Jantalia CP, Boddey RM, Urquiaga S (2017) Strategies for the use of urease and nitrification inhibitors with urea: Impact on N_2O and NH_3 emissions, fertilizer-^{15}N recovery and maize yield in a tropical soil. Agric Ecosyst Environ 247:54–62

Martins MR, Sarkis LF, Guareschi RF, Santos CA, Sant'Anna SAC, Zaman M, Jantalia CP, Alves BJR, Boddey RM, Araujo ES, Urquiaga S (2021a) A simple and easy method to measure ammonia volatilization: accuracy under field conditions. Pedosphere (in press)

Martins MR, Sarkis LF, Sant'Anna SAC, Santos CA, Araujo KE, Santos RC, Zaman M, Araujo EA, Jantalia CP, Boddey RM, Alves BJR, Urquiaga S (2021b) Optimizing the use of open chambers to measure ammonia volatilization in field plots amended with urea. Pedosphere (in press)

Mastepanov M, Christensen TR (2008) Bimembrane diffusion probe for continuous recording of dissolved and entrapped bubble gas concentrations in peat. Soil Biol Biochem 40:2992–3003

Minamikawa K, Tokida T, Sudo S, Padre A, Yagi K (2015) Guidelines for measuring CH_4 and N_2O emissions from rice paddies by a manually operated closed chamber method. National Institute for Agro-Environmental Sciences, Tsukuba, Japan

Misselbrook TH, Nicholson FA, Chambers BJ, Johnson RA (2005) Measuring ammonia emissions from land-applied manure: an intercomparison of commonly used samplers and techniques. Environ Pollut 135:389–397

Mosier A, Kroeze C, Nevison C, Oenema O, Seitzinger S, Van Cleemput O (1998) Closing the global N_2O budget: nitrous oxide emissions through the agricultural nitrogen cycle—OECD/IPCC/IEA phase ii development of IPCC guidelines for national greenhouse gas inventory methodology. Nutr Cycl Agroecosyst 52:225–248

Mosier AR (1989) Chamber and isotope techniques. In: Andreae MO, Schimel DS (eds) Exchange of trace gases between terrestrial ecosystems and the atmosphere. Wiley, New York, pp 175–187

Mosier AR (1990) Gas flux measurement techniques with special reference to techniques suitable for measurements over large ecologically uniform areas. In: Bouwman AF (ed) Soils and the greenhouse effect proceedings of the international conference soils and the greenhouse effect. International soil reference and information centre (ISRIC). Wiley, New York, pp 289–301

MSU—Montana State University (2019) Methodology (for Ammonia volatilization), accessed 29 May 2020

Müller C (2003) Plants affect the in situ N_2O emissions of a temperate grassland ecosystem. J Plant Nutr Soil Sci 166:771–773

Müller C, Stevens RJ, Laughlin RJ, Jäger HJ (2004) Microbial processes and the site of N_2O production in a temperate grassland soil. Soil Biol Biochem 36:453–461

Müller C, Laughlin RJ, Spott O, Rütting T (2014) Quantification of N_2O emission pathways via a ^{15}N tracing model. Soil Biol Biochem 72:44–54

Nara H, Tanimoto H, Tohjima Y, Mukai H, Nojiri Y, Katsumata K, Rella CW (2012) Effect of air composition (N_2, O_2, Ar, and H_2O) on CO_2 and CH_4 measurement by wavelength-scanned cavity ring-down spectroscopy: calibration and measurement strategy. Atmos Meas Technol 5(11):2689–2701

Neftel A, Blatter A, Schmid M, Lehman B, Tarakanov SV (2000) An experimental determination of the scale length of N_2O in the soil of a grassland. J Geophys Res 105:12095–12103

Nickerson N (2016) Evaluating gas emission measurements using minimum detectable flux (MDF). Eocense

Nichols KL, Del Grosso SJ, Derner JD, Follett RF, Archibeque SL, Delgado JA, Paustian KH (2018) Nitrous oxide and ammonia emissions from cattle excreta on shortgrass steppe. J Environ Qual 47:419–426

Nikodem M, Wysocki G (2012) Molecular dispersion spectroscopy—new capabilities in laser chemical sensing. Ann NY Acad Sci 1260:101–111

Nômmik H (1973) The effect of pellet size on the ammonia loss from urea applied to forest soil. Plant Soil 39:309–318

O'Connor D, Dobbins W (1958) Mechanism of reaeration in natural streams. Trans Am Soc Civ Eng 123:641–684

Oertel C, Herklotz K, Matschullat J, Zimmerman F (2012) Nitric oxide emissions from soils: a case study with temperate soils from Saxony, Germany. Enviro Earth Sci 66:2343–2351

Oertel C, Matschullat J, Zurba K, Zimmermann F (2016) Greenhouse gas emissions from soils—a review. Chem Erde 76:327–352

Outram FN, Hiscock KM (2012) Indirect nitrous oxide emissions from surface water bodies in a lowland arable catchment: a significant contribution to agricultural greenhouse gas budgets? Environ Sci Technol 46:8156–8163

Parkin TB, Venterea RT (2010) Chamber-based trace gas flux measurements. In: Follett RF (ed) Sample protocols GRACEnet. USDA, pp 1 − 39

Pathak H, Upadhyay RC, Muralidhar M, Bhattacharyya P, Venkateswarlu B (2013) Measurement of greenhouse gas emission from crop, livestock and aquaculture, NICRA manual series. Indian Agricultural Research Institute, New Delhi, p 101

Peltola O, Hensen A, Helfter C, Belelli Marchesini L, Bosveld FC, Van den Bulk WCM Elbers JA, Haapanala S, Holst J, Laurila T, Lindroth A, Nemitz E, Röckmann T Vermeulen AT, Mammarella I (2014) Evaluating the performance of commonly used gas analysers for methane eddy covariance flux measurements: the InGOS inter-comparison field experiment. Biogeoscience 11:3163–3186

Premaratne M, Clough TJ, Kelliher FM (2017) Determining the nitrous oxide transfer velocity and emission factor of an agricultural drain. NZ J Agric Res 60:277–286

Raman CV, Krishnan KS (1928) A new type of secondary radiation. Nature 121:501–502

Rapson TD, Dacres H (2014) Analytical techniques for measuring nitrous oxide. Trends Anal Chem 54:65–74

Raymond PA, Zappa CJ, Butman D, Bott TL, Potter J, Mulholland P et al (2012) Scaling the gas transfer velocity and hydraulic geometry in streams and small rivers. Limnol Oceanogr 2:41–53

Reay DS, Smith KA, Edwards AC (2003) Nitrous oxide emission from agricultural drainage waters. Glob Change Biol 9:195–203

Rochette P (2011) Towards a standard non-steady-state chamber methodology for measuring soil N_2O emissions. Anim Feed Sci Technol 166–167:141–146

Rochette P, Bertrand N (2003) Soil air sample storage and handling using polypropylene syringes and glass vials. Can J Soil Sci 83:631–637

Rochette P, Angers DA, Chantigny MH, Gasser MO, MacDonald JD, Pelster DE, Bertrand N (2013) Ammonia volatilization and nitrogen retention: how deep to incorporate urea? J Environ Qual 42:1635–1642

Roper JD, Burton DL, Madani A, Stratton GW (2013) A simple method for quantifying dissolved nitrous oxide in tile drainage water. Can J Soil Sci 93:59–64

Russell PSJ (2003) Photonic crystal fibers. Science 299(5605):358–362

Saggar S, Andrew RM, Tate KR, Hedley CB, Rodda NJ, Townsend JA (2004) Modelling nitrous oxide emissions from dairy-grazed pastures. Nutr Cycl Agroecosys 68:243–255

Saggar S, Hedley CB, Giltrap DL, Lambie SM (2007) Measured and modelled estimates of nitrous oxide emission and methane consumption from a sheep-grazed pasture. Agric Ecosyst Environ 122:357–365

Sander R (2015) Compilation of Henry's law constants (version 4.0) for water as solvent. Atmos Chem Phys 15:4399–4981

Schiel D, Richter W (1987) Use of Raman spectrometry in gas analysis. Fresenius Z Anal Chem 327(3–4):335–337

Schwarzenbach RP, Gschwend PM Imboden DM (1993) Environmental organic chemistry. Wiley, New York

Šimek M, Hynšt J, Šimek P (2014) Emissions of CH_4, CO_2, and N_2O from soil at a cattle overwintering area as affected by available C and N. Appl Soil Ecol 75(5):2–62

Sieburg A, Jochum T, Trumbore SE, Popp J, Frosch T (2017) Onsite cavity enhanced Raman spectrometry for the investigation of gas exchange processes in the Earth's critical zone. Analyst 142(18):3360–3369

Sieburg A, Knebl A, Jacob JM, Frosch T (2019) Characterization of fuel gases with fiber-enhanced Raman spectroscopy. Anal Bioanal Chem 411(28):7399–7408

Smith KA (2017) Changing views of nitrous oxide emissions from agricultural soil: key controlling processes and assessment at different spatial scales. Eur J Soil Sci 68:137–155

Smekal A (1923) Zur Quantentheorie der Dispersion. Naturwissenschaften 11(43):873–875

Thomas SM, Ledgard SF, Francis GS (2005) Improving estimates of nitrate leaching for quantifying New Zealand's indirect nitrous oxide emissions. Nutr Cycl Agroecosyst 73(2):13–226

Tohjima Y, Katsumata K, Morino I, Mukai H, Machida T, Akama I, Amari T, Tsunogai U (2009) Theoretical and experimental evaluation of the isotope effect of NDIR analyser on atmospheric CO_2 measurement. J Geophys Res 114:D13302

Turner PA, Griffis TJ, Lee X, Baker JM, Venterea RT, Wood JD (2015) Indirect nitrous oxide emissions from streams within the US Corn Belt scale with stream order. Proc Natl Acad Sci 112:9839–9843

UNFCCC The United Nations Framework Convention on Climate Change (2008) Report of the conference of the parties on its thirteenth session, held in Bali from 3 to 15 December 2007

Venterea RT, Parkin TB, Cardenas L, Petersen SO, Pedersen AR (2012) Air sample collection, storage and analysis. In: de Klein CAM, Harvey MJ (eds) Nitrous oxide chamber methodology guidelines. Global Research Alliance on agricultural greenhouse gases. Ministry for Primary Industries, New Zealand, pp 95–121

Venterea RT, Petersen SO, De Klein CAM, Pedersen AR, Noble ADL, Rees RM, Gamble JD, Parkin TB (2020) Global Research Alliance N_2O chamber methodology guidelines: Flux calculations. J Environ Qual 49:1141–1155

Volk GM (1959) Volatile loss of ammonia following surface application of urea to turf or bare soils. Agron J 51:746–749

Wagner S, Reicosky D, Alessi R (1997) Regression models for calculating gas fluxes measured with a closed chamber. Agron J 89(2):279

Wanninkhof R (1992) Relationship between gas exchange and wind speed over the ocean. J Geophys Res 97:7373–7381

Wanninkhof R (2014) Relationship between wind speed and gas exchange over the ocean revisited. Limnol Oceanogr Methods 12:351–362

Weiss RF, Price BA (1980) Nitrous oxide solubility in water and seawater. Mar Chem 8:347–359

Werle PW (2004) Diode-laser sensors for in-situ gas analysis. In: Hering P, Lay JP, Stry S (eds) Laser in environmental and life sciences. Springer, Berlin

Wilcock RJ (1988) Study of river aeration at different flow rates. J Environ Eng 114:91–105

Wilson JD, Catchpoole VR, Denmead OT, Thurtell GW (1983) Verification of a simple microm-eteorological method for estimating the rate of gaseous mass transfer from the ground to the atmosphere. Agric Meteorol 29:183–189

Xu L, Furtaw MD, Madsen RA, Garcia RL, Anderson DJ, McDermitt DK (2006) On maintaining pressure equilibrium between a soil CO_2 flux chamber and the ambient air. J Geophys Res Atmos 111:1–14

Yan D, Popp J, Pletz MW, Frosch T (2017a) Highly sensitive broadband raman sensing of antibiotics in step-index hollow-core photonic crystal fibers. ACS Photonics 4(1):138–145

Yan D, Popp J, Frosch T (2017b) Analysis of fiber-enhanced Raman gas sensing based on Raman chemical imaging. Anal Chem 89(22):12269–12275

Zaman M, Saggar S, Blennerhassett JD, Singh J (2009) Effect of urease and nitrification inhibitors on N transformation, gaseous emissions of ammonia and nitrous oxide, pasture yield and N uptake in grazed pasture system. Soil Biol Biochem 41:1270–1280

Zaman M, Nguyen ML, Šimek M, Nawaz S, Khan MJ, Babar MN, Zaman S (2012) Emissions of nitrous oxide (N_2O) and di-nitrogen (N_2) from the agricultural landscapes, sources, sinks, and factors affecting N_2O and N_2 ratios. In: Liu G (ed) Greenhouse gases—emissions, measurement and management. InTech, pp. 3–32

3

Automated Measurements of Greenhouse Gases: Methods and Techniques

M. Zaman, K. Kleineidam, L. Bakken, J. Berendt, C. Bracken, K. Butterbach-Bahl, Z. Cai, S. X. Chang, T. Clough, K. Dawar, W. X. Ding, P. Dörsch, M. dos Reis Martins, C. Eckhardt, S. Fiedler, T. Frosch, J. Goopy, C.-M. Görres, A. Gupta, S. Henjes, M. E. G. Hofmann, M. A. Horn, M. M. R. Jahangir, A. Jansen-Willems, K. Lenhart, L. Heng, D. Lewicka-Szczebak, G. Lucic, L. Merbold, J. Mohn, L. Molstad, G. Moser, P. Murphy, A. Sanz-Cobena, M. Šimek, S. Urquiaga, R. Well, N. Wrage-Mönnig, S. Zaman, J. Zhang, and C. Müller

Abstract Methods and techniques are described for automated measurements of greenhouse gases (GHGs) in both the laboratory and the field. Robotic systems are currently available to measure the entire range of gases evolved from soils including dinitrogen (N_2). These systems usually work on an exchange of the atmospheric N_2 with helium (He) so that N_2 fluxes can be determined. Laboratory systems are often

M. Zaman (✉) · L. Heng
Soil and Water Management & Crop Nutrition (SWMCN) Section, Joint FAO/IAEA Division of Nuclear Techniques in Food and Agriculture, International Atomic Energy Agency (IAEA), Vienna, Austria
e-mail: m.zaman@iaea.org; zamanm_99@yahoo.com

K. Kleineidam · C. Eckhardt · A. Jansen-Willems · G. Moser · C. Müller
Institute of Plant Ecology, Justus Liebig University Giessen, Giessen, Germany

L. Bakken
Norwegian University of Life Sciences (NMBU), Aas, Norway

J. Berendt · S. Fiedler · N. Wrage-Mönnig
University of Rostock, Rostock, Germany

C. Bracken
School of Agriculture and Food Science and Earth Institute, University College Dublin, Dublin, Ireland

K. Butterbach-Bahl
Institute of Meteorology and Climate Research, Atmospheric Environmental Research (IMK-IFU), Karlsruhe Institute of Technology, Karlsruhe, Germany

Z. Cai
School of Geography Sciences, Nanjing Normal University, Jiangsu, China

S. X. Chang
Department of Renewable Resources, University of Alberta, Edmonton, AB T6G 2E3, Canada

used in microbiology to determine kinetic response reactions via the dynamics of all gaseous N species such as nitric oxide (NO), nitrous oxide (N_2O), and N_2. Latest He incubation techniques also take plants into account, in order to study the effect of plant–soil interactions on GHGs and N_2 production. The advantage of automated in-field techniques is that GHG emission rates can be determined at a high temporal resolution. This allows, for instance, to determine diurnal response reactions (e.g. with temperature) and GHG dynamics over longer time periods.

Keywords GHG · N_2O · N_2 · Helium

3.1 Automated Laboratory Techniques

This section focuses on automated high-resolution determinations of anaerobic atmospheric trace gas (H_2, NO, N_2O, CO_2, CH_4) production kinetics, and nitrate (NO_3^-)-nitrite (NO_2^-) analysis systems designed for ex situ incubations. Typically, such systems combine a temperature-controlled sampling unit holding several septum vials (with or without stirring or shaking) under a fully programmable robotic arm of an autosampler, which is used to repeatedly pierce the septa of the incubation vessels to subsample the headspace for concentration measurements by gas chromatography and/or other dedicated gas analysers. Processes are studied in different

T. Clough
Department of Soil & Physical Sciences, Faculty of Agriculture & Life Sciences, Lincoln University, Lincoln, New Zealand

K. Dawar
Department of Soil and environmental Sciences, University of Agriculture, Peshawar, Pakistan

W. X. Ding
Institute of Soil Science, Chinese Academy of Sciences, Nanjing, China

P. Dörsch · L. Molstad
Faculty of Environmental Sciences and Natural Resource Management, Norwegian University of Life Sciences (NMBU), Aas, Norway

T. Frosch
Leibniz Institute of Photonic Technology, Technical University Darmstadt, Darmstadt, Germany

J. Goopy
International Livestock Research Institute (ILRI), Nairobi, Kenya

C.-M. Görres
Department of Soil Science and Plant Nutrition/Department of Applied Ecology, Hochschule Geisenheim University, Geisenheim, Germany

A. Gupta
Independent Consultant India, Mumbai, India

S. Henjes · M. A. Horn
Institute of Microbiology, Leibniz University Hannover, Hannover, Germany

matrices, ranging from repacked or suspended soils (Horn et al. 2006; Palmer et al. 2012, 2015, 2016; Zhu et al. 2013; Schmidt et al. 2016), plant material (Keppler et al. 2006), and water samples (McCrackin and Elser 2010) to extracted bacterial cells (Dörsch et al. 2012; Brenzinger et al. 2015) and cultures of isolated (Lycus et al. 2017) or type strain organisms (Bergaust et al. 2012; Horn et al. 2005) including knock-out mutants. Automated incubation setups are a refinement of classical incubation techniques still used to determine process rates of "respiration" (e.g. by soda-lime trapping of evolved CO_2) or "denitrification enzyme activity" (DEA; N_2O accumulation in the presence of acetylene). While automated batch incubations allow for estimating process potentials and apparent Michaelis–Menten parameters (V_{max}, K_m) under quasi-steady-state conditions in short-term assays with ample substrates, high-resolution incubations can also be used to study "gas kinetics" of selected processes with finite amounts of substrates (Palmer et al. 2012). Gaseous intermediates (e.g. in denitrification) evolve transiently before gaseous end products accumulate to amounts given by the added substrate. In this way, high-resolution gas kinetics can provide insights into the metabolic functioning under changing substrate conditions. Typically, incubations are initiated under oxic or anoxic conditions by flushing the headspace of the incubation bottles with various gas mixtures. The transition from oxic to anoxic metabolism can be studied by letting the sample consume a finite amount of O_2 while monitoring the evolution of gaseous intermediates (NO, N_2O, H_2), and final products (N_2, CH_4, H_2S) involved in anoxic metabolism (Schlüter et al. 2019). Notably, replacing the bottle atmosphere with He or a mixture of He and O_2 allows for quantifying N_2, the final product of denitrification which cannot be detected at ambient N_2 background levels (but see section on Raman spectroscopy

M. E. G. Hofmann
Picarro B.V., 's-Hertogenbosch, The Netherlands

M. M. R. Jahangir
Department of Soil Science, Bangladesh Agricultural University, Mymensingh, Bangladesh

D. Lewicka-Szczebak
Laboratory of Isotope Geology and Geoecology, Institute of Geological Sciences, University of Wrocław, Wrocław, Poland

G. Lucic
Picarro Inc., Santa Clara, CA, USA

L. Merbold
Mazingira Centre, International Livestock Research Institute (ILRI), Nairobi, Kenya

J. Mohn
Laboratory for Air Pollution & Environmental Technology, Empa Dübendorf, Dübendorf, Switzerland

P. Murphy
Environment & Sustainable Resource Management Section, School of Agriculture & Food Science, and UCD Earth Institute, University College, Dublin, Ireland

A. Sanz-Cobena
Research Center for the Management of Environmental and Agricultural Risks (CEIGRAM), ETSIAAB, Universidad Politécnica de Madrid, Madrid, Spain

Chap. 2, Sect. 2.10). Using appropriate inhibitors (Yoshinari and Knowles 1976; Bedard and Knowles 1989; Taylor et al. 2015), selected processes can be studied.

High-resolution gas kinetics are also the method of choice to study the induction of enzymatic processes involved in GHG turnover, which can be studied by sequential analysis of gene expression levels or other genetic markers (Vollack and Zumft 2001; Bergaust et al. 2010; Liu et al. 2014; Brenzinger et al. 2015). In this way, the eco-physiology of specific microbial groups or organisms can be studied in great detail and advance our understanding of regulatory aspects involved in GHG turnover (Bakken et al. 2012; Braker and Conrad 2011; Hink et al. 2017).

3.1.1 Technical Challenges

- **Headspace pressure**: Subsampling closed incubation bottles repeatedly implies a decrease in headspace pressure. The incubation setup should be able to replace the removed sample volume automatically by an inert gas (e.g. He, Ar) to maintain bottle pressure at ~1 atm. This can be achieved by reversing the sampling pump (if used) or by replacing the volume with an automatically operated syringe. Alternatively, bottles with elevated headspace pressure can be used (Palmer et al. 2010). In both cases, the resulting dilution has to be corrected when calculating rates between sampling increments (Molstad et al. 2007).
- **Sample loss**: Irrespective of how the sampled volume is replaced, high-resolution gas kinetics with finite amounts of substrates suffer from sampling loss, i.e. part of the substrates and evolving intermediates are lost during analysis. This has to be taken into account when calculating internal mass balances between substrates and gaseous products.

M. Šimek
Institute of Soil Biology, Biology Centre of the Czech Academy of Sciences, and Faculty of Science, University of South Bohemia, České Budějovice, Czech Republic

M. dos Reis Martins · S. Urquiaga
EMBRAPA Agrobiologia Seropédica, Brazilian Agricultural Research Corporation, Seropedica, RJ, Brazil

R. Well
Thünen Institute of Climate-Smart Agriculture, Braunschweig, Germany

S. Zaman
University of Canterbury, Christchurch, New Zealand

J. Zhang
School of Geography, Nanjing Normal University, Nanjing, China

K. Lenhart
Bingen University of Applied Sciences, Berlinstr. 109, Bingen 55411, Germany

- **Leakage**: When analysing accumulation or depletion of gaseous products (N_2) or substrates (O_2) which are abundant in the atmosphere, there will be leakage into the bottles or the sampling line. Incubation vessels, gas sampling access, sample transport, and admission to the measurement system must be designed and tested carefully to minimize the unavoidable leakage of O_2 and N_2, e.g. caused by repeatedly pierced septa. This can be avoided by programming the autosampler so that the point of piercing varies within a predefined radius (Molstad et al. 2007). Most of the leakage happens when the gas sample is entering the analysis unit. If a peristaltic pump is transporting sample to dedicated sampling loops on multi-port valves, the lining in the pump is made of flexible Marprene or Tygon, both of which allow some diffusion of O_2 and N_2 into the sampling stream. If using an automated syringe, there will always be some dead volume filled with ambient air. In any case, the sampling system should be backflushed continuously between sampling to keep the lines free of O_2 and N_2. Leakage rates should be determined in each analytical batch using He-filled dry bottles as part of the sampling sequence.
- **Gas analytics**: Analysing multi-gas kinetics from singular samples requires complicated column and detector setups, possibly combining different types of instruments (GC, chemoluminescence, tunable diode laser (TDL), etc.). A particular challenge is analysers which require a high flow-through of analyte (chemoluminescence, TDL). If the gaseous compound of interest is low in ambient air, air can be used to flush the analyser while the sample is spiked into the air stream using a multi-port sampling loop. In case it is undesirable to measure the component on top of atmospheric background, the sample has to be mixed with a span gas prior to analysis (cf. Picarro small sampling unit). In both cases, the compound experiences a considerable dilution, which has to be taken into account when considering detection limits.
- **Detection ranges and precision**: Multi-gas kinetics cover a large dynamic spectrum ranging from the ppb range (NO, N_2O) to the ppm range (CH_4, CO_2) to the vol.% range (O_2, N_2). Common detectors used in gas chromatography-based multi-gas detection (TCD, ECD, PDHID, FID) may cover most of the dynamic range expected to occur in closed bottle incubations, partly overlapping in their response to single compounds. For instance, ECD detectors have unmatched sensitivity for low N_2O concentrations but suffer from severe non-linearity in higher ppm ranges. On the other hand, TCD detectors are insensitive for N_2O but respond linearly over a wide concentration range. Hence, by combining ECD and TCD detectors, N_2O concentration can be analysed over a wide dynamic range. A possibility to cope with gas concentrations exceeding the dynamic range of an analyser is to use splits which divert a defined fraction of a compound to later injection as implemented, for instance, in case of NO detection by chemoluminescence (Molstad et al. 2007). It must also be noted that gas concentrations undergo strong changes during batch incubations as gases accumulate or deplete. This can create separation problems in gas chromatography if two adjacent peaks change concentrations in opposite directions. For example, CO_2 may accumulate to the vol.% concentrations, while N_2O is consumed (in denitrification). Both gases

typically eluate close to each other, and integration routines have to be designed to account for these dynamics. Likewise, it may be difficult to integrate small O_2 peaks next to large N_2 peaks.

- **Calibration**: Calibration should occur by internal standards, i.e. dry bottles filled with certified gas mixtures in He. By including such bottles into the sampling sequence, problems in sample admission, changes in detector sensitivity and leakage can be detected. Standard concentrations should be adopted carefully to the expected concentration ranges.
- **Temperature regulation and agitation**: Automated incubation setups measure concentrations in the headspace only, and dissolved concentrations must be inferred using temperature and Henry constants, assuming equilibrium between sample and headspace. This means that the incubation units should be temperature controlled during the entire incubation which is typically achieved by thermostatic water baths or heating blocks. To facilitate the exchange of gases between headspace and sample, it is often desirable to constantly stir samples in aqueous solutions (e.g. suspensions or slurries). This can easily be achieved by placing incubation bottles equipped with Teflon-coated magnetic stirrers on submersible stirrers in a water bath. Alternatively, samples may be shaken in between sampling using programmable shaking units.

In the following sections, two automatic incubation systems will be described.

3.1.2 System 1

The system for determining anaerobic atmospheric trace gases, namely, H_2, NO, N_2O, CH_4, and CO_2 comprises a gas chromatograph (7890B GC, Agilent Technologies Inc., Santa Clara, CA, USA) equipped with a split/splitless injector as well as a micro-packed column (Shin Carbon ST 80/100, 2 m, 0,5 mm ID, Part#: 19043, Restek Corporation, Bellefonte, PA, USA) and a capillary column (RT-Molecular Sieve 5A, 30 m, 0,53 mm ID, 50 μm, Cat#: 19723, Restek Corporation, Bellefonte, PA, USA; Plate 3.1). A pulsed discharge helium ionization detector (PDHID, V1D-3-I-HP-220, Valco Instruments Company Inc. VICI AG International, Houston, TX, USA) is coupled to the outlet of the ShinCarbon column via a restriction capillary, and a thermal conductivity detector (TCD, G3440B, Agilent Technologies Inc., Santa Clara, CA, USA) is coupled to the outlet of the molecular sieve 5A column. The ShinCarbon column acts as a water-insensitive molecular sieve where CO_2 and N_2O elute (which is essentially impossible for "regular" molecular sieve columns that trap CO_2). Baseline separation of O_2 and N_2 in the % range is achieved on the molecular sieve 5A after "heart cutting" of the combined peak eluting from the ShinCarbon column.

The TCD is used for analytes in the % range. The PDHID is a universal detector which offers high sensitivity in the higher ppb to lower % range without the need to fulfil the increasing requirements for operating a radionuclide-dependent electron

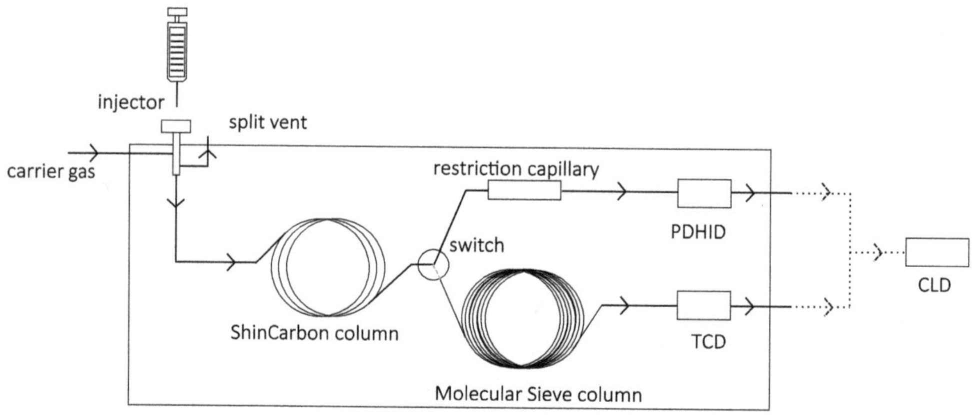

Plate 3.1 Setup of the gas chromatographic system. Switch represents a "Microfluidics Deans Switch" enabling "heart cutting" of peaks eluting from the ShinCarbon column

capture detector (ECD) (Forsyth 2004) or to purchase a rather cost-intensive mass sensitive detector (MSD). The gas chromatograph is operated with helium 6.0 as carrier gas that is purified by two sequential helium purifiers (V1HPM-220, Valco Instruments Company Inc. VICI AG International, Houston, TX, USA). One of the outlets of either the PDHID or the TCD can be connected to an NO analyser equipped with a chemiluminescence detector (CLD; Sievers NOA 280i, Zysense LLC, Weddington, NC, USA) to increase sensitivity for NO. The digital signal of the nitric oxide analyser is sent to an analogue-to-digital converter (A/D converter ADC 39500E, Agilent Technologies). The signals of all three detectors are recorded by the software Chemstation (Chemstation Rev. C.01.07.SR2 [255], Agilent Technologies Inc., Santa Clara, CA, USA) on a Windows computer, which is likewise used to control the gas chromatograph.

A robotized incubation system (Combi PAL-xt, CTC Analytics, Zwingen, Switzerland) is connected to the gas chromatograph to complete the setup (Plate 3.2). The Combi PAL-xt is equipped with two trays for 32 vials (10 or 20 ml) each. One of the trays can be temperature controlled in the range of 4 °C–70 °C. Furthermore, it comprises a temperature-controlled agitator (temperature range: 30 °C–200 °C; agitation speed: 250–750 rpm) shaking up to six samples at one time. A Y-cable (custom-made; Autosampler Guys, Alexandria, VA, USA) is used to enable the gas chromatograph to control both the incubation system and the A/D converter of the nitric oxide analyser. This has the advantage that all subordinate devices will only start a run when the gas chromatograph is "ready". Chronos xt 4.9 Master-Software for PAL samplers (Axel-Semrau, Sprockhövel, Germany) controls the robotized incubation system.

Altogether, this system allows for a cost-efficient online quantification of gas production and consumption during incubation experiments as well as the analysis of stored gas samples. A simultaneous quantification of H_2, O_2, N_2, NO, CO_2, CH_4, and N_2O is achieved. Although not in focus here, C_2H_2, C_2H_4, and C_2H_6 are likewise easily detectable.

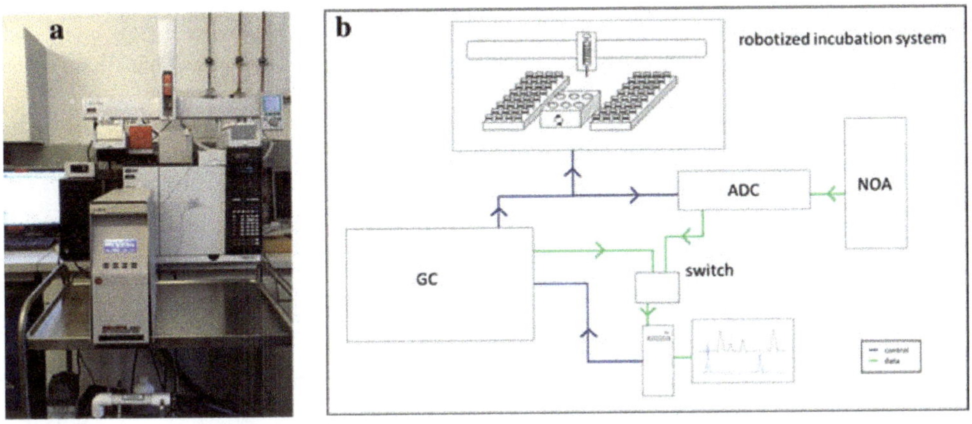

Plate 3.2 Photograph of the robotized incubation system (**a**) and configuration of the robotized incubation system, including the gas chromatograph (GC) and the nitric oxide analyser (NOA) with a chemiluminescence detector (CLD) connected via an A/D converter (ADC). "Switch" indicates an ethernet switch (**b**)

3.1.2.1 Measurements of Gas Samples

Analysis parameters are adjusted according to the gases to be measured as well as their expected concentrations. The most important consideration is whether the separation of a given analyte will be sufficient on the ShinCarbon column or whether "heart cutting" and further separation on the molecular sieve column is required. Specifically, baseline separation of O_2 and N_2 in the % range requires passage through the molecular sieve column. However, CO_2, N_2O, and NO are strongly retained or trapped, preventing detection. The choice of detector is generally dependent on the concentration of the gases with the PDHID being significantly more sensitive than the TCD. However, the TCD has a broader range of linear response than the PDHID. Multiple "heart cuts" (i.e. switches) between the molecular sieve—TCD path and the restrictor—PDHID path are possible during one run, enabling the quantification of trace concentrations of certain analytes in the presence of huge concentrations of others.

The following protocol provides good separation of the gases H_2, NO, CO_2, CH_4, N_2O, O_2, and N_2 (Table 3.1 and Fig. 3.1). The injection volume of gas samples is 250 μl. The system operates in constant pressure mode with a pressure of 379.21 kPa. The inlet is set to 160 °C, the total flow is 40.782 ml min^{-1}, and the split flow is 12.977 ml min^{-1}. The septum purge flow is 3 ml min^{-1}. Carrier gas is directed to the PDHID. A "heart cut" is performed via the Deans switch at 0.6 min to direct the flow to the TCD via the molecular sieve column, and it switches back at 0.76 min to redirect the flow to the PDHID. The outlet of the PDHID is connected to the CLD inside the nitric oxide analyser. The temperature protocol starts at 90 °C. This temperature is kept for 2 min before it is ramped at 50 °C min^{-1} to 120 °C and held for 2.1 min. Finally, the oven is further heated to 200 °C (50 °C min^{-1}), and the column is baked out for 2 min at this temperature. After 10 min, the gas chromatograph is

Table 3.1 Retention times, limits of detection (LOD), limits of quantification (LOQ), coefficient of variation (CV at 500 ppm apart from N_2), and linear range for different gases. Injection volume was 250 µl

Gases	Detector	Retention time [min]	LOD [ppm]	LOQ [ppm]	CV [%]	Linear Range [ppm]
H_2	PDHID	0.4	2	4	0.46	$4–10^4$
NO	CLD	0.8	0.05	0.1	n.d.[a]	$0.1–10^2$
CH_4	PDHID	1.7	2	4	0.35	$4–10^4$
O_2	TCD	1.9	n.d.	10^4	n.d.	$10^4–10^6$
N_2	TCD	2.2	n.d.	10^4	3.00[b]	$10^4–10^6$
CO_2	PDHID	3.2	n.d.	10^2	0.19	$10^2–10^4$
N_2O	PDHID	4.0	0.1	0.5	0.67	$1–5 \times 10^3$

[a]n.d., not determined
[b]Determined at 2% v/v N_2

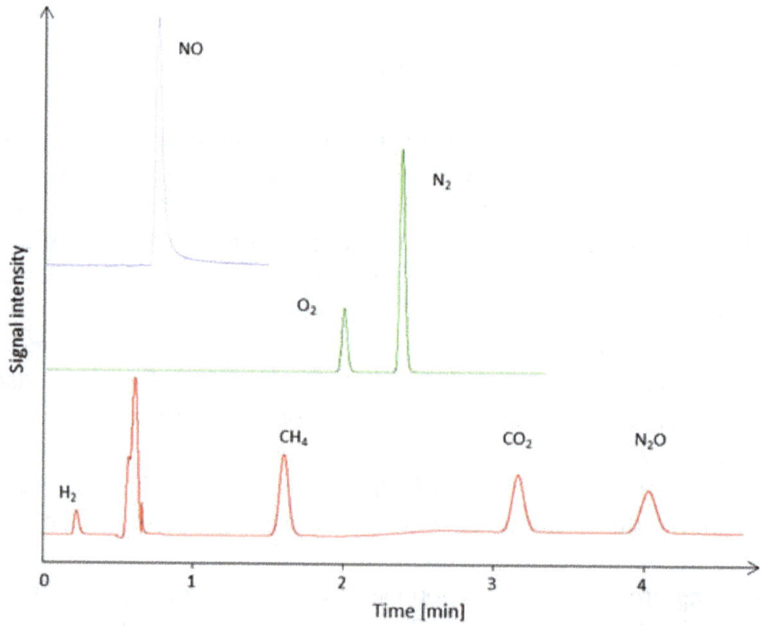

Fig. 3.1 Chromatograms of the three detectors connected to the GC system: CLD (blue), TCD (green), and PDHID (red)

ready to start a new run. H_2, CH_4, CO_2, and N_2O are detected on the PDHID, O_2 and N_2 on the TCD, and NO on the CLD (Fig. 3.1). The dynamic range for most of the tested gases includes four orders of magnitude with a single detector (Table 3.1).

The minimum change in concentration which can be regarded significant is 6.3 ppm for H_2, 5.0 ppm for CH_4, 2.9 ppm for CO_2, and 8.7 ppm for N_2O at an initial concentration of 500 ppm. For N_2, the minimum significant change in concentration is 0.2 vol.% (2000 ppm) at an initial concentration of 2%.

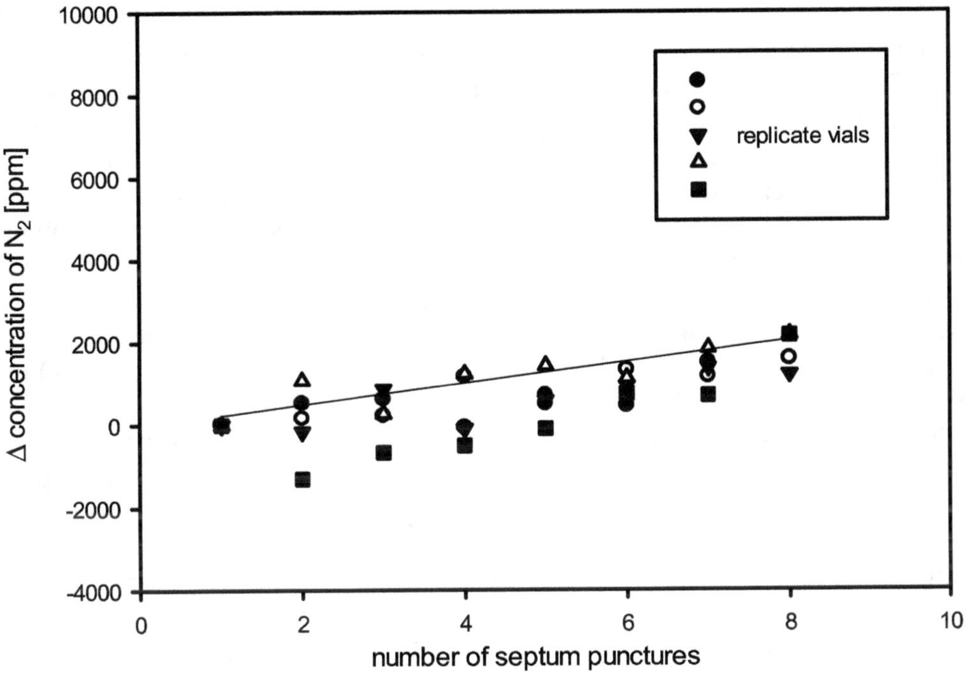

Fig. 3.2 Contamination of He-filled sample vials (500 mbar overpressure) with N_2 from air after multiple punctures

3.1.2.2 Possible Sample Containers

The system can be adapted to a variety of vial types. The sample seats in the provided autosampler trays accommodate 20 ml glass vials (75.5 mm × 22.5 mm with rounded bottom) or 10 ml vials with specific inserts (LEAP PAL Parts and Consumables LLC, Raleigh, NC, USA). Magnetic caps can be used to enable the autosampler to move the vials from the trays to the agitator if required. Acrylic glass blocks with milled sample seats (custom-made) can accommodate up to 100 exetainers (3 ml or 12 ml; Labco, Lampeter, UK) and other vial types offering the possibility for simple measurement of stored gas samples. Butyl septa (T.H. Geyer GmbH and Co. KG, Renningen, Germany) are sufficiently gastight after multiple punctures during incubation experiments (Fig. 3.2). However, the tightness of the septa should be carefully checked to ensure the reliability of the data obtained.

3.1.2.3 Semi-automated Determination of Nitrate and Nitrite

An analysis of the dissolved N-oxides NO_3^- and NO_2^- is required for the calculation of N and electron recoveries along with the determination of N-gases. We use a semi-automated, highly sensitive system based on reduction of NO_3^- and/or NO_2^- and analysis of the resulting nitric oxide. The lower detection limit for NO_3^- or NO_2^- in liquid samples is 100 nM.

The heart of the system is a nitric oxide analyser (Sievers 280i, Zysense LLC, Weddington, NC, USA) with a chemical reactor reducing N-oxides to NO ("purge vessel"). The purge vessel contains a reducing agent and is heated by a circulating water bath (Grant Instruments, Cambridge, UK). Up to 96 samples can be placed in a Combi PAL-xt autosampler (CTC Analytics, Zwingen, Switzerland) equipped with a large wash module and standard tray holders for 2 ml HPLC vials. A small volume of liquid sample (10–100 μl) is injected into the purge vessel. The evolving nitric oxide is purged out of the vessel by a stream of nitrogen and reaches the CLD inside the NOA. 50 mg of potassium iodide dissolved in 0.5 ml of double-distilled water and added to 4.5 ml of glacial acetic acid are used as reducing agent for NO_2^- in the purge vessel. Nitrate and NO_2^- are both reduced to NO when using 40 mg of vanadium chloride in 5 ml hydrochloric acid as reducing agent. The purge vessel is heated to 45 and 90 °C by a circulating water bath for NO_2^- only and combined NO_3^-/NO_2^- analyses, respectively. Water vapour is condensed before escaping the vessel by a second circulating water bath, set at 4 °C to cool the upper part of the vessel. In addition, a propylene filter is installed in the tubing which connects the purge vessel outlet to the nitric oxide analyser to prevent any liquids from reaching the CLD inside the NOA. A sodium hydroxide trap filled with 1 M NaOH is installed in line with the purge vessel in case of the combined NO_3^-/NO_2^- analysis to neutralize any HCl before it reaches the CLD (Fig. 3.2). The data is transferred to a Windows computer which also controls the autosampler via the software Cycle Composer Pal1 (CTC Analytics, Zwingen, Switzerland).

Before the analysis a calibration is performed using NO_2^- or NO_3^- standards in the range of 100 nM to 100 mM. The standards as well as the samples are placed in 2 ml HPLC glass vials (neoLab Migge GmbH, Heidelberg, Germany), fitting into the sample tray and are sealed with septa and caps. The autosampler can perform up to 50 injections before the reducing agent needs to be replaced manually. Each measurement takes 60 s, and the needle is flushed with double-distilled water twice in the large wash modules of the Combi PAL autosampler before taking the sample and after injecting.

The concentrations of NO_2^- can be calculated directly from the peak areas using the calibration curve. In the case of NO_3^-, the actual concentration is obtained by subtracting the NO_2^- concentration from the combined NO_3^-/NO_2^- concentrations obtained (Plate 3.3).

3.1.2.4 Application of the Automated Incubation System to Determine Denitrification Kinetics

The following protocol can be applied for the estimation of denitrification potentials by use of soil slurry incubations and simultaneous analysis of denitrification associated trace gases. The determination of N_2O production and N_2O consumption rates are used to derive NO_3^- dependent apparent Michaelis Menten parameters for soil denitrification.

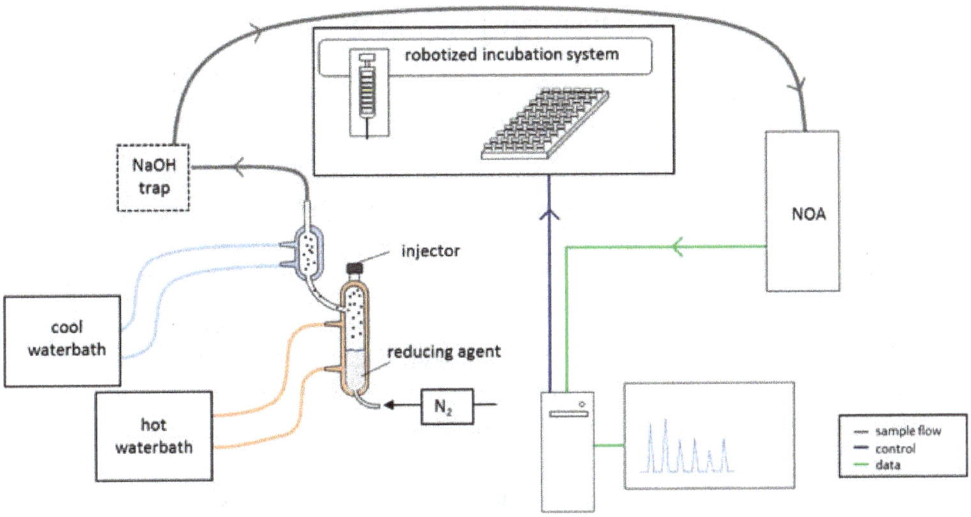

Plate 3.3 Schematic diagram of the semi-automated NO_3^- and NO_2^- analysis system

General Experimental Setup

Soil slurries are set up and pre-incubated under anoxic conditions to reduce the background of soil endogenous N-oxides. Pre-incubated soil slurries are supplemented with substrate and spiked with different concentrations of NO_3^-. Acetylene (10% v/v) is added to the gas phase of one of two subsets of samples to inhibit N_2O reduction (acetylene inhibition technique; Yoshinari and Knowles 1976). Subsequently, the concentration of N_2O in the headspace is monitored via gas chromatographic analyses while shaking the samples on a regular basis. The amount of N_2 which is produced in the non-inhibited samples can be estimated from the difference of the N_2O concentrations of the two subsets for low concentrations of supplemental NO_3^- or measured directly when high concentrations of NO_3^- in the mM range are supplemented.

Materials

- Duran glass bottles (DWK Life Sciences GmbH, Mainz, Germany)
- Butyl stopper (Ochs Glasgerätebau, Bovenden, Germany)
- Helium bottle (5.0, Linde AG, Munich, Germany) with tube connection
- Vacuum pump RV 3 with tube connection (Edwards, Burgess Hill, UK)
- 1 ml gastight syringe (Trajan Scientific and Medical, Ringwood, Australia)
- 100 μl micro-volume liquid syringe (Trajan Scientific and Medical, Ringwood, Australia)
- Disposable needles 0.45 mm (Braun GmbH, Kronberg, Germany)
- 1.5 ml tubes (Eppendorf, Hamburg, Germany)
- 20 ml glass vials with crimp top, rounded bottom, 75.5 × 22.5 mm with magnetic crimp caps and butyl septa (T.H. Geyer GmbH & Co. KG, Renningen, Germany)
- Crimping tool

- Filtropur $S0_2$ syringe filtration unit (Sarstedt AG and Co. KG, Nümbrecht, Germany)
- Acetylene (≥99.5%, Linde AG, Munich, Germany)
- Centrifuge Megafuge 8R (Heraeus, Hanau, Germany)
- Spectrophotometer Infinite M Plex (Tecan Austria GmbH, Grödig, Austria)
- Roller mixer (Ratek Instruments Pty Ltd, Boronia, Australia).

Experimental Procedure

For the pre-incubation, the soil is diluted 1:5 with distilled water in a Duran glass bottle. The bottle is closed with a butyl stopper, and the slurry is made anoxic by applying alternating cycles of vacuum and helium for five times.

A minor overpressure of helium (200 mbar) is applied to avoid mass flow of air into the bottle. The soil slurry is pre-incubated on a roller mixer until endogenous N-oxides are depleted. Incubation times are soil-dependent. 1 ml samples of the soil slurries are taken every 24 h using a helium-flushed syringe and needle to monitor combined NO_3^-/NO_2^- concentrations with the semi-automated NO_3^-/NO_2^- analysis system. The sample is centrifuged for 5 min at 16,000 × g and 4 °C before the supernatant is filtered through a 0.2 μm pore size membrane to remove microorganisms as well as any remaining particles. This filtrate is then used for NO_3^-/NO_2^- analysis (Plate 3.4). Alternatively, colorimetric methods can be used. The pre-incubation is stopped when NO_3^- and NO_2^- are depleted.

The pre-incubated soil slurry is aliquoted into 20 ml glass vials. Each vial receives 15 ml of the soil slurry via small funnels while shaking continuously by hand to avoid sedimentation. Acetate is added as C-source for denitrification as it is one of the most important fermentation products found in soils potentially fuelling denitrification. The vials are crimp-sealed with butyl septa and magnetic caps. To restore anoxic

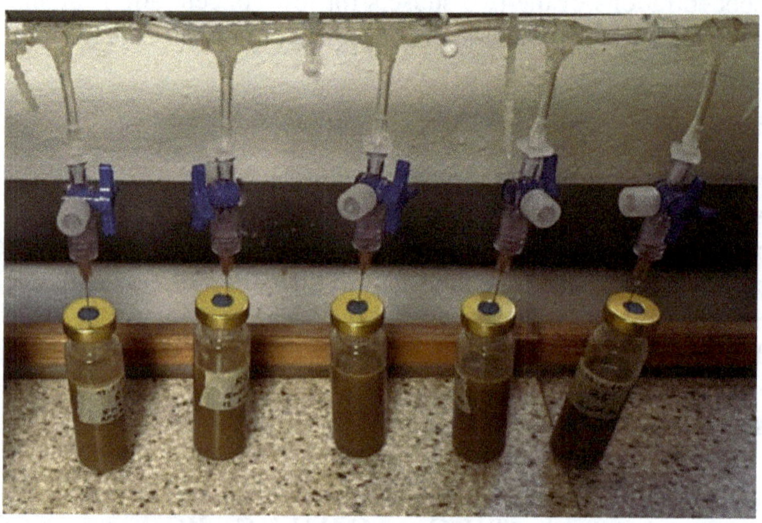

Plate 3.4 Flushing of incubation vials containing soil slurry

conditions in the samples, the septa are pierced by a thin needle and the headspace is evacuated and helium-flushed five times leaving an overpressure of 620 or 800 mbar in the subsets that will or will not receive acetylene, respectively. Those vials with 620 mbar overpressure receive 0.9 ml acetylene via a helium flushed gastight syringe with a thin needle, resulting in 800 mbar overpressure and 10% (v/v) acetylene in the headspace.

Finally, all vials receive the respective concentration of NO_3^- ranging from 20–1000 μM from anoxic stock solutions via a 100 μl micro-volume liquid syringe. After placing the vials in the temperature-controlled tray holders of the incubation system, the gas chromatographic analysis starts. Samples are transferred from the trays to the temperature-controlled agitator and vigorously shaken for 2 min before each analysis to allow for equilibration between slurry and headspace. The autosampler syringe is flushed with helium for 60 s before piercing the septum. It draws a 250 μl sample of the headspace and directly injects it into the gas chromatograph. Each vial is sampled 4 to 8 times during the incubation period depending on the initial NO_3^- concentration and the denitrification potential.

In addition to the gas analysis, NO_3^-, NO_2^- and ammonium can be analysed from t_0 as well as from t_{end} samples to check for N-recovery. For NO_3^- and NO_2^- analysis, the samples are centrifuged as stated above and analysed with the semi-automated NO_3^- and NO_2^- analyser (Plate 3.4). Ammonium binds to the cation exchange sites of soil particles. Thus, extraction with 2 M potassium chloride is performed to extract ammonium prior to quantification. 1 ml samples are mixed with 1 ml of 2 M potassium chloride in an Eppendorf tube before being centrifuged for 15 min at 16740 × g and 4 °C. The supernatant is then used for a colorimetric assay (Berthelot reaction). In brief, 100 μl of supernatant is mixed with 50 μl of 2% sodium phenoxide, 25 μl of 0.005% sodium nitroprusside, and 25 μl of sodium hypochlorite. The absorption of the formed indophenol is measured at 630 nm in the spectrophotometer after 30 min of incubation at 30 °C in the dark.

Data Evaluation

The gas chromatographic data is evaluated by firstly transforming peak areas into ppm values with the help of calibration curves. These ppm values are then corrected for the pressure in the vial headspace, which decreases with the number of samplings. It is important to account for the soluble fraction of CO_2 and N_2O in water. The temperature-dependent Bunsen coefficients of these gases can be used to estimate the physically dissolved fraction. For nitrous oxide, the Bunsen coefficient is 0.85 at 10 °C, 0.71 at 15 °C, 0.61 at 20 °C, and 0.52 at 25 °C. For carbon dioxide, it is 0.99 at 10 °C, 0.85 at 15 °C, 0.74 at 20 °C, and 0.65 at 25 °C (Blachnik 1998). Physically dissolved CO_2 reacts in a pH-dependent manner with water to H_2CO_3 that further reacts to bicarbonate and carbonate. This pH-dependent chemically dissolved fraction also needs to be considered.

Calculation of decreasing headspace pressure after sampling:

$$n_0 = \frac{p_0 \times V_h}{R \times T} \tag{3.1}$$

$$p_s = \frac{p_0 \times V_h}{V_h + V_s} \tag{3.2}$$

$$n_s = \frac{p_s \times V_s}{R \times T} \tag{3.3}$$

$$p_1 = \frac{(n_0 - n_s) \times R \times T}{V_h} \tag{3.4}$$

in the likely case that $p_1 = p_s$, this equation simplifies to:

$$p_1 = \frac{p_0 \times V_h}{V_s + V_h}$$

Symbol		Unit
p_0	Total pressure in the headspace before sampling	Pa
n_0	Amount of substance in the headspace before sampling	mol
V_h	Volume of the headspace	m^3
R	Gas constant	8.314 J K^{-1} mol^{-1}
T	temperature	K
V_s	Sample volume	m^3
p_s	Pressure during sampling	Pa
n_s	Amount of substance sampled	mol
p_1	Total pressure in the headspace after sampling	Pa

Calculation of amount of substance from ppm values:

$$n_{\text{total}} = n_{\text{gas}} + n_{\text{diss}} + n_{\text{chem}} \text{ (last term only if needed)} \tag{3.5}$$

$$n_{\text{gas}} = y_i \times \frac{p_h \times V_{\text{gas}}}{R \times T} \tag{3.6}$$

$$n_{\text{diss}} = \alpha \times \frac{n_{\text{gas}}}{V_{\text{gas}}} \times V_{\text{liq}} \tag{3.7}$$

$$n_{\text{chem}} = n_{\text{diss}} \times 10^{pH - pKa} \tag{3.8}$$

Symbol		Unit
n_{total}	Total amount of substance	mol
n_{gas}	Amount of substance in gaseous phase	mol
n_{diss}	Amount of substance in liquid phase	mol
n_{chem}	Chemically dissolved fraction	mol
V_{gas}	Headspace volume	m^3
V_{liq}	Slurry volume	m^3
y_i	Mole fraction	ppm $\times 10^{-6}$
p_h	Headspace pressure	Pa
R	Gas constant	8.314 J K^{-1} mol^{-1}
T	Temperature	K
α	Bunsen coefficient	

3.1.3 System 2

The described robotized gas analysis system is an improved version of the system described by Molstad et al. (2007). The current system has been published by Molstad et al. (2016). The system has replaced the microGC with a traditional GC (Agilent 7890A), equipped with various valves, columns, and detectors (TCD, FID, and ECD) to analyse all relevant gases (except NO) by a single injection. The setup has two PLOT columns: a Molesieve column for separation of N_2 and O_2, and a PLOTQ column to separate CH_4, N_2O, and CO_2. The system has three loop injectors–one for each PLOT column and one for injecting samples into the chemiluminescence NO detector. Short, packed Hayesep columns are used to retain and backflush water coming with the sample.

The use of ECD ensures sensitive measurements of N_2O at low concentrations, and the FID detects CH_4 in the sub-ppm range, thus allowing quantification of these gases at below ambient levels in the atmosphere (which are 0.35 ppmv for N_2O, 1.84 ppmv for CH_4). At high concentrations, N_2O can be quantified via the TCD signal, which is linear, ensuring accurate determination of N_2O over a wide concentration range.

The Gilson 222 (XL) autosampler of the old system has been replaced with the larger CTC GC-PAL (rebranded by Agilent), including a custom-made holder for a hypodermic needle. Up to 30 stirred and 16 non-stirred 120 ml serum flasks, incubated in a water bath, can now be monitored in one experiment (Plate 3.5).

To transfer a sample from the headspace to the three sampling loops (Plate 3.6), a peristaltic pump (Gilson Minipuls 3) is used. We now use Iso-Versinic Fluoroelastomer (IF) tubing (17 cm, ID = 2 mm; product nr 770130, Saint-Gobain or product nr F117744, John Morris) instead of Marprene tubing. This is because the diffusion of

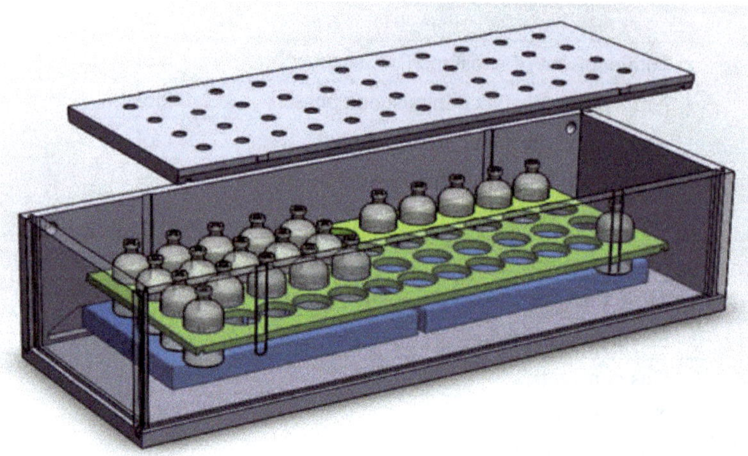

Plate 3.5 Water bath with stirrer plates (blue) and rack (green) that can hold up to 30 stirred and 16 non-stirred bottles. The lid fits onto the tops of the bottles. Other types of bottles may be supported by making other versions of the two plates that serve as rack and lid. The autosampler (not shown) is attached behind the bath. The bath, rack, and lid can be produced from CAD drawings with a CNC router, using seven PVC plates of 2 cm thickness. The CAD drawings are available upon request

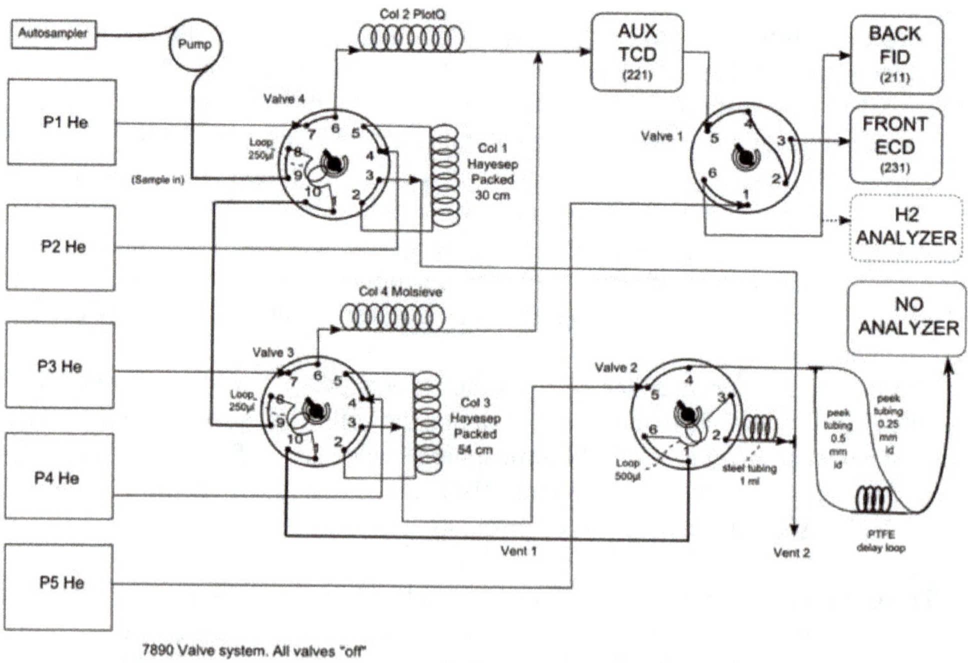

Plate 3.6 Architecture of the GC system. The GC exists in slightly different versions, equipped with one or two ovens (one for the PLOTQ column and one for the Molesieve column) which are held at constant temperature

N_2 and O_2 is slower through IF than Marprene, and because this replacement reduced the carryover of N_2O and CH_4 (see leaks, carryover, and dilution tests below).

In all other aspects, the new system is equal to the original (Molstad et al. 2007) regarding the repeated sampling by peristaltic pumping, replacement of the sampled gas by reversed pumping (injecting He), and transfer of gas to a chemiluminescence NO analyser. In short, samples are taken from 10–120 ml serum flasks using an autosampler connected to a peristaltic pump. The autosampler operates a needle which pierces the septa on the flasks (never twice at the same spot), and the pump takes approximately 1 ml headspace gas from the flasks into the sample loops of the GC and NO analyser. After injection (the turning of the valves leading to the gas in the loops into the GC columns and the NO analyser), the pump is reversed, pumping helium and the portion of the up-pumped gas in the pipelines back into the flasks, thus minimizing dilution of the gas in the flasks while maintaining atmospheric pressure.

The system monitors O_2, N_2, N_2O, NO, CO_2, and CH_4 (and various other gases if needed) by repeated sampling of the headspace, while sustaining the gas pressure near 1 atmosphere in the flasks due to replacement of sampled gas with He.

3.1.3.1 The GC

The flow scheme of the GC system is shown in Plate 3.6. Loop injection to the PLOT columns (Molesieve and PLOTQ) is done by Valve 3 and 4 via Hayesep backflush columns, which prevent water from entering the PLOT columns. For the Molesieve column, also CO_2 is backflushed. The third loop injection valve (Valve 2) sends samples to the NO analyser. Like in Sitaula et al. (1992), Valve 1 directs the outlet from the TCD to the FID most of the time to prevent oxidation of the detector, except for a brief period when N_2O is eluted from the PLOTQ column.

The flowchart for the valve positions during a run is shown in Plate 3.6, and a chromatogram of air is illustrated in Fig. 3.3. Figure 3.4 shows the valve positions, column back pressures, and substance elution times during a run. Further information about the GC is found in Table 3.2.

Slightly different versions of the system have been constructed with our help elsewhere. Currently, there are five systems operating in China and three in Norway. These systems are identical in principle, the main difference being that some of the GC's have the PLOTQ column in a separate oven. Our experience is now that this is not necessary for the compounds we are analysing. In one version, an H_2 analyser is attached in parallel to the FID. A split in front of the FID directs a minor amount of the column flow into a Peak Performer 1 RCP H_2 analyser.

3.1.3.2 The NO Analyser

For the analysis of nitric oxide, we use either a modified Teledyne 200 NO/NOx analyser or a Sievers NO 280i. The modification of the Teledyne analyser is held a trade secret by the Norwegian distributor of Teledyne (Industriell Måleteknikk, Oslo,

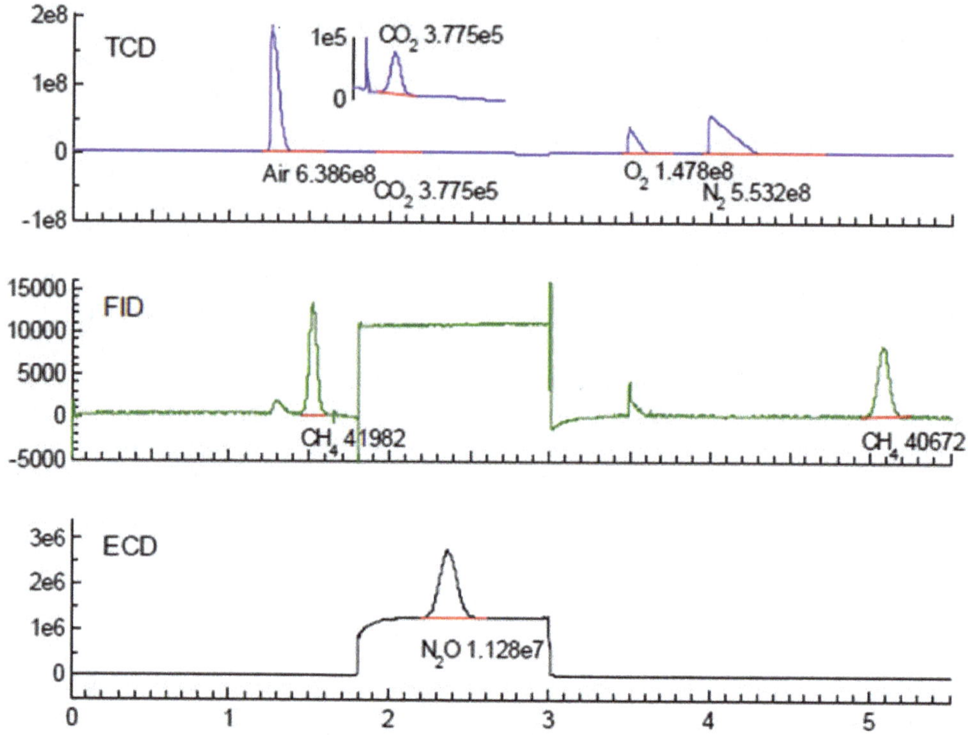

Fig. 3.3 Chromatogram of an air sample (5.5 min). The peaks before the 3rd minute come from the PLOTQ column, the others come from the Molesieve column. CO_2, N_2, and O_2 (at 2.1, 3.5, and 4 min) are detected on the TCD. Methane is detected, at 1.5 min from the PLOTQ column and at 5 min from the Molesieve column, on the FID, and also, for high concentrations, on the TCD. For a 1.2 min period, the gas from the TCD is switched from FID to ECD, allowing the quantification of low N_2O concentrations. At high concentrations, N_2O is also detected by the TCD, thus extending the linear range much beyond that of the ECD

<table>
<tr><td>Valve 1</td><td colspan="3">ON to FID</td><td>OFF to ECD</td><td colspan="2">ON to FID</td></tr>
<tr><td>Valve 2</td><td>OFF Loading loop 3</td><td colspan="4">ON Inject to NO analyzer</td><td>OFF</td></tr>
<tr><td>Valve 3</td><td>OFF Loading loop 1</td><td>ON inject to Hayesep and PLOTQ</td><td colspan="4">OFF Backflush</td></tr>
<tr><td>Valve 4</td><td>OFF Loading loop 2</td><td colspan="2">ON Inject to Hayesep and Molesieve</td><td colspan="3">OFF Backflush</td></tr>
<tr><td>P1</td><td colspan="4">16 psi</td><td colspan="2">10 psi</td></tr>
<tr><td>P3</td><td colspan="4">12 psi</td><td colspan="2">16 psi</td></tr>
<tr><td>TCD</td><td colspan="6">Air CH_4 CO_2 N_2O O_2 N_2 CH_4</td></tr>
<tr><td>FID</td><td colspan="6">CH_4 CH_4</td></tr>
<tr><td>ECD</td><td colspan="6">N_2O</td></tr>
<tr><td>Time (min)</td><td colspan="6">0 1 2 3 4 5</td></tr>
</table>

Fig. 3.4 Valve positions, column pressures, and eluted peaks during a single run (5 min)

Table 3.2 Characteristics of the GC

PLOTQ column	30 m × 0.53 mm × 40 μm, Agilent p/n 19094P-QO4
Molesieve column	30 m × 0.53 mm × 50 μm, Agilent p/n 19095P-MSO
Oven temp.	50 °C (If two ovens: 50 °C for the Molesieve and 35 °C for the PLOTQ)
TCD parameters	250 °C, 20 ml min^{-1} reference flow, 3 ml min^{-1} makeup flow (He 5.0)
FID parameters	250 °C, 40 ml min^{-1} H_2, 400 ml min^{-1} air, 20 ml/min makeup (N_2 5.0)
ECD parameters	340–375 °C (depending on the ECD), 25 ml min^{-1} makeup (90%Ar/10%CH4)
Carrier gas	He 5.0, approx. 24 ml min^{-1}

Norway) but involves an improvement of the response time. Without this modification, the Teledyne analyser has proven to be unsuitable for the incubation system, which injects a short pulse of 0.5 ml gas into the analyser. (An unmodified analyser gives a one-minute response to a 0.5 ml injection; depending on the NO concentration, this response can be either a broad flat-topped peak or a sharp peak followed by one minute of noise.) Based on this, it is recommended to either buy the Teledyne instrument from the Norwegian distributor or acquire a different NO analyser, like the (much more sensitive) Sievers NOA 280i, which is also supported by the software, when building the incubation system.

3.1.3.3 Leaks, Carryover, and Dilution of Headspace by Sampling

The new system has another type of injection (loop injection versus the solenoid-valve system of the microGC), and this leads to less dilution of the headspace per sampling and lower leaks. We have also replaced the Marprene tubing with Iso-Versinic Fluorelastomer tubing, which is more gastight, though less durable.

By sampling repeatedly from 120 ml flasks containing air (no liquid), we have found that the dilution of the headspace per sampling is 0.71%, which corresponds to replacing 0.85 ml of gas with helium. This test was conducted with active sampling to the NO analyser; without this (i.e. without turning Valve 2), the dilution per sampling was 0.36% or 0.43 ml. These volumes are lower than the sum of the volumes of the sample loops because they are heated to 100 °C. The theoretical loss for a 1 ml sample loop should be approximately 1.0 ml*293 K/373 K = 0.785 ml if the temperature in the bottle is 293 K. The dilution is slightly higher than this due to the mixing of sample gas with helium in the pump tubing and pipes.

We have also tested the leak of O_2 and N_2 into the flasks per sampling by repeated sampling from flasks with He (1 atm) at 20 °C. This was done with two types of tubing, Marprene (ID = 1.6 mm), and Iso-Versinic Fluorelastomer (ID = 2 mm).

Table 3.3 Carryover for N_2O and CH_4 in five consecutive helium samples after a standard containing 151 ppmv N_2O and 1% CH_4. Standard errors are, for N_2O: < 2ppbv; for CH_4: < 0.015ppmv (Two replicates for Marprene, three for Iso-Versinic)

Bottle no.		1	2	3	4	5
N_2O (ppb)	Marprene	125	43	27	14	14
	Iso-Versinic	55	20	11	7	6
CH_4 (ppm)	Marprene	1.0	0.4	0.2	0.1	0.1
	Iso-Versinic	0.15	0.05	0.03	0.03	0.01

The results for Marprene are

N_2: 10 ppmv/sampling = 50 nmol N_2/sampling
O_2: 5 ppmv/sampling = 25 nmol O_2/sampling

These leak estimates were obtained by running the pump at a speed resulting in a flow of 9 ml min^{-1}, sample pumping for 20 s = 3 ml total volume taken into the loops. If using slower pumping speed and longer times (3 ml min^{-1}, 60 s pumping), the leaks were approximately doubled.

The results for Iso-Versinic Fluorelastomer, pumping 12 ml min^{-1} (due to larger ID) for 15 s:

N_2: 2.3 ppmv/sampling = 11 nmol N_2/sampling
O_2: 1.2 ppmv/sampling = 6 nmol O_2/sampling

The leaks can be further reduced by enclosing the pump in a helium chamber. A simple plastic cover over the pump head, flushed with He (10 ml min^{-1}), reduced the leaks of N_2 and O_2 to 1.6 and 0.8 ppmv/sampling.

The carryover of N_2O and CH_4 from one sample to the next was measured by sampling from a standard containing 151 ppmv N_2O and 1 vol.% CH_4 before sampling in sequence five bottles containing only helium. The results (Table 3.3 and Fig. 3.5) show that, compared to Marprene, the Iso-Versinic tubing has less than half the carryover for N_2O, and less than one-fifth for CH_4. With Iso-Versinic, the carryover to the first helium sample yielded 0.055 ppmv N_2O (0.037% of the previous sample) and 0.15 ppmv CH_4 (0.0015% of the previous sample). The carryover of O_2 and N_2 was measured in a similar way by sampling from a helium bottle after sampling from a bottle with 20% O_2 and 79% N_2. The carryover of O_2 and N_2 was not detectable with Iso-Versinic; with Marprene, it yielded approximately 7 ppmv O_2 and 16 ppmv N_2. For CO_2, a bottle with 8.5% CO_2 yielded a carryover of 13.3 ppmv (0.016%) for Marprene and 7.3 ppmv (0.0086%) for Iso-Versinic

3.1.3.4 Software

The system is controlled by Python software which checks the readiness state of the chromatography software, controls the autosampler and the peristaltic pump, triggers the GC, records the output of the NO analyser, and sorts and tabulates the

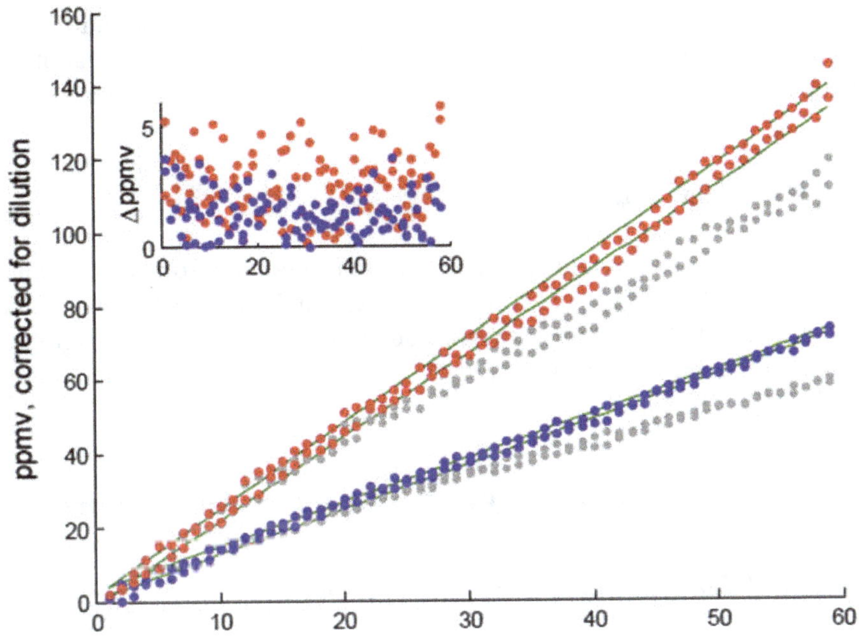

Fig. 3.5 Leak test: Results of repeated sampling from two 120 ml bottles containing helium. The O_2 (blue dots) and N_2 (red dots) concentrations increase on average by 1.2 ppmv and 2.3 ppmv per sampling, respectively. The insert shows the leaks, the difference between consecutive measured ppmv values. The leaks do not appear to increase over the 60 samplings (These measurements have been adjusted for the dilution caused by replacing the sampled gas with helium; unadjusted measurements are shown in grey)

results. The bottles may be given individual rates of sampling, and the sampling rate of each bottle may be easily changed during an experiment (The problem of collisions of sampling time is solved by putting the bottles in a priority queue and linking together bottles that must be sampled in sequence). It is also possible to let the sampling rates automatically depend on the measured concentrations during a run (for example, measuring more often when the oxygen concentration runs low), but this requires some programming from the experimenter. For the inspection of results during an experiment, a graphical user interface is provided which lets the experimenter quickly plot time series of raw data or estimated concentrations for chosen substances and bottles.

3.1.3.5 Data Handling and Analysis

The system produces large amounts of data, requiring adequate handling. We have developed software that compiles the data as a result matrix for each gas (vial nr * time), and calculation routines for estimation of gas transport between headspace and liquid, concentrations of the gases in the liquid, and the kinetics of transformations (Molstad et al. 2007). The software has been used in many studies of N_2O/N_2 gas kinetics, NO scavenging kinetics, analyses of oxic and anoxic respiration, etc. (see publication list of NMBU Nitrogen Group https://www.nmbu.no/en/research/groups/nitrogen/publications). The software is provided free of charge on request.

3.1.3.6 Documentation

This description of the robotic incubation system is modified from a technical report (Molstad et al. 2016) which is available on Researchgate. The report should be referred to by its DOI, together with Molstad et al. (2007), when publishing results obtained with this incubation system.

3.2 Automated Chamber Systems for Field Measurements

3.2.1 Field Techniques Using GC Systems

Accurate measurement of GHG flux from soils is challenging because of the high temporal and spatial variabilities in the processes contributing to GHG emissions (Butterbach-Bahl et al. 2016), which result in a variability of soil surface fluxes often spanning 2–3 magnitudes. Addressing these two components of variability remains a challenge.

Automated chamber systems are increasingly used to address temporal variability. These systems consist of a set of automated chambers, mostly the closed chamber type if fluxes of rather non-reactive trace gases such as CH_4, CO_2, or N_2O are measured, or of the dynamic chamber type, if soil–atmosphere fluxes of reactive gas species such as NO, NO_2, or NH_3 are measured (e.g. Butterbach-Bahl et al. 1997).

One of the first times an automated closed chamber system was used was for the quantification of CH_4 fluxes from rice paddy fields in Italy (Schütz et al. 1989). These authors showed that CH_4 fluxes vary strongly on diurnal to seasonal scales and, by using automated chamber systems with sub-daily measurements of fluxes, provided for the first time an accurate quantification of cumulative seasonal CH_4 fluxes. The development of automated measuring chambers continued over years, and Papen and Butterbach-Bahl (1999), Gasche and Papen (1999) and Butterbach-Bahl et al. (2002) provided examples of how such systems can be used for quantifying the soil–atmosphere exchange of N_2O, NO, NO_2, and CH_4 continuously over entire years (Plate 3.7). These measurements revealed that even during freeze–thaw periods forest soils are extremely strong sources for the atmospheric GHG N_2O (Papen and Butterbach-Bahl 1999) and that in years during which such freeze–thaw N_2O fluxes occur, the annual flux magnitude might be a factor of 2–10 times higher than in years without freeze–thaw events (Plate 3.7).

Only by using automated chamber systems, Wolf et al. (2010) were able to accurately quantify N_2O fluxes during freeze–thaw events in steppe ecosystems of Inner

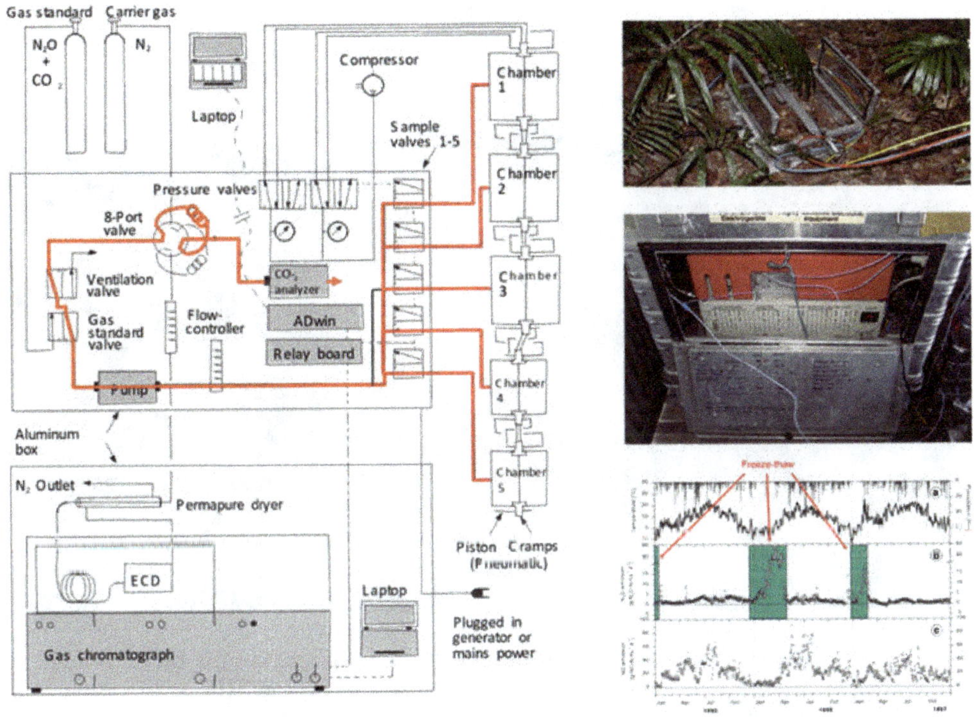

Plate 3.7 Schematic drawing of an automated chamber system for measurements of soil GHG fluxes with five chambers and a gas chromatograph equipped with an Electron Capture detector (for N_2O analysis) and Flame Ionization Detector (for analysis of CH_4 plus CO_2 [the latter requires use of a methanizer]) (right panel). The right panel show a picture of an automated chamber with pistons to close/open chambers by compressed air (top), a gas chromatograph fixed to a small box (middle), and a dataset of mean daily N_2O and NO fluxes as obtained by the use of an automated closed chamber system (N_2O) and an automated dynamic chamber system. Further details describing these measurements are provided in Papen and Butterbach-Bahl (1999) and Gasche and Papen (1999)

Mongolia. In this study, not only high seasonal, but also diurnal flux variations were observed. Meanwhile, to the knowledge of the authors, approx. 20–30 groups worldwide are using automated chamber systems, i.e. still in a minority of GHG studies are such systems used for quantifying GHG emissions from upland and lowland agricultural, forest, or grassland ecosystems, while installations are spanning various climate zones, i.e. from the tropics to alpine systems. Most recently, those automated closed chamber systems were combined with automated lysimeter systems, so that environmental N and C losses can be quantified at the same time (Fu et al. 2017, Plate 3.8). At the same time, the use of laser spectroscopy instead of gas chromatography allowed to lower the flux detection limit significantly, so that closing times for chambers could be reduced (Savage et al. 2014).

But how important are high-frequency measurements of GHG emissions from soils? Does this indeed significantly improve the accuracy of annual flux estimates? Or do weekly or monthly measurements suffice to conclude on annual exchange rates

Plate 3.8 Automated GHG flux measurements at a grassland site in Bavaria by the Institute of Meteorology and Climate Research, Atmospheric Environmental Research (IMK-IFU), Garmisch-Partenkirchen, Germany, using a closed chamber robot system in combination with lysimeters. For this installation, undisturbed soil monoliths (1 m diameter, 1.8 m depth) were taken at various locations along a climate gradient and re-installed at a common site. This setup allows to monitor short- and long-term effects of climate change (changes in rainfall and temperature) on ecosystem CO_2, N_2O, and CH_4 fluxes as well as on groundwater recharge and leaching of nitrogen compounds (DON, NO_3^-, NH_4^+). Photo: courtesy of Dr. Ralf Kiese, who is also leading this research programme at IMK-IFU, which is part of the German Helmholtz TERENO initiative

of GHG between soils and the atmosphere? These questions were addressed in a study by Barton et al. (2015) using 28 published datasets of sub-daily N_2O fluxes from a variety of different terrestrial ecosystems across the globe. The aim of the study was (a) to provide guidelines for sampling frequency to estimate annual N_2O fluxes using manual chambers and (b) to quantify the uncertainty if the sampling frequency is lowered to 2–3 times per week, weekly, or monthly or if the sampling frequency is varied, with intensive measurements during episodes with expected high fluxes (e.g. following fertilization events or first rainfalls following a dry period), but lower frequency measurements during periods with expected low fluxes (e.g. dry periods). The analyses by Barton et al. (2015) show that "Nitrous oxide emissions need to be measured daily to accurately determine annual N_2O flux in environments where data has not previously been collated", and that daily sampling was required to achieve annual N_2O fluxes within 10% accuracy compared to sub-daily measurements with automated chambers. Lowering the measuring frequency to weekly, two weekly or monthly measurements result in an under- or over-estimation of annual N_2O fluxes by up to 93.5%. Overall, the authors suggested that the current uncertainty in global terrestrial N_2O budgets associated with the upscaling of field-based datasets can be decreased significantly using adequate sampling frequencies (daily, every second day, or by using automated measuring systems). For further reading the paper of the Global Research Alliance by Grace et al. (2020) on automated techniques is recommended.

3.2.2 Combination of Automatic Chamber System and CRDS Analyser for Field GHG Flux Measurements

The LI-8100A Automated Soil CO_2 Flux System (Licor, Inc.) is designed to measure CO_2 efflux with an IR gas analyser (IRGA) from soils and low vegetation using up to 16 automated chambers with the closed non-steady-state transient approach controlled by a multiplexer unit LI-8150. This system allows the combination with a CRDS analyser (see Chap. 2, Sect. 2.7), capable of measuring other trace gases, including CH_4, N_2O, and NH_3. They can be interfaced with the LI-8100A System and the SoilFluxPro data processing software (freely downloadable from the LI-COR website) to compute fluxes for the additional gases. There is an application note available that describes how to integrate a third-party analyser like the Picarro G2508, which can measure CO_2, CH_4, N_2O, NH_3, and H_2O simultaneously, into the system and the procedure used to compute fluxes of additional gases in SoilFluxPro software (Licor 2016).

There are three options of combining the two systems (see Plates 3.9 and 3.10); the first two options (Plate 3.9a, b) connecting the CRDS analyser to the primary sampling loop or as a subsampling loop are described in the application note and need adjustments of the flow and a calibration of the IRGA (see application note, Licor 2016). The third mode of gas flow path does not need these adjustments, and the LI-8100A might also be used without the CRDS analyser without the need for readjustments.

If the system is used in humid and cold regions, a water trap or nafion tube to dry the measurement gas flow is recommended to protect the CRDS analyser from condensed water. In the LI-COR control software, LI-8100A the total gas volume of the system (chamber, tubes, analyser internal tubing, water trap) has to be defined for correct flux calculations. With the LI-8100A software, the valve system of the Multiplexer and the measurement sequence of the 16 automated chambers can be controlled.

It is important to synchronize the system clocks of the IRGA and the Picarro frequently (daily–weekly); otherwise, the SoilFluxPro software cannot match the LI-COR and the Picarro dataset for flux calculations properly. Matching of the LI-COR and the Picarro dataset in the SoilFluxPro software is also explained in the application note (Licor 2016).

For long-term field measurement campaigns, the Picarro Analyser should be placed in a water proof box. If this system is used in a hot environment, the housing box for the Picarro Analyser has to be cooled so that temperature in the box stays

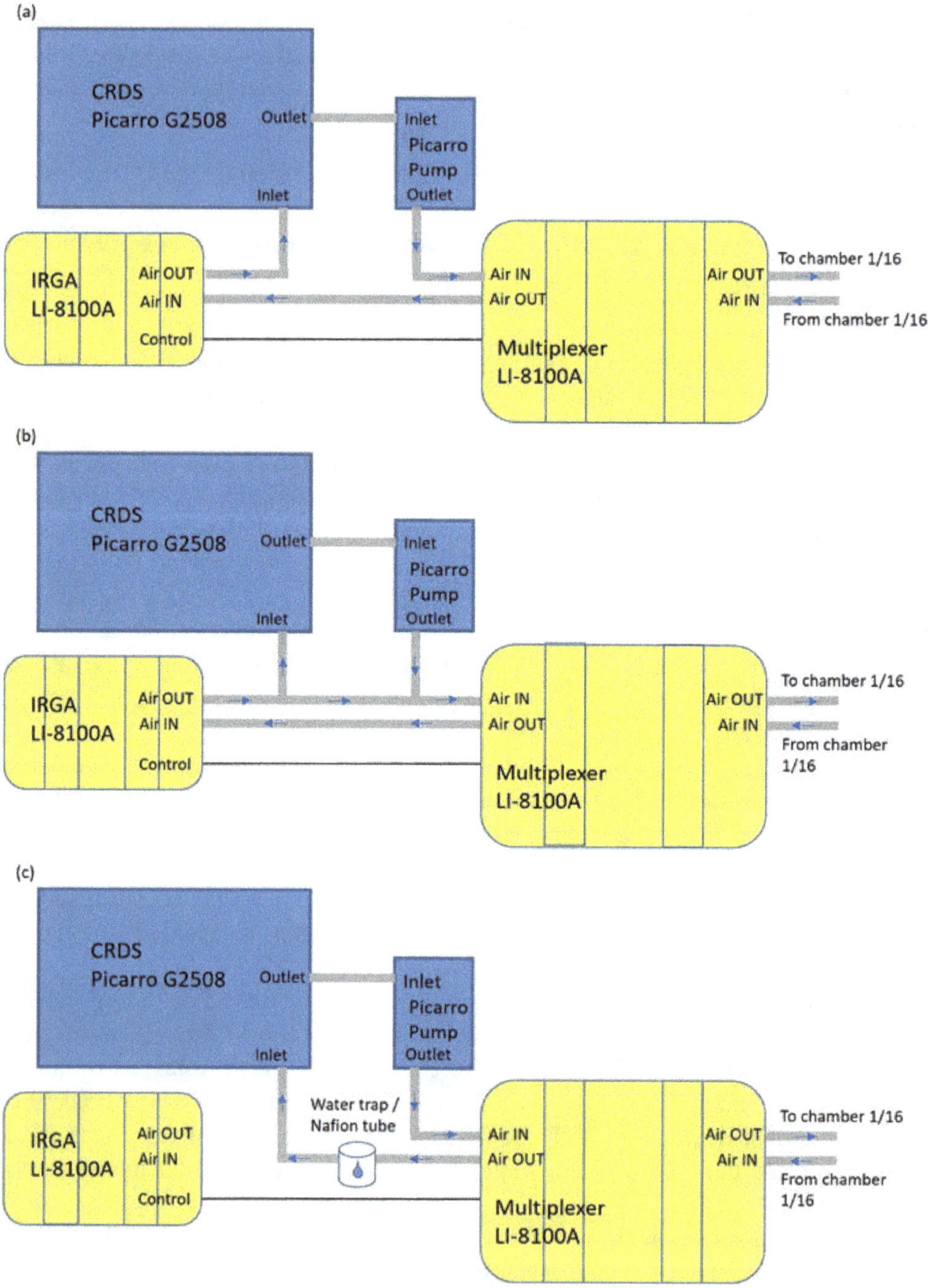

Plate 3.9 Schematic modes of combining the LI-8100A automated chamber system including up to 16 clear or opaque chambers with a CRDS Picarro five-gas analyser G2508 for simultaneous flux measurement of CO_2, CH_4, N_2O, NH_3, and H_2O: **a** entire measurement gas flow enters the IRGA first, then the CRDS analyser, **b** only part of the measurement gas flow from the IRGA Outlet enters the CRDS analyser, and **c** the IRGA unit is only used to control the pump and valve system of the multiplexer but not for CO_2 flux measurement. The multiplexer has 16 ports and can be connected to clear and opaque chambers, to soil air probes or aboveground sampling ports

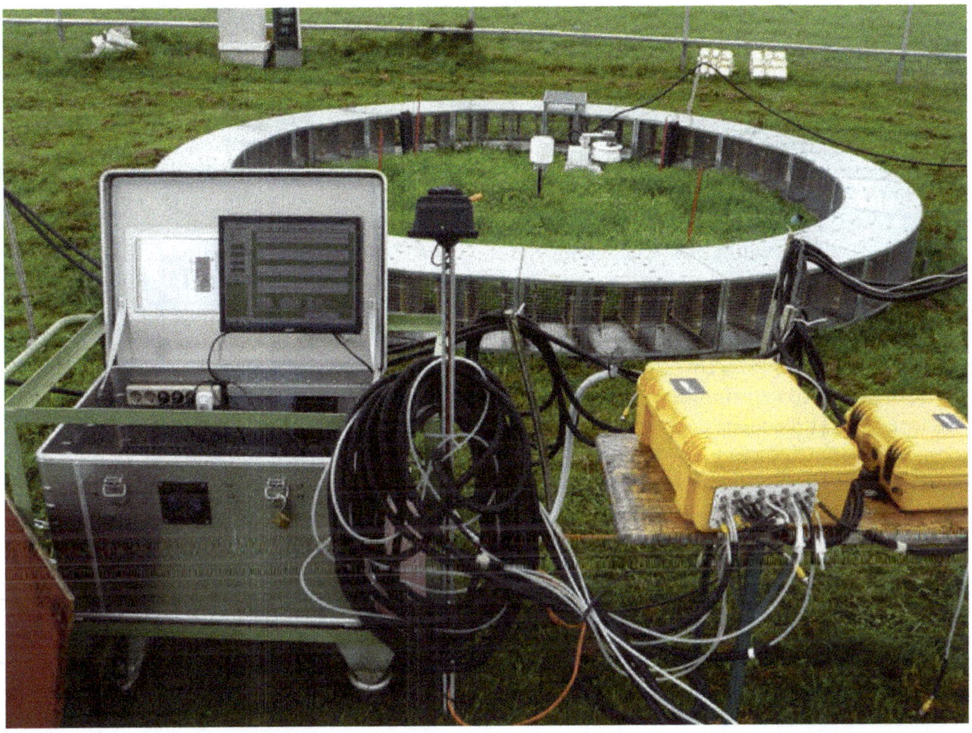

Plate 3.10 The combined automatic chamber system with CRDS analyser for field GHG-flux measurement in the Giessen climate change experiment combining air Temperature warming of + 2 °C and free-air CO_2 Enrichment of +20% (T-FACE). In the left-hand box, the Picarro analyser is installed including its monitor and pump. The LI-8150 Multiplexer (big yellow box) is connected to seven automated chambers and eight soil air probes and switches every 5 to 30 min from one port to the next controlled by the LI-8100A Analyzer Control Unit (small yellow box). The screen (lower left side) shows stable N_2O concentrations (zero flux, 1st panel above), CH_4 uptake (2nd panel), stable H_2O concentrations (3rd panel), and ecosystem CO_2 respiration (4th panel on the screen)

below 35 °C; otherwise, the analyser will go into error. A cooling solution with a small air condition unit is possible and is described in the Picarro Application Note AN041 (Anthony and Silver 2020).

This combination of an automatic chamber system and CRDS Analyser allows to set up a high-frequency measurement sequence including up to 16 ports. A combination of clear and opaque chambers, soil air probes (e.g. Accurel© tubes installed in different soil depths), and aboveground air sampling ports is possible to create detailed data on GHG concentrations in different soil depths and aboveground and the resulting gas fluxes. The LI-8100A system allows to install and measure soil temperature and moisture probes at each automated chamber to record the top soil conditions during flux measurement. The result of such data sets allows a better understanding of hot moments of GHG emission peaks during long-term measurement campaigns and the hotspots of GHG production within the soil profile.

References

Anthony TL, Silver WL (2020) Field deployment of Picarro G2508 and Eosense eosAC/eosMx in perennially or periodically flooded soils. Picarro Application Note (AN041)

Bakken LR, Bergaust L, Liu B, Frostegård Å (2012) Regulation of denitrification at the cellular level: a clue to the understanding of N_2O emissions from soils. Philos Trans R Soc B Biol Sci 367:1226–1234

Barton L, Wolf B, Rowlings D, Scheer C, Kiese R, Grace P, Stefanova K, Butterbach-Bahl K (2015) Sampling frequency affects estimates of annual nitrous oxide fluxes. Sci Rep 5:16912

Bedard C, Knowles R (1989) Physiology, biochemistry, and specific inhibitors of CH_4, NH_4+, and CO oxidation by methanotrophs and nitrifiers. Microbiol Rev 53:68–84

Bergaust L, Mao Y, Bakken L, Frostegard A (2010) Denitrification response patterns during the transition to anoxic respiration and posttranscriptional effects of suboptimal pH on nitrogen oxide reductase in Paracoccus denitrificans. Appl Environ Microbiol 76:6387–6396

Bergaust L, van Spanning RJM, Frostegård Å, Bakken LR (2012) Expression of nitrous oxide reductase in Paracoccus denitrificans is regulated by oxygen and nitric oxide through FnrP and NNR. Microbiol 158:826–834

Blachnik R (1998) Taschenbuch für Chemiker und Physiker Ed. 4 Vol 3. Springer, Heidelberg

Braker G, Conrad R (2011) Chapter 2- Diversity, Structure, and Size of N_2O-Producing Microbial Communities in Soils—What Matters for Their Functioning? In, Laskin AI, Sariaslani S, Gadd GMBT-A. in A.M. (eds). Academic Press, pp 33–70

Brenzinger K, Dörsch P, Braker G (2015) pH-driven shifts in overall and transcriptionally active denitrifiers control gaseous product stoichiometry in growth experiments with extracted bacteria from soil. Front Microbiol 6

Butterbach-Bahl K, Gasche R, Breuer L, Papen H (1997) Fluxes of NO and N_2O from temperate forest soils: impact of forest type, N deposition and of liming on the NO and N_2O emissions. Nutr Cycl Agroecosyst 48:79–90

Butterbach-Bahl K, Rothe A, Papen H (2002) Effect of tree distance on N_2O^- and CH_4^- fluxes from soils in temperate forest ecosystems. Plant Soil 240:91–103

Butterbach-Bahl K, Sander B, Pelster D, Díaz-Pinés E (2016) Quantifying Greenhouse Gas Emissions from Managed and Natural Soils. Methods for Measuring Greenhouse Gas Balances and Evaluating Mitigation Options in Smallholder Agriculture, pp 71–96

Dörsch P, Braker G, Bakken LR (2012) Community specific pH response of denitrification: experiments with cells extracted from organic soils. FEMS Microbiol Ecol 79:530–541

Forsyth DS (2004) Pulsed discharge detector: theory and applications. J Chromatogr A 1050:63–68

Fu J, Gasche R, Wang N, Lu H, Butterbach-Bahl K, Kiese R (2017) Impacts of climate and management on water balance and nitrogen leaching from montane grassland soils of S-Germany. Environ Pollut 229:119–131

Gasche R, Papen H (1999) A 3-year continuous record of nitrogen trace gas fluxes from un- treated and limed soil of a N-saturated spruce and beech forest ecosystem in Germany: 2. NO and NO_2 fluxes. J Geophys Res 104:18505–18520

Grace PR, Van der Weerden TJ, Rowlings DW, Scheer C, Brunk C, Kiese R, Butterbach-Bahl K, Rees RM, Robertson GP, Skiba U, (2020) Global Research Alliance N_2O chamber methodology guidelines: Considerations for automated flux measurement. J Environ Qual 49:1126–1140

Hink L, Lycus P, Gubry-Rangin C, Frostegård Å, Nicol GW, Prosser JI, Bakken LR (2017) Kinetics of NH_3-oxidation, NO-turnover, N_2O-production and electron flow during oxygen depletion in model bacterial and archaeal ammonia oxidisers. Env Microbiol 19:4882–4896

Horn MA, Ihssen J, Matthies C, Schramm A, Acker G, Drake HL (2005) *Dechloromonas denitrificans* sp. nov., *Flavobacterium denitrificans* sp. nov., *Paenibacillus anaericanus* sp. nov. and *Paenibacillus terrae* strain MH72, N_2O-producing bacteria isolated from the gut of the earthworm *Aporrectodea caliginosa.* Int J Syst Evolut Microbiol 55(3): 1255–1265

Horn MA, Mertel R, Gehre M, Kästner M, Drake HL (2006) In vivo emission of dinitrogen by earthworms via denitrifying bacteria in the gut. Appl Environ Microbiol 72(2):1013–1018

Keppler F, Hamilton JTG, Braß M, Röckmann T (2006) Methane emissions from terrestrial plants under aerobic conditions. Nature 439:187–191

Liu B, Frostegard A, Bakken LR (2014) Impaired reduction of N_2O to N_2 in acid soils is due to a posttranscriptional interference with the expression of nosZ. mBio 5: 1383–1314

Licor (2016) Capturing and processing soil GHG fluxes using the LI-8100A—Picarro. Application Note. Licor, Linoln, Nebraska

Lycus P, Lovise Bøthun K, Bergaust L, Peele Shapleigh J, Reier Bakken L, Frostegård Å (2017) Phenotypic and genotypic richness of denitrifiers revealed by a novel isolation strategy. ISME J 11:2219–2232

McCrackin ML, Elser JJ (2010) Atmospheric nitrogen deposition influences denitrification and nitrous oxide production in lakes. Ecology 91:528–539

Molstad L, Dörsch P, Bakken LR (2007) Robotized incubation system for monitoring gases (O_2, NO, N_2O, N_2) in denitrifying cultures. J Microbiol Methods 71:202–211

Molstad L, Dörsch P, Bakken L (2016) Improved robotized incubation system for gas kinetics in batch cultures. ResearchGate

Palmer K, Drake HL, Horn MA (2010) Association of novel and highly diverse acid-tolerant denitrifiers with $N_{(2)}O$ fluxes of an acidic fen. Appl Environ Microbiol 76(4):1125–1134

Palmer K, Biasi C, Horn MA (2012) Contrasting denitrifier communities relate to contrasting N_2O emission patterns from acidic peat soils in arctic tundra. ISME J 6(5):1058–1077

Palmer K, Horn MA (2015) Denitrification activity of a remarkably diverse fen denitrifier community in Finnish Lapland is N-oxide limited. PLoS ONE 10(4):e0123123–e0123123

Palmer K, Kopp J, Gebauer G, Horn MA (2016) Drying-rewetting and flooding impact denitrifier activity rather than community structure in a moderately acidic fen. Front Microbiol 7:727

Papen H, Butterbach-Bahl K (1999) A 3-year continuous record of nitrogen trace gas fluxes from untreated and limed soil of a N-saturated spruce and beech forest ecosystem in Germany, 1. N_2O emissions. J Geophys Res 104:18487–18503

Savage K, Phillips R, Davidson E (2014) High temporal frequency measurements of greenhouse gas emissions from soils. Biogeosci 10:2709–2720

Schlüter S, Zawallich J, Vogel HJ, Dörsch P (2019) Physical constraints for respiration in microbial hotspots in soil and their importance for denitrification. Biogeosci 16:3665–3678

Schmidt O, Hink L, Horn MA, Drake HL (2016) Peat: home to novel syntrophic species that feed acetate- and hydrogen-scavenging methanogens. ISME J 10:1954–1966

Schütz H, Holzapfel-Pschorn A, Conrad R, Rennenberg H (1989) A 3-year continues record on the influence of daytime, season, and fertilizer treatment on methane emission rates from an Italian rice paddy. J Geophys Res 941:16405–16416

Sitaula BK, Luo J, Bakken LR (1992) Rapid analysis of climate gases by wide bore capillary gas chromatography. J Environ Qual 21(3):493–496

Taylor AE, Taylor K, Tennigkeit B, Palatinszky M, Stieglmeier M, Myrold DD, Schleper C, Wagne M, Bottomley PJ (2015) Inhibitory effects of C_2 to C_{10} 1-alkynes on ammonia oxidation in two Nitrososphaera species. Appl Environ Microbiol 81:1942–1948

Vollack KU, Zumft WG (2001) Nitric Oxide Signaling and Transcriptional Control of Denitrification Genes in Pseudomonas stutzeri. J Bacteriol 183:2516–2526

Wolf B, Zheng X, Brüggemann N, Chen W, Dannenmann M, Han X, Sutton MA, Wu H, Yao Z, Butterbach-Bahl K (2010) Grazing-induced reduction of natural nitrous oxide release from continental steppe. Nature 464:881–884

Yoshinari T, Knowles R (1976) Acetylene inhibition of nitrous oxide reduction by denitrifying bacteria. Biochem Biophys Res Commun 69:705–710

Zhu J, Mulder J, Solheimslid SO, Dörsch P (2013) Functional traits of denitrification in a subtropical forest catchment in China with high atmogenic N deposition. Soil Biol Biochem 57:577–586

4

Measurement of GHG Emissions: Micrometeorological Techniques

M. Zaman, K. Kleineidam, L. Bakken, J. Berendt, C. Bracken, K. Butterbach-Bahl, Z. Cai, S. X. Chang, T. Clough, K. Dawar, W. X. Ding, P. Dörsch, M. dos Reis Martins, C. Eckhardt, S. Fiedler, T. Frosch, J. Goopy, C.-M. Görres, A. Gupta, S. Henjes, M. E. G. Hofmann, M. A. Horn, M. M. R. Jahangir, A. Jansen-Willems, K. Lenhart, L. Heng, D. Lewicka-Szczebak, G. Lucic, L. Merbold, J. Mohn, L. Molstad, G. Moser, P. Murphy, A. Sanz-Cobena, M. Šimek, S. Urquiaga, R. Well, N. Wrage-Mönnig, S. Zaman, J. Zhang, and C. Müller

Abstract Micrometeorological techniques are useful if greenhouse gas (GHG) emissions from larger areas (i.e. entire fields) should be integrated. The theory and the various techniques such as flux-gradient, aerodynamic, and Bowen ratio as well as Eddy correlation methods are described and discussed. Alternative methods also used

---

M. Zaman (✉) · L. Heng
Soil and Water Management & Crop Nutrition (SWMCN) Section, Joint FAO/IAEA Division of Nuclear Techniques in Food and Agriculture, International Atomic Energy Agency (IAEA), Vienna, Austria
e-mail: m.zaman@iaea.org; zamanm_99@yahoo.com

K. Kleineidam · C. Eckhardt · A. Jansen-Willems · G. Moser · C. Müller
Institute of Plant Ecology, Justus Liebig University Giessen, Giessen, Germany

L. Bakken
Norwegian University of Life Sciences (NMBU), Aas, Norway

J. Berendt · S. Fiedler · N. Wrage-Mönnig
University of Rostock, Rostock, Germany

C. Bracken
School of Agriculture and Food Science and Earth Institute, University College Dublin, Dublin, Ireland

K. Butterbach-Bahl
Institute of Meteorology and Climate Research, Atmospheric Environmental Research (IMK-IFU), Karlsruhe Institute of Technology, Karlsruhe, Germany

Z. Cai
School of Geography Sciences, Nanjing Normal University, Jiangsu, China

S. X. Chang
Department of Renewable Resources, University of Alberta, Edmonton, AB T6G 2E3, Canada

T. Clough
Department of Soil & Physical Sciences, Faculty of Agriculture & Life Sciences, Lincoln University, Lincoln, New Zealand

are Eddy correlation, mass balance techniques, and tracer-based methods. The analytical techniques with current state-of-the-art approaches as well as the calculation procedures are presented.

Keywords Eddy correlation · Flux gradient · Energy balance · Mass balance

4.1 Introduction

Gas transport takes place through the eddying motion of the atmosphere, which transfers sections of air from one level to another; this is used as the basic concept of the micrometeorological approaches used to measure GHG fluxes to or from the soil surface (Denmead 1983). Viscous forces reduce turbulence in distances less than 1 mm, and molecular diffusion is crucial for transport within these very short distances. The basic transportation of gas throughout the free atmosphere, up to within approximately 1 mm of the emitting or absorbing surface, is caused by the turbulent diffusion, where individual eddies are relocated. Therefore, in practically measured distances, the overriding approach is turbulent transport. In the easiest micrometeorological methods, it is possible to measure the GHG flux through sensing the

K. Dawar
Department of Soil and environmental Sciences, University of Agriculture, Peshawar, Pakistan

W. X. Ding
Institute of Soil Science, Chinese Academy of Sciences, Nanjing, China

P. Dörsch · L. Molstad
Faculty of Environmental Sciences and Natural Resource Management, Norwegian University of Life Sciences (NMBU), Aas, Norway

T. Frosch
Leibniz Institute of Photonic Technology, Technical University Darmstadt, Darmstadt, Germany

J. Goopy
International Livestock Research Institute (ILRI), Nairobi, Kenya

C.-M. Görres
Department of Soil Science and Plant Nutrition/Department of Applied Ecology, Hochschule Geisenheim University, Geisenheim, Germany

A. Gupta
Independent Consultant India, Mumbai, India

S. Henjes · M. A. Horn
Institute of Microbiology, Leibniz University Hannover, Hannover, Germany

M. E. G. Hofmann
Picarro B.V., 's-Hertogenbosch, The Netherlands

M. M. R. Jahangir
Department of Soil Science, Bangladesh Agricultural University, Mymensingh, Bangladesh

velocities and concentrations of components of the turbulence (Fowler and Duyzer 1989).

With the exception of the mass balance method (see below), micrometeorological methods are based on the assumption that the GHG flux to or from the surface is identical to the vertical flux measured at the reference level some distance above the surface (Eugster and Merbold 2015). These two quantities may differ as a consequence of three processes: (i) chemical reaction within the air column between the measurement level and the surface; (ii) changes in concentration with time and therefore changes in the storage of the trace gas within the air column; and (iii) horizontal gradients in air concentration leading to advection (Fowler and Duyzer 1989).

Additionally, prerequisites of these methods include large uniform surface areas, uniform flux to the surface in the upwind area that effects the sample point (fetch), and the atmospheric conditions must be unremitting throughout each measurement period. Measuring the flux at the selected sampling points, on flat homogeneous terrain, provides the average vertical flux over the upwind fetch. The constant flux layer is at a range extending vertically to approximately 0.5% of the upwind uniform fetch. Therefore, the constant flux layer, on a uniform 200 m field, is located around 1 m deep at the downwind edge. The height of the sampling points should be placed within the height range where this vertical flux is constant (Monteith 1973; Monteith

D. Lewicka-Szczebak
Laboratory of Isotope Geology and Geoecology, Institute of Geological Sciences, University of Wrocław, Wrocław, Poland

G. Lucic
Picarro Inc., Santa Clara, CA, USA

L. Merbold
Mazingira Centre, International Livestock Research Institute (ILRI), Nairobi, Kenya

J. Mohn
Laboratory for Air Pollution & Environmental Technology, Empa Dübendorf, Dübendorf, Switzerland

P. Murphy
Environment & Sustainable Resource Management Section, School of Agriculture & Food Science, and UCD Earth Institute, University College, Dublin, Ireland

A. Sanz-Cobena
Research Center for the Management of Environmental and Agricultural Risks (CEIGRAM), ETSIAAB, Universidad Politécnica de Madrid, Madrid, Spain

M. Šimek
Institute of Soil Biology, Biology Centre of the Czech Academy of Sciences, and Faculty of Science, University of South Bohemia, České Budějovice, Czech Republic

M. dos Reis Martins · S. Urquiaga
EMBRAPA Agrobiologia Seropédica, Brazilian Agricultural Research Corporation, Seropedica, RJ, Brazil

R. Well
Thünen Institute of Climate-Smart Agriculture, Braunschweig, Germany

and Unsworth 1990). It is evident that experimental areas need to be quite large and uniform before these techniques can be used in accordance with theory.

Two general micrometeorological techniques are used to measure trace gas flux density: eddy correlation and flux gradient (gradient diffusion). The description of these methods is derived primarily from Baldocchi et al. (1988), Denmead (1983), and Fowler and Duyzer (1989). While the overall principle of micrometeorological measurements has not changed, there have been advancements in analysers capable of measuring other greenhouse gases such as CH_4 and N_2O besides CO_2 and H_2O in relevant temporal resolution to allow continuous field-scale observations of these gases. Similarly, standardization approaches have been implemented during the past decade, allowing comparison of measurements across sites, ecosystem types, and climatic regions (Baldocchi 2014; Eugster and Merbold 2015).

4.2 Flux-Gradient Method

Though there are differences, the flux-gradient theory assumes that molecular diffusion is the same process as the turbulent transfer of gas. However, molecular diffusion occurs due to the random motion of molecules, while turbulent transfer, or eddy diffusion, occurs when parcels of air move from one level to another. This results in eddy diffusion generally being several times greater than molecular diffusion. Wind speed, height above surface (i.e. soil, plant canopy, or water), the vertical temperature gradient, and the aerodynamic roughness of the surface are what determine the actual magnitudes. The gradient is positive by convention if the concentration increases towards the surface, and vice versa. Additionally, the turbulent flux is proportionate to the product of an eddy diffusion coefficient and the average vertical concentration gradient of the gas.

The basic data for numerous methods of measuring vertical fluxes over large surfaces are stipulated by measurements within the constant flux layer of gradients with height of wind velocity, concentration of trace gases, and air temperature. The Bowen ratio (energy balance) and the aerodynamic techniques are two of the more popular methods of calculating vertical fluxes.

S. Zaman
University of Canterbury, Christchurch, New Zealand

J. Zhang
School of Geography, Nanjing Normal University, Nanjing, China

K. Lenhart
Bingen University of Applied Sciences, Berlinstr. 109, Bingen 55411, Germany

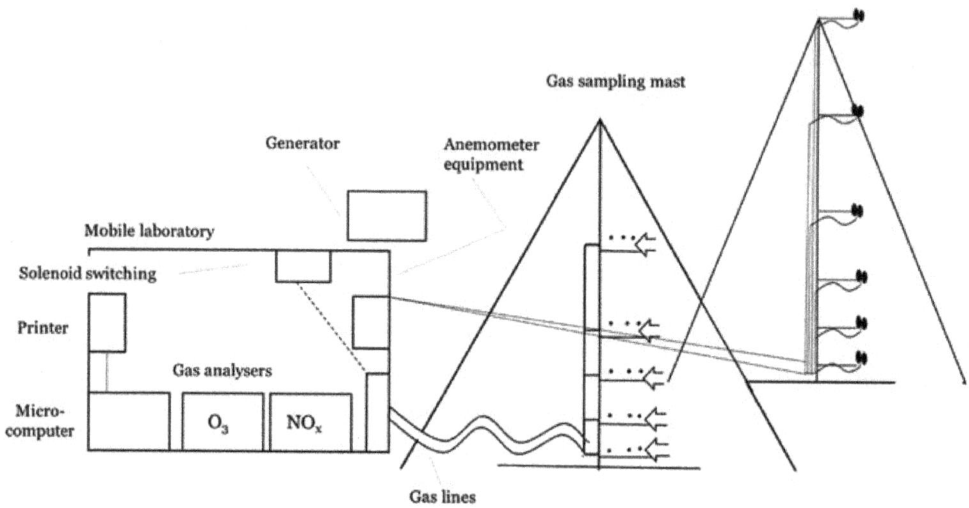

Plate 4.1 Typical instrumentation for flux-gradient measurement in the field

4.3 Aerodynamic Method

The aerodynamic method is based on the relationship between the momentum flux equation and the wind speed gradient. This technique requires the measurement of wind speed at two or more heights and the concentration gradient of the gas at these heights. The trace gas flux is influenced by vertical air density gradients due to water vapour and heat fluxes, and stability corrections must be made (Fowler and Duyzer 1989).

Flux-gradient measurements can be used to estimate emissions from field-size areas of land: 1–100 ha. Plate 4.1 shows the system employed for gases which can be analysed directly in a flowing stream. Gas streams are pumped through sampling tubes from different heights on a gas sampling mast, and wind speed and temperature at different heights are also measured.

4.4 Bowen Ratio (Energy Balance Method)

This procedure is based on the energy balance at the surface, and measurements of the wind velocity profile are not necessary. The inward net radiation is divided among the soil heat flux, sensible heat flux, and the latent heat flux. The Bowen ratio consists of the ratio of sensible to latent heat fluxes. The measurements required for this method are the vertical gradients of gas concentration, humidity, and temperature. Without introducing the uncertainty and difficulty that occurs with stability corrections, which are necessary for the aerodynamic method, these measurements specify approximations of the fluxes of sensible heat, the trace gas, and water vapour. Furthermore, this method allows for the calculation of the rate that water evaporates (Denmead 1983).

The largest and most important drawback for trace gas fluxes is the substantial net radiation fluxes required by energy balance methods. In cloudy, night, or winter conditions, the available net radiation is frequently too small to permit satisfactory flux estimates (Fowler and Duyzer 1989). A further night-time problem is that condensation of dew on radiation instruments leads to erroneous measurements. The employment of both aerodynamic and energy balance methods would seem advisable at any time, but particularly so when the diurnal pattern of trace gas loss is being investigated (Denmead 1983).

4.5 Eddy Correlation Approach

Using the eddy correlation technique, a trace gas flux density (vertical transport of gas past a point in the atmosphere) is obtained by correlating the instantaneous vertical wind speed at a point with the instantaneous concentration of that gas. In the natural environment, the eddies which are important in the transport process occur with frequencies extending up to 5 or 10 Hz. Therefore, a rapid response detector is required (Denmead 1983; Aubinet et al. 2012). With a fast response detector, eddy correlation methods offer many advantages. In particular, they require a minimum number of assumptions about the nature of the transport process. The method requires real time, continuous measurement of wind speed, and gas concentration at only one height above the surface. The technique can also be used inside plant canopies, above soils, and at night (Eugster and Merbold 2015). The method does require fast responding computing facilities to cope with the rapid data acquisition rates and rather large storage capacity. Moreover, the correction of the trace gas density is required with respect to water vapour and heat transfer. These corrections become quite large for some gases (Denmead 1987; Burba et al. 2008; Webb et al. 1980). Applications of both eddy correlation and flux-gradient approaches are limited to situations in which the air analysed has passed over a homogeneous exchange surface for a long distance so that profiles of gas concentration in the air are in equilibrium with the local rates of exchange. The methods assume that horizontal concentration gradients are negligible as a standard requirement to apply the eddy correlation technique (Baldocchi et al. 1988, 2001). With the development of easy-to-deploy quantum cascade laser absorption spectrometers QCLAS (i.e. Mohn et al. 2008, 2013; McManus et al. 2010), field applications of these instruments to continuously measure CH_4 and N_2O exchange above ecosystems are possible. Previous tunable diode laser (TDL) absorption techniques (Thurtell et al. 1991) were often difficult to maintain in the field over a long period. This was particularly the case of the TDL technique which is based on infrared (IR) absorption spectrometry, whereby the extent of absorption depends upon path length, line strength, and absorber concentration. Liquid nitrogen temperature diodes are commercially available to cover the IR spectrum from about 2 to 10 pm, the region where most trace gases have absorption spectra. The diode laser is mounted in a liquid nitrogen-cooled dewar. A heater mounted inside the dewar controls precisely the laser temperature in the 78–110 K region. The centre frequency of the laser emission is controlled by the cold head temperature. Each diode is temperature tunable over about 100 cm^{-1}. With the newly developed and by now commercially available instruments, difficult setups including liquid nitrogen supply in the field are no longer required. Nevertheless, the instruments capable of measuring CH_4 and N_2O at high temporal resolution as needed for the application in eddy correlation setups still need to be handled with more care than instruments capable of measuring CO_2 and H_2O concentration. A recent overview paper highlights the necessary steps to set up reliable and long-lasting measurements of CH_4 and N_2O with the eddy covariance technique (see below).

Standardization of GHG flux measurements was already initiated two decades ago (Aubinet et al. 2012). With the establishment of environmental research infrastructures in Europe, the US and Australia (i.e. Integrated Carbon Observation Systems–ICOS, National Ecological Observatory Network–NEON, and the Terrestrial Ecosystem Research Network–TERN), further steps towards standardized GHG flux measurements were made. These go beyond the standardization of eddy flux measurements and further include GHG chamber measurements as well as standardized collection of ancillary data to interpret GHG flux data. Most of the best practices concerning eddy covariance (EC) flux measurements, including the actual setup of EC towers, the data collections as well as calculation procedures have been summarized in a recent Special Issue in International Agrophysics (Franz et al. 2018). While the approaches explained in this special issue are largely focusing on the highest quality of data, modified and often simpler approaches including instruments that may be used in more remote environments where grid power may not be available are also covered and referred to for more details.

4.6 Alternative Micrometeorological Methods

4.6.1 Eddy Accumulation

The eddy accumulation technique has been proposed as a possible means of measuring the fluxes of constituents, e.g. N_2O. The technique involves the collection of upward and downward transported material in two separate containers at rates proportional to the vertical wind speed. The flux is evaluated as the difference in concentration accumulated over the sampling period. Unfortunately, there are substantial practical difficulties in measuring small concentration differences in the accumulators and sampling the air according to vertical wind velocity (Baldocchi et al. 1988).

4.6.2 *Mass Balance Technique*

The mass balance method has been used to measure fluxes of ammonia from small fields (Denmead et al. 1977; Wilson et al. 1983). Gas flux density is related to the horizontal distance from the upwind edge of the field and the top of the air layer influenced by the emission of the gas. The method assumes that the mean horizontal turbulent flux is much smaller than the mean horizontal advective flux. The top of the air layer influenced by the emission of the gas is a function of stability and surface roughness which can be estimated (Denmead 1983). It is recommended that concentrations of the gas and wind speeds are measured at five or more heights above the surface. The horizontal distance from the upwind edge must be known. Thus, to minimize the effect of changing wind direction on this horizontal distance, it is recommended that experiments be conducted in a circular plot with the instrument array in the centre (Baldocchi et al. 1988).

4.7 Non-isotopic Tracer Release and Measurement Methods

Methods that have been used to consider an assortment of atmospheric transport and diffusion problems include non-isotopic tracer techniques (e.g. using halogenated gases such as sulphur hexafluoride (SF_6)) (Benner and Lamb 1985). Procedures exercising tracer release and measurement techniques have been developed to approximate flux rates from area and point sources. These tracer release and measurement techniques can be utilized when a definite plume of the target gas, methane, for example, can be readily detected in the ambient environment. When these conditions are present, based on analysis of downwind and upwind samples, the plume of dispersed emissions can be located. Next, the tracer gas (e.g. SF_6) should be released, at a known rate, in a configuration similar to that of the target gas. For example, if release of the target gas is at several distinct points, the release of the tracer gas should be at those same points. To ensure that the locations and shapes of the target and tracer plumes match one another, downwind and upwind samples are analysed. The known release rate of the tracer and the ratio of the tracer and target gas concentration are used to calculate the flux rate of the target gas. The ideal conditions for this technique are with steady meteorological surroundings, which allow the target and tracer plumes to comprehensively mix. Further, the flux rates need to be sizeable enough to permit the detection of the target and tracer gases at distances that are great enough to enable thorough mixing. To determine the desired flux rate if there is insufficient knowledge about the pattern of the target gas release, it is possible that single-point release of the tracer can be used and an analysis of the ratio of the crosswind integral of the tracer and target gas plumes.

It is essential to conduct immediate, on-site, analysis to achieve the most success. Chapter 2 describes the preferred equipment for continuous mobile gas analysis.

References

Aubinet M, Vesala T, Papale D (2012) Eddy covariance—a practical guide to measurement and data analysis. Springer, Heidelberg

Baldocchi D (2014) Measuring fluxes of trace gases and energy between ecosystems and the atmosphere—the state and future of the eddy covariance method. Glob Change Biol 20(12):3600–3609

Baldocchi D, Falge E, Gu L et al (2001) FLUXNET: a new tool to study the temporal and spatial variability of ecosystem-scale carbon dioxide, water vapor, and energy flux densities. B Am Meteorol Soc 82(11):2415–2434

Baldocchi DD, Hicks BB, Meyers TP (1988) Measuring biosphere-atmosphere exchanges of biologically related gases with micrometeorological methods. Ecology 69:1331–1340

Benner RL, Lamb B (1985) A fast response continuous analyzer for halogenated atmospheric tracers. J Atmos Oceanic Technol 2:582–589

Burba GG, McDermitt DK, Grelle A, Anderson DJ, Xu L (2008) Addressing the influence of instrument surface heat exchange on the measurements of CO_2 flux from open-path gas analyzers. Glob Change Biol 14:1854–1876

Denmead OT (1983) Micrometeorological methods for measuring gaseous losses of nitrogen in the field. In: Freney JR, Simpson JR (eds) Gaseous loss of nitrogen from plant-soil systems. Developments in Plant and Soil Sciences, vol 9. Springer, Dordrecht

Denmead OT (1987) Notes on measuring trace gas fluxes with enclosures. In: Workshop on measurement of surface exchange and flux divergence of chemical species in the global atmosphere. Columbia University, New York

Denmead OT, Simpson JR, Freney JR (1977) A direct field measurement of ammonia emission after injection of anhydrous ammonia. Soil Sci Soc Am J 41:1001–1004

Eugster W, Merbold L (2015) Eddy covariance for quantifying trace gas fluxes from soils. Soil 1:187–205

Fowler D, Duyzer JH (1989) Micrometeorological techniques for the measurement of trace gas exchange. In: Andreae MO, Shimel DS (eds) Exchange of trace gases between terrestrial ecosystems and the atmosphere. J Wiley and Sons, New York, pp 189–207

Franz et al (2018) Towards long-term standardised carbon and greenhouse gas observations for monitoring Europe´s terrestrial ecosystems: a review. Int Agrophys 32(4):439–455

McManus JB, Zahniser MS, Nelson DD, Shorter JH, Herndon S, Wood E, Wehr R (2010) Application of quantum cascade lasers to high-precision atmospheric trace gas measurements. Opt Eng 49:111–124

Mohn J, Steinlin C, Merbold L, Emmenegger L, Hagedorn F (2013) N_2O emissions and source processes in snow-covered soils in the Swiss Alps. Isotopes Environ Health Stud 49:520–531

Mohn J, Zeeman MJ, Werner RA, Eugster W, Emmenegger L (2008) Continuous field measurements of ^{13}C-CO_2 and trace gases by FTIR spectroscopy. Isotopes Environ Health Stud 44:241–251

Monteith JL (1973) Principles of environmental physics. Edward Arnold, London, p 241

Monteith JL, Unsworth MH (1990) Principles of environmental physics, 2nd edn. Edward Arnold, London, p 291

Thurtell GW, Kidd GE, King KM, Edwards GC (1991) Measurement of N_2O and CH_4 fluxes from natural ecosystems using a TDL-based monitor and micrometeorological techniques. In: IGBP focus 7 meeting, Boulder, CO

Webb EK, Pearman GI, Leuning R (1980) Correction of flux measurements for density effects due to heat and water vapour transfer. QJR Meteorol Soc 106:85–100

Wilson JD, Catchpoole VR, Denmead OT, Thurtell GW (1983) Verification of a simple micrometeorological method for estimating ammonia loss after fertilizer application. Agric Meteorol 29:183–189

5

Soil GHG Fluxes: Role of Fungi, Plants and Fauna

M. Zaman, K. Kleineidam, L. Bakken, J. Berendt, C. Bracken, K. Butterbach-Bahl, Z. Cai, S. X. Chang, T. Clough, K. Dawar, W. X. Ding, P. Dörsch, M. dos Reis Martins, C. Eckhardt, S. Fiedler, T. Frosch, J. Goopy, C.-M. Görres, A. Gupta, S. Henjes, M. E. G. Hofmann, M. A. Horn, M. M. R. Jahangir, A. Jansen-Willems, K. Lenhart, L. Heng, D. Lewicka-Szczebak, G. Lucic, L. Merbold, J. Mohn, L. Molstad, G. Moser, P. Murphy, A. Sanz-Cobena, M. Šimek, S. Urquiaga, R. Well, N. Wrage-Mönnig, S. Zaman, J. Zhang, and C. Müller

Abstract Soils harbour diverse soil fauna and a wide range of soil microorganisms. These fauna and microorganisms directly contribute to soil greenhouse gas (GHG) fluxes via their respiratory and metabolic activities and indirectly by changing the physical, chemical and biological properties of soils through bioturbation, fragmentation and redistribution of plant residues, defecation, soil aggregate formation,

M. Zaman (✉) · L. Heng
Soil and Water Management & Crop Nutrition (SWMCN) Section, Joint FAO/IAEA Division of Nuclear Techniques in Food and Agriculture, International Atomic Energy Agency (IAEA), Vienna, Austria
e-mail: m.zaman@iaea.org; zamanm_99@yahoo.com

K. Kleineidam · C. Eckhardt · A. Jansen-Willems · G. Moser · C. Müller
Institute of Plant Ecology, Justus Liebig University Giessen, Giessen, Germany

L. Bakken
Norwegian University of Life Sciences (NMBU), Aas, Norway

J. Berendt · S. Fiedler · N. Wrage-Mönnig
University of Rostock, Rostock, Germany

C. Bracken
School of Agriculture and Food Science and Earth Institute, University College Dublin, Dublin, Ireland

K. Butterbach-Bahl
Institute of Meteorology and Climate Research, Atmospheric Environmental Research (IMK-IFU), Karlsruhe Institute of Technology, Karlsruhe, Germany

Z. Cai
School of Geography Sciences, Nanjing Normal University, Jiangsu, China

S. X. Chang
Department of Renewable Resources, University of Alberta, Edmonton, AB T6G 2E3, Canada

herbivory, and grazing on microorganisms and fungi. Based on recent results, the methods and results found in relation to fauna as well as from fungi and plants are presented. The approaches are outlined, and the significance of these hitherto ignored fluxes is discussed.

Keywords Soil fauna · Fungi · Microorganisms · GHG

5.1 Greenhouse Gases from Soil Fauna

5.1.1 Introduction

Soils harbour a diverse group of fauna. Based on their size and the resulting occupied soil space, soil animals can be grouped into microfauna (<200 μm: protists, some nematodes), mesofauna (0.2–2 mm, e.g. nematodes, microarthropods, enchytraeids, molluscs), and macrofauna (>2 mm, e.g. earthworms and other worms, ants, beetles, termites, spiders, molluscs) (Lavelle et al. 2006). These animals directly contribute to soil GHG fluxes via their respiratory and metabolic activities and indirectly by changing the physical, chemical and biological properties of soils through bioturbation, fragmentation and redistribution of plant residues, defecation, soil aggregate

T. Clough
Department of Soil & Physical Sciences, Faculty of Agriculture & Life Sciences, Lincoln University, Lincoln, New Zealand

K. Dawar
Department of Soil and environmental Sciences, University of Agriculture, Peshawar, Pakistan

W. X. Ding
Institute of Soil Science, Chinese Academy of Sciences, Nanjing, China

P. Dörsch · L. Molstad
Faculty of Environmental Sciences and Natural Resource Management, Norwegian University of Life Sciences (NMBU), Aas, Norway

T. Frosch
Leibniz Institute of Photonic Technology, Technical University Darmstadt, Darmstadt, Germany

J. Goopy
International Livestock Research Institute (ILRI), Nairobi, Kenya

C.-M. Görres
Department of Soil Science and Plant Nutrition/Department of Applied Ecology, Hochschule Geisenheim University, Geisenheim, Germany

A. Gupta
Independent Consultant India, Mumbai, India

S. Henjes · M. A. Horn
Institute of Microbiology, Leibniz University Hannover, Hannover, Germany

formation, herbivory, and grazing on microorganisms and fungi. Additionally, microhabitats are created that can support greater microbial activity than bulk soil does (Brown et al. 2000; Lubbers et al. 2013). Thus, soil fauna can substantially influence the spatial and temporal variability of GHG fluxes in the field. Additionally, climate, abiotic soil conditions, land management and interactions within the soil food web modify the abundance, activity and vertical distribution of fauna in soils. The magnitude of the effect of soil fauna on soil GHG fluxes remains poorly quantified. Most of our current knowledge comes from laboratory experiments, while field data are scarce, which are often controversial and have been limited to only a few regions and species. Filser et al. (2016) provided the latest review and an extensive literature list on soil fauna and its effects on soil organic matter (SOM) turnover and nutrient cycling. The following sections provide a broad overview of the current knowledge of direct and indirect effects of soil animals on soil GHG fluxes and summarize available field methods for quantifying these effects.

5.1.2 Overview of Fauna on GHG Emissions

5.1.2.1 Carbon Dioxide (CO_2)

Spatial clustering of soil fauna can result in an increased spatial variability of soil CO_2 emissions due to the creation of CO_2 point sources (Ohashi et al. 2007). Positive correlations between soil CO_2 emissions and faunal biomass have been observed (Binet et al. 1998), and faunal respiration can contribute between 2 and 40% of total

M. E. G. Hofmann
Picarro B.V., 's-Hertogenbosch, The Netherlands

M. M. R. Jahangir
Department of Soil Science, Bangladesh Agricultural University, Mymensingh, Bangladesh

D. Lewicka-Szczebak
Laboratory of Isotope Geology and Geoecology, Institute of Geological Sciences, University of Wrocław, Wrocław, Poland

G. Lucic
Picarro Inc., Santa Clara, CA, USA

L. Merbold
Mazingira Centre, International Livestock Research Institute (ILRI), Nairobi, Kenya

J. Mohn
Laboratory for Air Pollution & Environmental Technology, Empa Dübendorf, Dübendorf, Switzerland

P. Murphy
Environment & Sustainable Resource Management Section, School of Agriculture & Food Science, and UCD Earth Institute, University College, Dublin, Ireland

A. Sanz-Cobena
Research Center for the Management of Environmental and Agricultural Risks (CEIGRAM), ETSIAAB, Universidad Politécnica de Madrid, Madrid, Spain

soil respiration (Briones et al. 2004; Lubbers et al. 2013). Ignoring these soil CO_2 flux hotspots might result in substantial errors in ecosystem carbon (C) balances (Ohashi et al. 2007). However, indirect effects of the presence of soil fauna on total soil respiration often seem to be more important than direct respiratory contributions.

The most studied soil fauna groups with respect to soil CO_2 fluxes are earthworms, potworms (*Enchytraeids*), termites and beetles. Earthworms are divided into three functional groups based on their feeding and movement patterns in soils. Anecic earthworms live in deeper soil zones where they ingest moderate amounts of mineral soil as well as litter which they drag down from the soil surface into their burrows. Earthworms living and feeding mainly in the rhizosphere are called endogeic. These earthworms ingest substantial amounts of mineral soil. Epigeic earthworms live and feed preferentially in the litter zone above the mineral soil (Horn et al. 2006). Soil properties are altered by its passage through the earthworm gut, and earthworm casts, burrows and middens provide microhabitats for smaller soil animals and microorganisms (Brown et al. 2000; Lubbers et al. 2013). Interactions of functional earthworm groups seem to lead to greater mean soil CO_2 emissions compared to soils where only one functional earthworm group is present (Speratti and Whalen 2008). Increased soil CO_2 emissions through earthworm activity seem to be transient, though in the long-term earthworms might increase C storage in soils through stabilization of SOM in stable micro-aggregates (Lubbers et al. 2013; Six and Paustian 2014).

Earthworm abundances increase and species compositions change when moving from conventional tillage systems to systems with reduced or no tillage (Lubbers et al. 2013). In conventionally tilled fields, enchytraeids seem to be more important for organic matter (OM) mineralization than earthworms (van Vliet et al. 2004). Like earthworms, enchytraeids are burrowing animals that can ingest large amounts of soil, including fungi, bacteria, algae and even dead bodies of larger invertebrates (Briones et al. 2004; van Vliet et al. 2004).

M. Šimek
Institute of Soil Biology, Biology Centre of the Czech Academy of Sciences, and Faculty of Science, University of South Bohemia, České Budějovice, Czech Republic

M. dos Reis Martins · S. Urquiaga
EMBRAPA Agrobiologia Seropédica, Brazilian Agricultural Research Corporation, Seropedica, RJ, Brazil

R. Well
Thünen Institute of Climate-Smart Agriculture, Braunschweig, Germany

S. Zaman
University of Canterbury, Christchurch, New Zealand

J. Zhang
School of Geography, Nanjing Normal University, Nanjing, China

K. Lenhart
Bingen University of Applied Sciences, Berlinstr. 109, Bingen 55411, Germany

Termites are primarily concentrated in tropical grasslands and forests, and based on their nesting habits are classified as mound-building, wood-nesting or subterranean termites (Jamali et al. 2011a). Termite mounds are point sources of CO_2 emissions (Brümmer et al. 2009) which can show large seasonal patterns due to seasonal population dynamics of termites inhabiting the mounds, as well as changes in gas diffusivity of the mound walls (Jamali et al. 2011b, 2013). These patterns are controlled by changes in temperature, moisture, food quantity and quality, as well as the termites' life cycle, and can be highly species-specific (Jamali et al. 2011b). Although total mound CO_2 emissions are mainly comprised of termite respiration, microbial respiration within the mound walls can also be a significant contributor. Suitability of mound walls as habitats for microorganisms depends on properties such as mound bulk density and wall thickness that again can vary widely between termite species (Jamali et al. 2013). Termite CO_2 emissions might be negligible on an ecosystem scale, but uncertainties in CO_2 flux estimates are high (Brümmer et al. 2009; Jamali et al. 2013). These uncertainties are due to a lack of field studies for termites that do not construct mounds, uncertainties in the global estimates of total termite biomass and number of nests, and a lack of process understanding of gaseous exchange between termites and the atmosphere (Jamali et al. 2011a).

In recent years, another group of arthropods has received increasing attention regarding its effects on soil GHG fluxes–dung beetles. On farmland with grazing animals, dung pats are GHG flux hotspots. Dung beetles are important contributors to dung composition and can significantly alter the temporal pattern of CO_2 emissions from dung pats. Dung beetles can either enhance or suppress CO_2 emissions from dung pats based on the species and their feeding behaviours (Penttilä et al. 2013; Piccini et al. 2017).

5.1.2.2 Methane (CH_4)

Soil fauna groups known to emit CH_4 are termites, scarab beetles, millipedes and cockroaches. Emissions have been observed from tropical as well as temperate species, but not from all tested species within a group, and within-group variability is high (Egert et al. 2005; Sustr and Šimek 2009; Brune 2010; Kammann et al. 2017). Methane is produced by methanogenic microorganisms in the insect gut which either live on intestinal surfaces or as endosymbionts inside gut-inhabiting protozoa (Brune 2010). Like termite respiration, termite CH_4 fluxes scale with termite biomass (Jamali et al. 2011b). Methane emitted by termites can significantly affect the CH_4 balance of an ecosystem, offsetting part of the CH_4 sink of the surrounding soil (Jamali et al. 2011a). However, not all CH_4 produced by termites reaches the atmosphere because part of it is oxidized to CO_2 by methanotrophic microorganisms during its diffusive passage through mound walls and the soil surrounding termite nests (Sugimoto et al. 1998; Jamali et al. 2011a, 2013). Termites can also directly enhance CH_4 oxidation as their burrowing activities increase soil diffusivity, and the increasing soil CH_4 concentration supports a larger and more active methanotrophic community (Bender and Conrad 1995). Methane emissions from other soil-dwelling insects have thus far been regarded as negligible at the ecosystem scale (Egert et al. 2005; Sustr and Šimek 2009). However, recent studies suggest that CH_4 emitted by Scarab beetle larvae have the potential to significantly enhance soil CH_4 oxidation and thus affect the soil CH_4 balance in well-aerated upland soils (Kammann et al. 2017). However, the observations on dung beetles are controversial. Methane emissions from dung pats, especially if they are from grazing dairy cows, can be extremely high, being able to switch a field from a net CH_4 sink to a net CH_4 source. Depending on species, dung beetles have been shown to either drastically reduce or increase dung pat CH_4 fluxes (Penttilä et al. 2013; Piccini et al. 2017). Adult scarab beetles can also be CH_4 sources (Bijnen et al. 1996). Earthworms are generally not considered as CH_4 emitters (Drake et al 2006), but recently, CH_4-emitting earthworms were discovered in Brazil (Schulz et al. 2015).

5.1.2.3 Nitrous Oxide (N_2O)

Earthworms are the only faunal group for which a considerable amount of literature on their effects on soil N_2O fluxes is available (Kuiper et al. 2013). Earthworms themselves as well as their casts and burrow walls can be N_2O emission hotspots, able to increase soil N_2O emissions by more than 40% (Brown et al. 2000; Lubbers et al. 2013). In contrast to CH_4-emitting soil fauna, earthworms do not have a quantitatively significant indigenous microbial biome in their guts. Instead, N_2O emissions are due to the activation of nitrate- and nitrite-reducing bacteria in the ingested material during its gut passage. Conditions favourable for denitrifiers inside the earthworm gut include anoxia, an ample supply of C as well as nitrate and nitrite, a suitable pH, and a high moisture content (Drake and Horn 2006; Horn et al. 2006). Gut passage does not only affect soil denitrifiers, but the overall soil bacterial community composition. This is due to selective digestion of bacteria by the earthworms and the mixing of ingested material with mucus, an aqueous secretion rich in organic molecules (Drake et al. 2006). However, the impact of earthworms on functional bacterial soil communities remains unclear (Chapuis-Lardy et al. 2010).

The impact of earthworms on soil N_2O emissions depends on soil type, environmental conditions, earthworm functional groups and species compositions, but the emissions themselves do not seem to scale with earthworm biomass (Chapuis-Lardy et al. 2010). The impact of earthworms on soil structure, gas diffusion, and N and C availability seems to have a much larger effect on soil N_2O fluxes than direct N_2O emissions from the earthworms (Bertora et al. 2007; Kuiper et al. 2013). Increase in N_2O emissions in the presence of earthworms has only been observed for soils to which crop residues or fertilizers were applied (Chapuis-Lardy et al. 2010; Speratti and Whalen 2008). Earthworms initially speed up residue mineralization and thus increase inorganic N availability. However, as already seen for CO_2 emissions, this seems to be a transient phenomenon and in the long-term earthworm presence may lead to lower N_2O emissions, compared to fields without an abundant earthworm population (Bertora et al. 2007; Chapuis-Lardy et al. 2010). The earthworm effect on N_2O fluxes due to residue incorporation can be partly replaced by ploughing in agricultural systems (Bertora et al. 2007). Other soil invertebrate fauna can also significantly modify soil N_2O fluxes, both positive and negative, with the effect being dependent on the fauna's ability to modify soil structure and its relative biomass in the soil fauna food web (Kuiper et al. 2013; Penttilä et al. 2013; Piccini et al. 2017). However, for soil fauna groups other than earthworms, studies have mainly focused on N mineralisation, rather than on direct gaseous N fluxes (van Vliet et al. 2004; Kuiper et al. 2013).

5.1.3 Field Methodology

Soil fauna can substantially influence the spatial and temporal variability of GHG fluxes in the field through direct and indirect effects, but it is often logistically or methodologically not possible to quantify them in situ. Nevertheless, a basic knowledge of the local soil fauna can significantly aid the interpretation and upscaling of GHG flux data. Thus, in the planning phase of a GHG flux study, one should consider the following questions:

(i) What are the key species or functional groups of the soil fauna at the study site and to which degree do they have to be considered in order to reach the aim of the study?
(ii) Are the measurement plots representative for the selected land management types or ecosystems with regard to the soil fauna?
(iii) Do temporal changes in the behaviour of key faunal species or their functional groups have to be considered in the measurement schedule?

There are no standard field methods and measurement protocols available specifically for quantifying soil fauna GHG fluxes since the vast majority of studies have been in the laboratory or in field mesocosms under controlled and simplified conditions. However, all of the measurement methods described in this book can potentially be adapted to quantify soil fauna GHG fluxes in situ, and examples of such adaptations are listed in the following paragraphs.

Measurements of net GHG fluxes with the chamber or eddy covariance method (Chap. 4) include all soil fauna effects at the plot scale regardless of the available knowledge on these animals at a particular field site. Abundances and total biomass of key species and functional groups can be quantified in accompanying soil sampling campaigns. Abundance, weight and vertical distribution in the soil are the easiest parameters to obtain for soil fauna characterization. Greenhouse gas emissions of excavated animals can be directly estimated in the field by incubating them in hermetically sealed vessels, preferably glass vessels. Vessel size (e.g. Exetainers, test tubes, glass jars) and incubation time depend on animal size and expected emission strength (Sustr and Šimek 2009). Figure 5.1 shows a freshly excavated Scarab beetle larva (2.6 g) which was incubated in a 110 ml test tube sealed with a rubber stopper for an hour.

Fig. 5.1 Freshly excavated Scarab beetle incubated in a 110 mL test tube

At the end of the incubation period, 25 ml of gas were extracted from the test tube with a syringe for gas concentration analysis. The GHG emission rates are then calculated from the increase of gas concentrations in the headspace air of the incubation vessel over time. Vessels should be kept in the shade or a cooler to prevent a large temperature increase relative to the soil layer from which the animals were excavated. For incubation times over an hour, wet pieces of paper towel should be added to the vessels to keep the animals from drying out. Each incubation series should include blank measurements (=empty sealed vessels). In some studies, individuals were rinsed with water before incubating them to remove attached soil particles (e.g. Horn et al. 2006). Each vessel should only be used once in the field as earthworms and scarab beetle larvae can defecate during the incubation. Any emissions from the faeces can be determined by sealing the vessel again for at least an hour after the animal has been removed. These incubations can also be performed on soil fauna which is kept in the laboratory under controlled conditions. However, one has to keep in mind that GHG emissions may strongly depend on the available food source and thus, laboratory and field quantifications might not be directly comparable.

Chambers can also be used to measure directly net GHG fluxes by soil fauna clusters like dung pats or termite mounds (Penttilä et al. 2013; Jamali et al. 2011a, b). The flexibility of the chamber and collar design permits it to fit as precisely as possible to the size of the clusters, thus reducing the relative contribution of soil adjacent to the clusters to the measured GHG flux. However, when targeting soil faunal effects on GHG fluxes with the chamber method, there are two additional considerations which have to be taken into account during construction and installation of permanent collars. First, permanent collar installation can constrain the horizontal movement of soil fauna and, as a result, change their behaviour. Depending on the targeted soil fauna groups, it might not be possible to find a collar insertion depth that adheres to the guidelines for airtight chamber measurements and permits unconstrained soil fauna movement at the same time. Second, the chosen collar material has to be durable enough to withstand any fragmentation attempts by the soil fauna, especially the mandibles of macro-arthropods (e.g. scarab beetle larvae).

Net GHG fluxes at soil fauna clusters may also be quantified using soil gas sampling probes (see Chap. 2, Sect. 2.3.) and the flux gradient technique. Jamali et al. (2013) installed nylon tubes in termite mounds for measuring CO_2 and CH_4 concentrations. Net mound CO_2 and CH_4 fluxes measured with chambers were significantly correlated to the internal mound CO_2 and CH_4 concentrations, respectively. There was also a significant relationship between net mound CH_4 fluxes and internal mound CO_2 concentrations. However, all these relationships were highly species-specific, and using internal mound gas concentrations as a proxy for determining net mound fluxes results in higher uncertainties in the flux estimates compared with direct chamber measurements. Nevertheless, in areas where chamber measurements on termite mounds are not possible, this method can be used with great caution to derive at least rough estimates of net mound CH_4 and CO_2 emissions (Jamali et al. 2013).

Stable isotope methods provide possibilities to measure GHG fluxes in situ without disturbing the soil system and are especially helpful to estimate gross fluxes, such as the contribution of soil fauna. Sugimoto et al. (1998) estimated the proportion of CH_4 oxidized by diffusion through the walls of termite mounds from the difference in $\delta^{13}C\text{-}CH_4$ between the CH_4 produced inside the mounds and the CH_4 captured outside during chamber measurements (natural abundance approach). An isotope pool dilution technique developed by von Fischer and Hedin (2002) allows the simultaneous estimation of gross CH_4 production and gross CH_4 oxidation rates in soils. Originally developed for incubating intact soil cores in hermetically sealed vessels, this technique can also be used in combination with chambers (Yang and Silver 2016). Immediately after a chamber is placed airtight onto a collar to start a flux measurement, $^{13}C\text{-}CH_4$ is injected into the chamber headspace to reach an isotopic enrichment of 2–10 atom % $^{13}C\text{-}CH_4$. However, headspace CH_4 concentration should remain at the ambient level and not increase by more than 0.1 ppm. The chamber headspace can additionally be labelled with trace amounts of SF_6 (~10 ppb) to quantify diffusional losses of the $^{13}C\text{-}CH_4$ label from the chamber and then correct the measurement results accordingly. After labelling, the chamber measurement proceeds as usual, taking gas samples either manually with a syringe or automatically with an attached gas analyser (Yang and Silver 2016). Section 5.3 provides a practical example of the analysis of manually collected discrete gas samples with a stable carbon isotope analyser. Gross CH_4 production and gross CH_4 oxidation rates are estimated by fitting equations for the change of the amount of labelled CH_4, the total amount of CH_4, and the isotope ratio over time (von Fischer and Hedin 2002). This technique has the potential to be used to quantify in situ CH_4 production of soil-dwelling Scarab beetle larvae and other CH_4-emitting macro-arthropods in well-aerated upland soils, although it is yet to be field-tested for this purpose. The various CH_4 production and oxidation processes can also be analysed via a stable isotope tracing model similar to *Ntrace*$_{Gas}$ described in Chap. 7, Sect. 7.5.7.

Another way to use stable isotopes in soil fauna GHG flux studies is to add ^{13}C and/or ^{15}N labelled plant material to soils. Methane production by scarab beetle larvae in well-aerated upland soils feeding on the residues could be estimated by quantifying the emission of ^{13}C-CH_4 from the soil by using chambers. Measuring ^{15}N-N_2O soil emissions using chambers allows estimation of the plant material contribution to the overall N_2O emissions. Excavated earthworms can be freeze-dried, ball-milled and dried at 60 °C for ^{15}N determination. This information can untangle which of the earthworm species is most active in digesting plant material residues (Giannopoulos et al. 2010).

5.2 Greenhouse Gases from Fungi and Plants

Many experiments focus on the greenhouse gas balance of an ecosystem. Therefore, CH_4, N_2O and CO_2 fluxes from the surface to the atmosphere are measured, either with chamber techniques on a number of plots or–on larger scale–via eddy covariance techniques. With these techniques, the net fluxes of GHG between the ecosystem and the atmosphere can be quantified and the sink or source strength can be determined. Nevertheless, this approach neglects simultaneously occurring production and consumption processes of CH_4 and N_2O plants that are part of the soil–plant–atmosphere continuum (SPAC).

Currently, when investigating greenhouse gas fluxes and emission pathways, the focus lies on microbial-derived greenhouse gas emissions from soils. Recently, it has been shown for a broad range of species that eukaryotic organisms–namely, algae (Scranton and Brewer 1977), fungi (Lenhart et al. 2012), cryptogams (Lenhart et al. 2015), animal cells (Ghyczy et al. 2008) and higher plants (Keppler et al. 2006)—release considerable amounts of CH_4 and N_2O to the atmosphere. From those organisms, only fungi are known to produce N_2O from soil. In addition, fungi can also emit CH_4 (Lenhart et al. 2012). Plants can either be a source of CH_4 (Keppler et al. 2006) and N_2O (Lenhart et al. 2019) themselves, or they can have a "chimney" function, transporting dissolved CH_4 and N_2O via the transpiration stream. Depending on ecosystem and environmental conditions, this will lead to an over- or underestimation of GHG fluxes (Machacova et al. 2016). Lichens and mosses will be of minor importance in most agricultural ecosystems due to their low abundance in managed ecosystems. Nevertheless, depending on cryptogamic biomass and metabolic activity, cryptogam-derived emissions are of relevance in natural ecosystems.

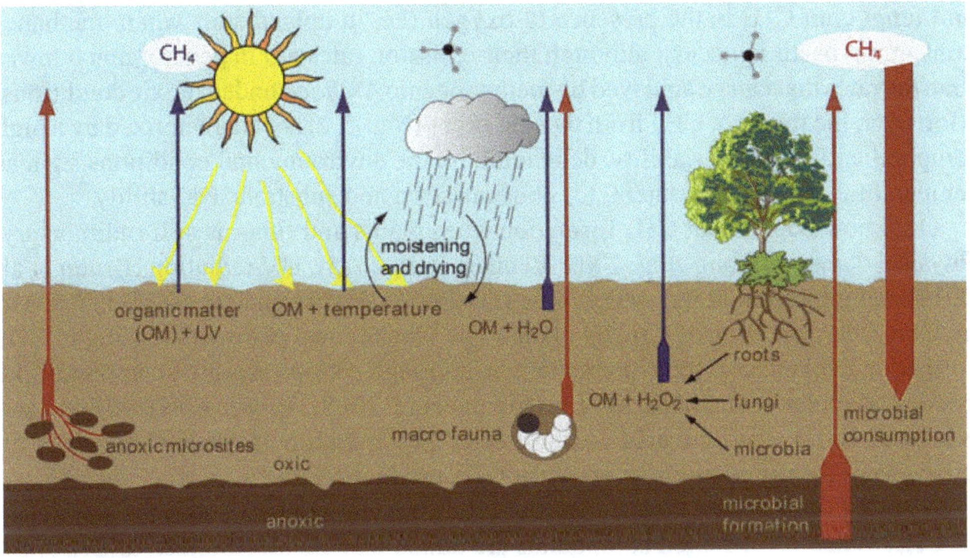

Fig. 5.2 Conceptual scheme of sources and sinks of soil air CH_4 in an aerated soil ecosystem (Jugold et al. 2012). Methane uptake from the atmosphere is the dominant process, biotic CH_4 formation (red arrows) under oxic conditions in plants and fungi, biotic CH_4 formation under anoxic conditions in free-living Archaea and larvae containing methanogenic Archaea in their gut and chemical formation of CH_4 (blue arrows) from soil organic matter are sources of soil air CH_4

5.2.1 *Methane (CH_4)*

Investigating CH_4 production in aerated soil in the field is difficult because it occurs simultaneously to CH_4 consumption. For aerated ecosystems (e.g. upland soils), CH_4 consumption usually exceeds CH_4 production by several orders of magnitude; thus, it is extremely difficult to detect and quantify small CH_4 production rates. Methane consumption by methanotrophic bacteria is the only biological CH_4 sink of atmospheric CH_4 and causes–due to the arising concentration gradient–flux of CH_4 from the atmosphere into the soil. Besides atmospheric CH_4, which is usually the predominant CH_4 source in aerated soils, several biotic or abiotic sources are known in the plant–soil system (Wang et al. 2013). An overview is given in Fig. 5.2.

Abiotic CH_4 formation from soil organic matter is triggered by solar radiation, temperature and wetting–drying cycles. For a detailed description of abiotic greenhouse gas sources, we refer to Wang et al. (2017). Biotic formation of CH_4 occurs in anoxic microsites and in the gut of soil macrofauna by methanogenic archaea. Plants release CH_4 from roots to soil air and from aboveground biomass directly to the atmosphere. Due to their strong connection via mycorrhiza, it is virtually impossible to distinguish between root- and fungi-derived CH_4 in nature.

Methane formation of methanogenic archaea is restricted to anoxic conditions. Thus, free-living archaea occur in deeper soil layers, water-saturated soils, anoxic microsites or within the gut of soil animals (Hackstein and Stumm 1994). In contrast to *Archaea,* plants and fungi are not restricted to anoxic conditions. Thus, plants and fungi emit CH_4 in the presence of oxygen (i.e. in upland soils where methane-consuming bacteria occur), although their emission rates are in general much lower than the rates that can be achieved by methanogenic *Archaea* under anoxic conditions. Moreover, the fluxes of CH_4 from these sources (Fig. 5.2) are characterized by a high temporal and spatial variability, depending on the environmental conditions such as temperature, pH, water content, O_2 concentration and substrate availability.

Our knowledge about CH_4 formation from plants and fungi is still rudimentary. Physical damage (Wang et al. 2009; Lenhart et al. 2015), UV-radiation (Bruhn et al. 2014) and inhibition of the cytochrome C oxidase (Wishkerman et al. 2011) leads to an increased formation of CH_4 by plants. The biochemical pathways leading to CH_4 formation in plants and fungi are unknown, although various organic compounds like pectins (Keppler et al. 2008), lignins (Vigano et al. 2008), hydrocarbons (Etiope and Klusman 2002), ascorbic acid (Althoff et al. 2010) and methionine (Lenhart et al. 2015) were identified as a precursor for CH_4 production.

By using isotopic labelling techniques (^{13}C), it is possible to identify the precursors of CH_4. When ^{13}C labelled compounds are added that are metabolized to CH_4, also the ^{13}C label in headspace-CH_4 will be labelled. Moreover, using position-specific components (i.e. compounds where only one atom is ^{13}C labelled), it is even possible to determine a specific atom or functional group of a molecule that is converted to CH_4. One example of position-specific labelling is the amino acid methionine, where only the sulphur bound methyl group of methionine is ^{13}C labelled (S-$^{13}CH_3$). In a laboratory approach, it was shown that sterile cultivated plants are able to convert this methyl group to CH_4 (Lenhart et al. 2015). In a similar experiment where fungi were supplemented with acetate, and where either the carboxy group ($HOO^{13}C\text{-}CH_3$) or the methyl group ($HOOC\text{-}^{13}CH_3$) was ^{13}C labelled (Fig. 5.3), it was shown that fungi can convert the methyl group of acetate to CH_4.

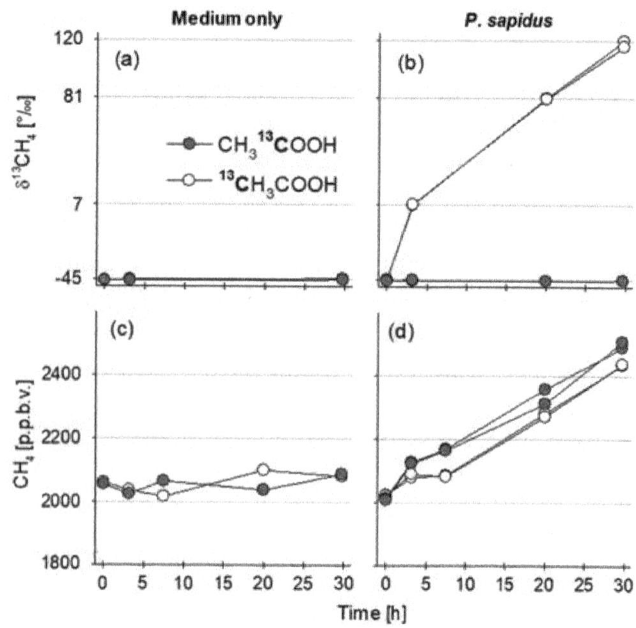

Fig. 5.3 CH_4 production form ^{13}C labelled acetate in sterile culture of *Pleurotus sapidus* (n = 3) at 25 °C in the dark. $\delta^{13}C$ (**a**, **b**) and concentration (**c**, **d**) of headspace CH_4 are presented separately for the control (**a**, **c**) "medium only" and *P. sapidus* (**b**, **d**)

5.2.2 A Laboratory Approach to Study CH_4 Production from Plants and Fungi

Traditional approaches that measure greenhouse gas fluxes from the plant–soil system often measure net fluxes; determination of gross rates is rare. Only one-field study of plant-derived CH_4 exists (Wang et al. 2008). However, to get insights into the complex system of soil air CH_4 sources and sinks occurring in the plant–soil system, we suggest investigating CH_4 production processes under controlled laboratory conditions. We are aware that transfer of laboratory results to the field site is very complex and not always reasonable. On the other hand, with our current knowledge, it is not possible to conduct the following experiments in the field.

To quantify CH_4 formation, a *closed chamber system* (Fig. 5.4) must be used to achieve a sufficient CH_4 enrichment. For open flux chamber systems, the CH_4 (and N_2O) production rates of plants and fungi are probably too low to be detected. This might cause artificial effects, e.g. by affecting gas diffusion (including O_2 availability) in the soil. When measuring CH_4 production in the plant–soil system, methane oxidation by methanotrophic bacteria must be inhibited. This can be achieved with the gaseous inhibitors Difluormethane (DFM, 1 ml l^{-1}) or Acetylene (1 ml l^{-1}). Both substances are reversible inhibitors of the enzyme methane monooxygenase, resulting in a nearly complete inhibition of CH_4 consumption by methanotrophic bacteria.

Fig. 5.4 Example of a closed chamber system that can be used to measure trace gas emissions from organisms–also under sterile conditions. If the lid is equipped with two Festo tube connectors, the flask can be connected to a gas analyser for automated analysis (see Sect. 3.2.2)

5.2.3 *Measuring Procedure*

Field-fresh soil samples should be sieved (2–6 mm) to increase the diffusivity of the inhibitors in the soil. When collecting soil cores with vegetation (e.g. grassland soil), the efficiency of inhibitors is limited due to the lower diffusivity compared to sieved soil.

Insert the field-fresh sample into an airtight flask or chamber equipped with a septum to collect gas samples or connect it to an online system for trace gas measurements (e.g. Picarro system, Plate 3.10). Add either DFM or Acetylene. To accelerate the diffusion of the inhibitor into the soil, we recommend pressure fluctuations, e.g. by "pumping" air in and out with a syringe (=volume change).

Attention: Those inhibitors often cause interferences with a laser-based system → check before the start of the incubation. It is therefore recommended to collect gas samples and measure the concentration with a GC system.

The time to collect gas samples depends on the concentration change, which in turn depends on the amount of sample, the volume of the flask/chamber and the production rate (mostly temperature dependent). For a temperate grassland soil, we obtained good results with 2.7 l headspace, 300 g samples (soil + vegetation), and 25 °C and 15 h incubation. A pre-experiment is recommended to determine the emission rates of CH_4, N_2O and CO_2 under "normal" conditions before working with inhibitors. Depending on the outcome of this pre-experiment, the system must be adjusted (volume, temperature, amount of sample, sampling time). For all incubations, we add three control flasks where only water was added and no changes in CH_4 concentration over time can be expected. In those flasks, gas concentration should remain constant.

5.3 Measuring Discrete Gas Samples with a Cavity Ring-Down Spectrometer for CO_2 and CH_4 Concentration and Carbon Isotope Analysis

In greenhouse gas flux studies, it is common practice to perform the gas concentration and isotope analysis with a gas chromatograph and mass spectrometer, respectively. Some commercially available optical gas analysers are capable to combine both analyses, thus simplifying sample collection and sample processing for the user. The gas analyses to quantify the concentrations of GHG emissions and the stable isotope compositions as described in Sects. 5.1 and 5.2. were predominantly carried out with an optical analyser. One such analyser is the Picarro G2201-*i* cavity ring-down spectrometer. The Picarro G2201-*i* is a field-deployable analyser capable of simultaneous concentration and $\delta^{13}C$ measurements for both CO_2 and CH_4. It can be directly connected to a chamber for continuous closed-loop measurements or employed for the analysis of discrete gas samples, which is the focus here. The following sections address sample injection modes, carrier gas stream and sample volume choice, as well as sample processing times specifically for the Picarro G2201-*i*, but large parts of the information are either directly or in a slightly modified version transferable to other commercially available optical gas analysers.

Sample injection mode

An important gas analyser property to consider when designing an experiment is its sample injection mode since this has implications on the maximum number of processible samples and the method of sample storage prior to analysis. The Picarro G2201-*i* has three possible sample injection modes: (a) injecting each sample manually with a syringe directly into the carrier gas stream, (b) injecting each sample manually into a Picarro Small Sample Introduction Module (SSIM) which automatically passes the samples on into the carrier gas stream, or (c) equipping the SSIM with a manifold to completely automate the injection of multiple samples (see Fig. 5.5). For all three discrete injection modes, the basic requirement is the supply of the Picarro G2201-*i* with a continuous carrier gas stream of either dry zero air (i.e. air with <1 ppm CO_2 and CH_4 and <10 ppm H_2O) or dry standard air with ambient trace gas concentrations (Dickinson et al. 2017) at a pressure of >3 psi (0.2 bar) and <8 psi (~0.5 bar). The slight overpressure at the sample inlet ensures that the analyser can draw in the required ~25 ml min^{-1} into its optical cavity.

An example of a custom-made direct manual sample injection setup has been described by Dickinson et al. (2017). It was tested with different Picarro gas analysers, but the principle of operation can also be transferred to other optical gas analysers. In general, the preferred tubing material for the carrier gas stream line is stainless steel due to its gas tightness and chemical inertness, but PTFE and FEP tubing can also be used, with FEP being the least expensive material. In this case, FEP tubing was equipped with two Luer lock three-way valves. The valve closest to the analyser is the designated sample injection port. Regardless of the sample injection mode, the tubing between the sample injection port and the analyser's sample inlet should be minimized to decrease potential dead volumes, mixing and lag times between sample injection and the actual measurement inside the analyser. The other valve is used for controlling the carrier gas stream. Samples were delivered to the carrier

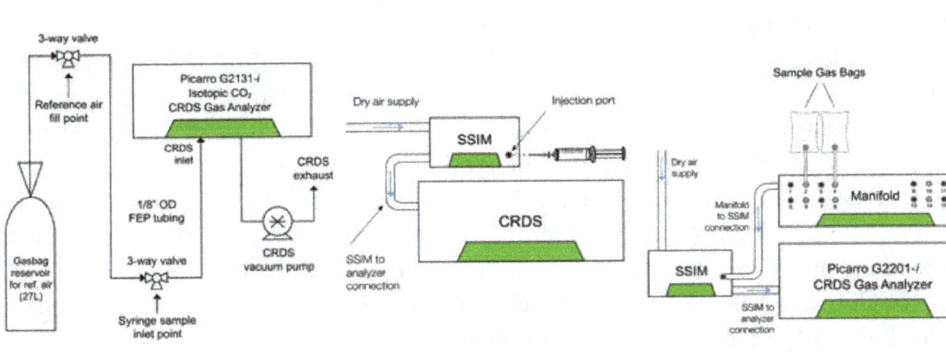

Fig. 5.5 Comparison of different discrete sample injection modes: *left* direct manual injection (figure modified after Dickinson et al. 2017), *center* injection via Small Sample Introduction Module (SSIM) and *right* automated injection of multiple samples via a 16-port-manifold. The manifold allows to attach up to 8 sample bags (every second inlet is capped off for purging cycles)

gas stream with a gastight syringe equipped with a push-button valve and Luer lock fitting. Once the syringe is connected to the sample injection port, the three-way valve is closed manually to stop the carrier gas stream. Then the syringe valve is opened, and the gas sample drawn steadily into the analyser without any further assistance since the analyser operates with a vacuum inside the optical cavity. After complete evacuation of the syringe, the carrier gas stream is manually switched back on. A sample volume of 50 ml was estimated to be the minimum volume for reliable operation of this injection mode. Larger volumes can be administered as well but would require some adjustments in the data processing routine which are described in Dickinson et al. (2017). This custom-made injection setup requires a well-cleaned and silicone-lubricated glass syringe to ensure smooth sample evacuation. Any friction between the plunger and the syringe wall can cause the plunger to jump during sample evacuation which can create small pressure fluctuations inside the analyser's optical cavity and thus increase measurement noise.

Injecting discrete gas samples via the SSIM avoids the problem of possible pressure variations. The SSIM is composed of a 20 ml sample chamber with a sample injection port, a solenoid valve system, an internal pressure sensor and an external vacuum pump. It is inserted into the carrier gas stream line as close as possible to the analyser's sample inlet. The solenoid valve system can shut off the sample cell from the carrier gas stream, but the carrier gas stream into the analyser is never interrupted. At the beginning of a discrete sample measurement, the SSIM is purged with carrier gas and subsequently evacuated several times to remove residues of previous samples. The last pre-measurement step is always a sample chamber evacuation. Injections of discrete sample volumes between 20 and 25 ml can be performed actively with a syringe, or by attaching larger sample containers (e.g. Teflon bags) to the SSIM sample injection port from which samples can be drawn in by the sample chamber vacuum. Once sample injection into the SSIM sample chamber is complete, the sample injection port is closed and the solenoid valve connecting the sample chamber with the carrier gas stream is opened to release the sample into the carrier gas stream. As the pressure in the sample chamber decreases, the outlet valve of the analyser slowly closes to maintain a constant pressure in the cavity. In this way, the flow rate through the cavity is reduced to ensure that the residence time of the sample in the cavity is maximized for a longer peak integration. It is also possible to inject less than 20 ml of gas into the SSIM sample chamber. In that case, the remaining sample chamber volume is filled up with carrier gas prior to releasing the sample to the analyser, resulting in sample dilution. This process is later in this case study referred to as automatic zero air dilution. When working in the dilution mode, one has to make sure that the CO_2 and/or CH_4 concentrations are within the dynamic range of the analyser. The standard specification range for the G2201-*i* is 380–2000 ppm for CO_2 and 1.8–1000 ppm for CH_4 (unless the analyser was upgraded for low and/or high CO_2 concentration measurements).

To automate discrete sample injection, the SSIM can be equipped with a 16-port manifold. The 16-port manifold is a rotary valve that allows to attach up to 8 discrete samples, while the other 8 ports must be closed off so that they can be used for purging between measurements to reduce memory effects due to sample carry-overs. Also, when attaching the manifold to the SSIM, it is important to keep the tubing as short as possible to reduce dead volume effects which might degrade the accuracy for concentration measurements. The sequence of sample and standard measurements can be defined in the SSIM control software (referred to as 'Coordinator' software) also allowing for repeated sample injection. After defining the measurement sequence and other measurement parameters, the SSIM performs a purge and pump step and then asks the user to attach all sample containers to the manifold keeping the valves to the container closed. The Coordinator software then starts another purge and pump cycle for all selected input ports, and once completed, the user has to manually open all sample container valves before the actual unattended sample measurement starts. The duration of a single measurement takes 12 or 8 min in the standard or fast measurement mode, respectively (Fig. 5.5).

Effect of carrier gas and sample volume on the measurement result

The Picarro G2201-*i* continuously measures gas concentrations and $\delta^{13}C$ in its optical cavity at a rate of approximately three to five seconds depending on whether one focuses only on the stable carbon isotopes of one gas species or the simultaneous quantification of both $\delta^{13}C$-CO_2 and $\delta^{13}C$-CH_4. The carrier gas stream provides a baseline measurement on which discrete sample measurements are superimposed as peaks. Choosing dry standard air instead of zero air as carrier gas is foremost an economical choice, but it has important implications for the data processing. Adding zero air to a discrete gas sample dilutes it resulting in lower CO_2 and CH_4 concentrations. However, the isotopic signature of the sample remains unchanged since no additional ^{12}C and ^{13}C are added to the sample. There are three possible reasons for diluting discrete gas samples: (a) it was not possible to collect sufficiently large gas sample volumes during an experiment, (b) CO_2 and/or CH_4 concentrations in the samples are too high and exceed the analyser's operational range, or (c) the samples contain contaminants (e.g. ammonia, hydrogen sulphide, ethylene) in high enough concentrations to negatively affect the spectroscopic measurements. When the Picarro G2201-*i* is coupled with the SSIM, the analyser's software automatically calculates the isotopic signature for each discrete gas sample regardless of whether it has been diluted with zero air in the SSIM or not, and no further data processing by the user is required. However, there is a limit to sample dilution. If the absolute concentrations of ^{12}C and ^{13}C in the sample approach the lower end of the analyser's operational range, instrument noise can increase to a degree that a reliable estimate of the isotopic signature becomes impossible.

Under certain conditions, the Picarro G2201-*i* is also able to provide reliable CO_2 and CH_4 concentration measurements (Picarro Application Note AN038). As previously mentioned, zero air addition to a discrete gas sample dilutes its CO_2 and CH_4 concentrations and ultimately biases the measurement result towards the carrier gas. Biases can also be introduced by sample carry-overs. Sample carry-overs are significantly reduced in the SSIM by the purging and evacuation cycles taking place immediately prior to sample injection. To physically limit zero air dilution of the sample, it is important to minimize the path length between the SSIM injection port and the sample, e.g. by using a septum port, and to inject at least 20 ml of gas into the SSIM sample chamber. If less sample is injected, a bias in the concentration data is unavoidable and has to be corrected by post-processing the data.

Dickinson et al. (2017) have developed data post-processing routines to correct for biases in the analyser's output data, both for the concentration and the $\delta^{13}C$ values. Post-processing of the $\delta^{13}C\text{-}CO_2$ and $\delta^{13}C\text{-}CH_4$ is required when dry standard air is used as carrier gas since it already contains CO_2 and CH_4 at ambient levels with a specific stable carbon isotopic signature. Another important point to consider when opting for dry air as carrier gas is that its CO_2 and CH_4 concentrations and their isotopic signatures have to be significantly different from the expected values in the discrete gas samples because otherwise no unambiguous peak identification in the analyser's continuous data output can be achieved.

Sample processing time

The processing time for a single discrete gas sample depends on the injection mode, the sample volume and required analyser precision. For the direct injection of 50 ml samples into the carrier gas stream, the sample-to-sample time is ~5 min. Half of this time is needed for complete sample evacuation from the syringe. If one likes to use this injection mode with sample volumes >50 ml, sample-to-sample time will increase accordingly (Dickinson et al. 2017). For the combination of the Picarro G2201-*i* with the SSIM, the user has the possibility to choose between a standard and a fast measurement mode, which have a sample-to-sample time of 15 and 10 min, respectively. This includes 3 min for the purge and evacuation cycle to clean the SSIM and the analyser between samples. The actual measurement times for the $\delta^{13}C$ and concentration values are 4 min and 9 min for the fast and standard mode, respectively. The analyser's software reports the discrete sample measurement results as averages with their respective standard deviations for these time intervals. Longer measurement time per sample increases measurement precision, but significantly reduces the amount of processible samples per day. The sample-to-sample time is independent of the gas sample volume injected into the SSIM sample cell since the SSIM adds carrier gas to the sample cell if less than 20 ml have been injected. As a result, the SSIM always releases the same gas sample volume into the carrier gas stream. However, if a user is not accustomed yet to manually injecting discrete samples into the SSIM sample cell or if there is a problem during the injection, significant time can be lost during this step and sample-to-sample times can increase up to 20 min. Daily sample throughput can be maximized by automating the entire sample injection process, i.e. equipping the SSIM with a programmable manifold. This does not decrease sample-to-sample times, but samples can also be processed during laboratory off-hours since no personnel has to be present for the sample injection process. For the calculation of the total number of processible samples per day, one does not only have to take into account the number of discrete gas samples obtained from an experiment, but also allocate time for instrument warm-up periods and instrument maintenance as well as the regular analysis of gas standards.

A practical example of discrete gas sample measurements with a Picarro G2201-i

This example demonstrates step-by-step workflow for discrete gas sample measurements with a Picarro G2201-*i* equipped with an SSIM. The focus of this example is solely on the physical handling of the gas samples and the analyser setup. Post-processing of the measurement data to account for biases will not be discussed but we refer to the considerations described by Dickinson et al. (2017). The gas samples were from a field study using chambers to measure soil CO_2 and CH_4 fluxes. 45 ml gas samples were taken from the chambers and stored in 20 ml screw-capped glass vials which were sealed with pierceable grey chlorobutyl rubber septa. The overpressure serves as protection against sample contamination during transport in case of minor leakages, and to ensure the extraction of ~20 ml of gas from the vials for injection into the SSIM to minimize sample dilution.

Figure 5.6 shows an example of how a sample measurement plan for the analysis of discrete gas samples with a Picarro G2201-*i* can look like. The first row identifies the experiment and the day on which the gas samples were obtained. The second row lists the names of the files in which the measurement data is stored on the analyser. As soon as the analyser is operational after being switched on, it automatically starts recording all data it collects regardless of whether the user will later use the data for analysis or not. These continuous data are stored in dat files at a sampling rate of approximately three to five seconds depending on the chosen measurement mode. In this example, the dat file name includes the date and the instrument time at which it was generated. To operate the SSIM, the CRDS coordinator software has to be installed on the analyser and launched, as well as a software monitoring the pressure inside the SSIM sample cell. The coordinator software automatically detects the peaks and integrates the signals to provide average $\delta^{13}C$ and gas concentration values and their uncertainties for each discrete sample injected via the SSIM. The discrete data are stored in csv files, and the file names also include the date and the instrument time at which it was generated as well as the abbreviation SSIM. When using the concentration data, e.g. for the calculation of soil chamber fluxes, one has to make sure to use the dry concentration values, not the uncorrected wet concentration values as this could lead to flux underestimation.

Sampling campaign:		Mesocosms	27.09.2017				
File ID:	CFIDS2102-20180124-112533Z-DataLog_User.dat			CFIDS2102-SSIM-20180124-103635.csv			
Vial ID	**Sample ID**	**SSIM Run Num**	**ml injected**	**SSIM pressure injec.**	**SSIM pressure dilut.**	**Injection time**	**Comments**
zero air	Continuously measuring zero air with the Picarro Analyzer via the SSIM, start time: 10:40						
zero air		1	0.0	7.8	978.2	11:30	
lab air		2	20.0	926.8	983.6	11:40	
lab air		3	15.0	702.6	981.7	11:50	
lab air		4	10.0	472.1	982.5	12:00	
lab air		5	5.0	250.7	980.9	12:10	
lab air		6	20.0	921.1	979.5	12:20	
245	Mesocosm_2b_t1	7	19.0	943.5	984.4	12:30	
246	Mesocosm_2b_t2	8	19.5	977.6	981.6	12:40	
249	Mesocosm_2b_t3	9	19.0	968.7	985.9	12:50	
247	Mesocosm_2b_t4	10	18.5	923.8	987.8	13:00	
177	Mesocosm_7b_t2	11	19.0	959.3	984.5	13:11	
248	Mesocosm_8b_t0	12	19.5	981.7	982.9	13:21	
250	Mesocosm_8b_t1	13	18.5	946.0	986.9	13:31	
251	Mesocosm_8b_t2	14	19.0	958.8	985.1	13:41	
252	Mesocosm_8b_t3	15	18.5	936.4	988.6	13:51	
253	Mesocosm_8b_t4	16	17.5	891.9	989.7	14:01	
lab air		17	20.0	925.8	985.9	14:11	
zero air		18	0.0	8.0	988.7	14:21	

Fig. 5.6 Example of a measurement plan for a Picarro G2201-*i* equipped with an SSIM

The measurement plan in Fig. 5.6 contains columns for sample descriptions (vial and sample id), a running number which is taken from the csv file (SSIM Run Num), the injected sample volume (ml injected), the pressure inside the SSIM sample cell after sample injection (SSIM pressure injec.) and after automatic zero air dilution (SSIM pressure dilut.), the time at which each sample was injected into the SSIM (injection time), and comments. The measurement schedule consists of three different phases–a start-up phase, a leak testing and calibration phase, and the actual measurement phase. The start-up phase includes all the necessary steps to get the analyser ready for discrete sample injection. If the analyser is not running yet, but still has to be switched on, it needs at least 45 min to warm up and achieve the necessary vacuum in its optical cavity, after which it should run for about two more hours just measuring room air for laser stabilization. This is to ensure the best possible analyser performance. The next step is to connect the carrier gas stream to the SSIM and the SSIM to the analyser (Fig. 5.7). Before conducting any discrete sample measurements, the analyser should be purged with the carrier gas for about an hour to ensure the absence of memory effects. A well-established carrier gas reading is also essential for any required data post-processing due to memory effects and data biases (Dickinson et al. 2017). These preparatory steps can also be conducted the day before the actual discrete measurements. However, if the carrier gas is zero air, it

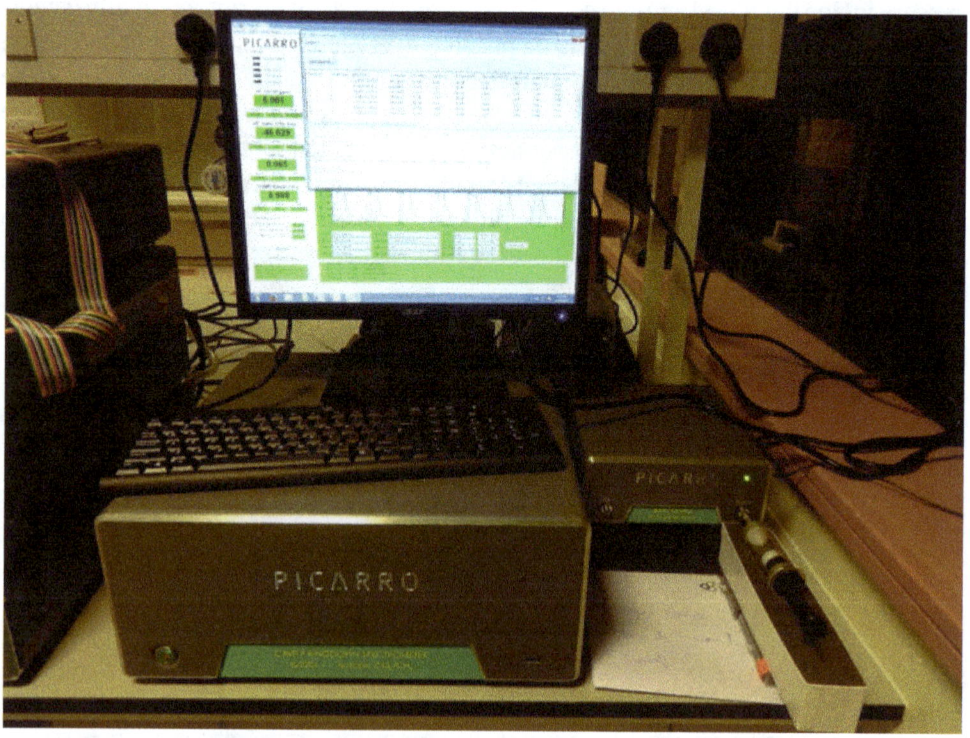

Fig. 5.7 A Picarro G2201-*i* (large box) equipped with an SSIM (small box to the right) during measurements. A syringe containing a discrete sample can be seen attached to the SSIM sample injection port

is not recommended to have the analyser measure the gas stream for several hours as the uninterrupted absence of CO_2 and CH_4 can cause the analyser to drift and offset the analyser's calibration.

Once the analyser is ready for the injection of discrete gas samples, it is very important to perform leak tests and any calibration checks deemed necessary before measuring unknown samples. The most common location for leakages is the sample injection port. In this example, the SSIM injection port was equipped with a so-called septum injector nut with ¼" GC septum (VICI International, Schenkon, Switzerland). The septum used was a 6-mm-thick EC grade, high-temperature silicone septum (Trajan Scientific Europe Ltd, Milton Keynes, United Kingdom). A septum injector nut provides the shortest possible injection pathway from the sample to the SSIM sample cell. It is screwed onto the SSIM injection port, but it is very important not to overtighten the nut. Nut overtightening can significantly compress the septum, thus reducing the septum's flexibility and resealing capability, ultimately resulting in leakage. Additionally, a tightly compressed septum is more difficult to pierce with a needle.

To check the overall system's gas tightness, one should monitor the water concentration and the SSIM pressure. When the dry carrier gas stream is connected to the analyser, the water concentration should drastically drop below 10 ppm and then very slowly continue to decrease throughout the measurement day. The slow continuous decrease is due to the fact that it takes a significant amount of time to purge an analyser completely of all moisture as water vapour attached to the tubing walls will only slowly detach and enter the carrier gas stream. To specifically check the gas tightness of the SSIM, one should first perform a discrete sample injection without actually injecting a sample (Fig. 5.6, SSIM Run Num 1). The normal injection routine for the SSIM consists of six steps: (1) The SSIM coordinator prompts the user to attach a syringe with a vacuum-proof valve to the sample injection port with the syringe valve closed. (2) The sample cell including the tubing to the injection port (and thus the syringe needle) is purged with carrier gas and subsequently evacuated to remove any gas residues. (3) The user is prompted to open the syringe valve, and the sample is sucked into the sample chamber by the vacuum as indicated by a slowly inward moving plunger. As the vacuum in the sample cell is diminished by the inflowing sample, the plunger movement will slow down, and the user has to press in the plunger completely to finish the sample injection. Once the sample has been injected, the syringe can be removed. (4) The sample cell is filled up with carrier gas if necessary (=automatic zero air dilution). (5) The gas in the sample cell is released into the carrier gas stream.(6) The SSIM is purged with carrier gas and evacuated several times to prepare it for the next sample. When no sample is injected during step 3, the pressure in the sample cell should remain stable at ~0 Torr (Fig. 5.6, SSIM pressure injec.). If the injection port is not gastight, the pressure in the sample cell will slowly increase. This test should be regularly repeated since the septum wears down with time and then has to be changed. If no sample is injected, the sample cell is filled with ~20 ml of carrier gas in step 4. The total amount depends on the actual pressure of the carrier gas stream which is depicted by "SSIM pressure dilute" in Fig. 5.5. The user's manual reading of this value from the analyser screen will vary a bit because this is only a short pressure spike in the sample cell before the sample is released to the analyser. However, the exact value can be retrieved later from the dat file if necessary. Apart from the stable vacuum during step 3, the discrete carrier gas sample measurement should have a water vapour concentration only slightly above the reading of the continuous carrier gas stream if the injection port is gastight.

Apart from leakage checks, the SSIM pressure can be used to estimate the volume for a gas sample if the user does not know it. The relationship between the gas volume and the pressure in the SSIM sample cell is theoretically perfectly linear. The first six measurements in Fig. 5.5 were all injections of known gas sample volumes which were used to construct a simple linear regression model. The resulting equation was

$$\text{SSIM pressure injec.}\,[Torr] = 45.576 Torr/mL * Volume_{injected}[mL] + 15.130 Torr$$

with $R^2 = 1.000$. If the user is not able to establish this linear relationship with an R^2 value of 1, this is also indicative of leakages in the system. This equation was used to later calculate the gas sample volumes for the unknown samples (Fig. 5.6, SSIM Run Num 7–16). These samples were extracted from the 20-ml screw-capped glass vials. The syringe used was a 25-ml 1025 SL Hamilton SampleLock Syringe for 22 gauge needles. To retrieve a discrete sample from a glass vial, the syringe needle was inserted into the vial and the syringe plunger pulled out all the way to the 25 ml mark. The plunger was held in this position for about five seconds to assure that a pressure equilibrium was reached between the vial and the syringe as the air needed some time to flow through the needle. Then the syringe valve was closed, and the plunger could be released. When transferring the sample from the vial to the syringe, the sample gas will equilibrate between these two volumes and only a part of the actual sample gas can be used for analysis, the remainder stays behind in the vial. Thus, as mentioned before, it is mandatory to have an overpressure in the vial of at least the volume which one wishes to inject into the analyser. After the sample was secured in the syringe, the plunger was gently pressed into the syringe until it met a slight resistance and then released again. The mark at which the plunger finally stopped was an indicator of the sample volume retrieved from the vial, albeit a very rough one. This value was noted down in the "ml injected" column; however, the precise SSIM pressure reading during sample injection was later used to derive a better sample volume estimate as described above.

At the end, a small remark regarding the syringe needle. We recommend the usage of side port needles to prevent septum coring. Septum coring means that the needle detaches a piece of silicone from the septum while piercing it. Usually, this silicone piece gets stuck inside the needle completely blocking it. A sure sign for coring is that the syringe plunger does not start to move when the syringe is subjected to the SSIM sample cell vacuum. However, needle blockage also means that no sample can be lost when coring occurs. One can simply close the syringe valve, remove the syringe, clean the needle, and repeat the injection. Needles which are open at the tip are much more prone to coring than side port needles. Side port needles are more expensive than other needle types; however, in the long run, frequent septum coring requires the septum in the injector nut to be changed more often and one also loses valuable measurement time.

References

Althoff F, Jugold A, Keppler F (2010) Methane formation by oxidation of ascorbic acid using iron minerals and hydrogen peroxide. Chemosphere 80:286–292

Bender M, Conrad R (1995) Effect of CH_4 concentrations and soil conditions on the induction of CH_4 oxidation activity. Soil Biol Biochem 27:1517–1527

Bertora C, van Vliet PCJ, Hummelink EWJ, van Groenigen JW (2007) Do earthworms increase N_2O emissions in ploughed grassland? Soil Biol Biochem 39:632–640

Bijnen FGC, Harren FJM, Hackstein JHP, Reuss J (1996) Intracavity CO laser photoacoustic trace gas detection: cyclic CH_4, H_2O and CO_2 emission by cockroaches and scarab beetles. Appl Opt 35:5357–5368

Binet F, Fayolle L, Pussard M (1998) Significance of earthworms in stimulating soil microbial activity. Biol fertil Soil 27:79–84

Briones MJI, Poskitt J, Ostle N (2004) Influence of warming and enchytraeid activities on soil CO_2 and CH_4 fluxes. Soil Biol Biochem 36:1851–1859

Brown GG, Barois I, Lavelle P (2000) Regulation of soil organic matter dynamics and microbial activity in the drilosphere and the role of interactions with other edaphic functional domains. Eur J Soil Biol 36:177–198

Bruhn D, Mikkelsen TN, Rolsted MMM, Egsgaard H, Ambus P (2014) Leaf surface wax is a source of plant methane formation under UV radiation and in the presence of oxygen. Plant Biol 16:512–516

Brümmer C, Papen H, Wassmann R, Brüggemann N (2009) Fluxes of CH_4 and CO_2 from soil and termite mounds in south Sudanian savannah of Burkina Faso (West Africa). Glob Biogeoch Cyl 23: GB1001

Brune A (2010) Methanogens in the digestive tract of termites. In: Hackstein JHP (ed) (Endo)symbiotic Methanogenic Archaea (Microbiology Monographs). Springer, Berlin, pp 81–100

Chapuis-Lardy L, Brauman A, Bernard L, Pablo AL, Toucet J, Mano MJ, Weber L, Brunet D, Razafimbelo T, Chotte JL, Blanchart E (2010) Effect of the endogeic earthworm *Pontoscolex corethrurus* on the microbial structure and activity related to CO_2 and N_2O fluxes from a tropical soil (Madagascar). Appl Soil Ecol 45:201–208

Dickinson D, Bodé S, Boeckx P (2017) System for $\delta^{13}C$–CO_2 and xCO_2 analysis of discrete gas samples by cavity ring-down spectroscopy. Atmosph Meas Tech 10:4507–4519

Drake HL, Horn MA (2006) Earthworms as a transient heaven for terrestrial denitrifying microbes: a review. Engl Life Sci 6:261–265

Drake HL, Schramm A, Horn MA (2006) Earthworm gut microbial biomes: Their importance to soil microorganisms, denitrification, and the terrestrial production of the greenhouse gas N_2O. In: König H, Varma A (eds): Intestinal microorganisms of termites and other invertebrates. Soil Biology, vol 6. Springer, Berlin, Heidelberg

Egert M, Stingl U, Dyhrberg Bruun L, Pommerenke B, Brune A, Friedrich MW (2005) Structure and topology of microbial communities in the major gut compartments of Melolontha melolontha larvae (Coleoptera: Scarabaeidae). Appl Environ Microbiol 71:4556–4566

Etiope G, Klusman RW (2002) Geologic emissions of methane to the atmosphere. Chemosph 49:777–789

Filser J, Faber JH, Tiunov AV, Brussaard L, Frouz J, De Deyn G, Uvarov AV, Berg MP, Lavelle P, Loreau M, Wall DH, Querner P, Eijsackers H, Jimenez JJ (2016) Soil fauna: key to new carbon models. Soil 2:565–582

Ghyczy M, Torday C, Kaszaki J, Szabo A, Czobel M, Boros M (2008) Hypoxia-induced generation of methane in mitochondria and eukaryotic cells—an alternative approach to methanogenesis. Cell Physiol Biochem 21:251–258

Giannopoulos G, Pulleman MM, van Groenigen JW (2010) Interactions between residue placement and earthworm ecological strategy affect aggregate turnover and N_2O dynamics in agricultural soil. Soil Biol Biochem 42:618–625

Hackstein JH, Stumm CK (1994) Methane production in terrestrial arthropods. PNAS 91(12):5441–5445

Horn MA, Mertel R, Gehre M, Kästner M, Drake HL (2006) In vivo emission of dinitrogen by earthworms via denitrifying bacteria in the gut. Appl Environ Microbiol 72:1013–1018

Jamali H, Livesley SJ, Dawes TZ, Hutley LB, Arndt SK (2011a) Termite mound emissions of CH_4 and CO_2 are primarily determined by seasonal changes in termite biomass and behaviour. Oecologia 167:525–534

Jamali H, Livesley SJ, Grover SP, Dawes TZ, Hutley LB, Cook GD, Arndt SK (2011b) The importance of termites to the CH_4 balance of a tropical savanna woodland of Northern Australia. Ecosyst 14:698–709

Jamali H, Livesley SJ, Hutley LB, Fest B, Arndt SK (2013) The relationships between termite mound CH_4/CO_2 emissions and internal concentration ratios are species specific. Biogeosci 10:2229–2240

Jugold A, Althoff F, Hurkuck M, Greule M, Lenhart K, Lelieveld J, Keppler F (2012) Non-microbial methane formation in oxic soils. Biogeosci 9(12):5291–5301

Kammann C, Ratering S, Görres CM, Guillet C, Müller C (2017) Stimulation of methane oxidation by CH_4-emitting rose chafer larvae in well-aerated grassland soil. Biol fertil Soil 53:491–499

Keppler F, Hamilton JTG, Brass M, Röckmann T (2006) Methane emissions from terrestrial plants under aerobic emissions. Nature 439:187–191

Keppler F et al (2008) Methoxyl groups of plant pectin as a precursor of atmospheric methane: evidence from deuterium labelling studies. New Phytol 178:808–814

Kuiper I, de Deyn GB, Thakur MP, van Groenigen JW (2013) Soil invertebrate fauna affect N_2O emissions from soil. Glob Chang Biol 19:2814–2825

Lavelle P, Decaens T, Aubert M, Barot S, Blouin M, Bureau F, Margerie P, Mora P, Rossi JP (2006) Soil invertebrates and ecosystem services. Eur J Soil Biol 42:3–15

Lenhart K, Bunge M, Ratering S, Neu TR, Schuttmann I, GreuleM Kammann C, Schnell S, Müller C, Zorn H et al (2012) Evidence for methane production by saprotrophic fungi. Nat Commun 3:1046

Lenhart K, Weber B, Elbert W, Steinkamp J, Clough T, Crutzen P, Pöschl U, Keppler F (2015) Nitrous oxide and methane emissions from cryptogamic covers. Glob Chang Biol 21:3889–3900

Lenhart K, Behrendt T, Greiner S, Steinkamp J, Well R, Giesemann A, Keppler F (2019) Nitrous oxide effluxes from plants as a potentially important source to the atmosphere. New Phytol 221:1398–1408

Lubbers IM, van Groenigen KJ, Fonte SJ, Six J, Brussaard L, van Groenigen JW (2013) Greenhouse-gas emissions from soils increased by earthworms. Nat Clim Chang 3:187–194

Machacova K, Back J, Vanhatalo A, Halmeenmaki E, Kolari P, Mammarella I, Pumpanen J, Acosta M, Urban O, Pihlatie M (2016) Pinus sylvestris as a missing source of nitrous oxide and methane in boreal forest. Sci Rep 6:23410

Ohashi M, Kume T, Yamane S, Suzuki M (2007) Hot spots of soil respiration in an Asian tropical rainforest. Geophys Res Lett 34:L08705

Penttilä A, Slade EM, Simojoki A, Riutta T, Minkkinen K, Roslin T (2013) Quantifying beetle-mediated effects on gas fluxes from dung pats. PLoS ONE 8:e71454

Picarro Inc (2018) Application Note AN038. Measuring small volume gas concentrations with the SSIM, pp 1–3

Piccini I, Arnieri F, Caprio E, Nervo B, Pelissetti S, Palestrini C, Roslin T, Rolando A (2017) Greenhouse gas emissions from dung pats vary with dung beetle species and with assemblage composition. PLoS ONE 12:e0178077

Schulz K, Hunger S, Brown GG, Tsai SM, Cerri CC, Conrad R, Drake HL (2015) Methanogenic food web in the gut contents of methane-emitting earthworm *Eudrilus eugeniae* from Brazil. ISME J 9:1778–1792

Scranton MI, Brewer PG (1977) Occurrence of methane in the near-surface waters of the western subtropical North Atlantic. Deep Sea Res 24:127–138

Six J, Paustian K (2014) Aggregate-associated soil organic matter as an ecosystem property and a measurement tool. Soil Biol Biochem 68:4–9

Speratti AB, Whalen JK (2008) Carbon dioxide and nitrous oxide fluxes from soil as influenced by anecic and endogeic earthworms. Appl Soil Ecol 38:27–33

Sugimoto A, Inoue T, Kirtibutr N, Abe T (1998) Methane oxidation by termite mounds estimated by the carbon isotopic composition of methane. Glob Biogeoch Cyl 12:595–605

Sustr V, Šimek M (2009) Methane release from millipedes and other soil invertebrates in Central Europe. Soil Biol Biochem 41:1684–1688

van Vliet PCJ, Beare MH, Coleman DC, Hendrix PF (2004) Effects of enchytraeids (Annelida: Oligochaeta) on soil carbon and nitrogen dynamics in laboratory incubations. Appl Soil Ecol 25:147–160

Vigano I et al (2008) Effect of UV radiation and temperature on the emission of methane from plant biomass and structural components. Biogeosci 5:937–947

von Fischer JC, Hedin LO (2002) Separating methane production and consumption with a field-based isotope pool dilution technique. Glob Biogeoch Cyl 16:10

Wang ZP, Han XG, Wang GG, Song Y, Gulledge J (2008) Aerobic methane emission from plants in the Inner Mongolia steppe. Environ Sci Tech 42:62–68

Wang S et al (2009) Methane emissions by plant communities in an alpine meadow on the Qinghai-Tibetan Plateau: a new experimental study of alpine meadows and oat pasture. Biol Lett 5:535–538

Wang ZP, Chang SX, Chen H, Han XG (2013) Widespread non-microbial methane production by organic compounds and the impact of environmental stresses. Earth Sci Rev 127:193–202

Wang ZP, Han SJ, Li HL, Deng FD, Zheng YH, Liu HF, Han XG (2017) Methane production explained largely by water content in the heartwood of living trees in upland forests. J Geophys Res-Biogeo 122:2479–2489

Wishkerman A et al (2011) Enhanced formation of methane in plant cell cultures by inhibition of cytochrome c oxidase. Plant, Cell Environ 34:457–464

Yang WH, Silver WL (2016) Net soil-atmosphere fluxes mask patterns in gross production and consumption of nitrous oxide and methane in a managed ecosystem. Biogeosci 13:1705–1715

6

Ruminant Livestock Animals and Methane Emissions

M. Zaman, K. Kleineidam, L. Bakken, J. Berendt, C. Bracken, K. Butterbach-Bahl, Z. Cai, S. X. Chang, T. Clough, K. Dawar, W. X. Ding, P. Dörsch, M. dos Reis Martins, C. Eckhardt, S. Fiedler, T. Frosch, J. Goopy, C.-M. Görres, A. Gupta, S. Henjes, M. E. G. Hofmann, M. A. Horn, M. M. R. Jahangir, A. Jansen-Willems, K. Lenhart, L. Heng, D. Lewicka-Szczebak, G. Lucic, L. Merbold, J. Mohn, L. Molstad, G. Moser, P. Murphy, A. Sanz-Cobena, M. Šimek, S. Urquiaga, R. Well, N. Wrage-Mönnig, S. Zaman, J. Zhang, and C. Müller

Abstract Agriculture is a significant source of GHGs globally and ruminant livestock animals are one of the largest contributors to these emissions, responsible for an estimated 14% of GHGs (CH_4 and N_2O combined) worldwide. A large portion of GHG fluxes from agricultural activities is related to CH_4 emissions from ruminants.

M. Zaman (✉) · L. Heng
Soil and Water Management & Crop Nutrition (SWMCN) Section, Joint FAO/IAEA Division of Nuclear Techniques in Food and Agriculture, International Atomic Energy Agency (IAEA), Vienna, Austria
e-mail: m.zaman@iaea.org; zamanm_99@yahoo.com

K. Kleineidam · C. Eckhardt · A. Jansen-Willems · G. Moser · C. Müller
Institute of Plant Ecology, Justus Liebig University Giessen, Giessen, Germany

L. Bakken
Norwegian University of Life Sciences (NMBU), Aas, Norway

J. Berendt · S. Fiedler · N. Wrage-Mönnig
University of Rostock, Rostock, Germany

C. Bracken
School of Agriculture and Food Science and Earth Institute, University College Dublin, Dublin, Ireland

K. Butterbach-Bahl
Institute of Meteorology and Climate Research, Atmospheric Environmental Research (IMK-IFU), Karlsruhe Institute of Technology, Karlsruhe, Germany

Z. Cai
School of Geography Sciences, Nanjing Normal University, Jiangsu, China

S. X. Chang
Department of Renewable Resources, University of Alberta, Edmonton, AB T6G 2E3, Canada

T. Clough
Department of Soil & Physical Sciences, Faculty of Agriculture & Life Sciences, Lincoln University, Lincoln, New Zealand

Both direct and indirect methods are available. Direct methods include enclosure techniques, artificial (e.g. SF_6) or natural (e.g. CO_2) tracer techniques, and micrometeorological methods using open-path lasers. Under the indirect methods, emission mechanisms are understood, where the CH_4 emission potential is estimated based on the substrate characteristics and the digestibility (i.e. from volatile fatty acids). These approximate methods are useful if no direct measurement is possible. The different systems used to quantify these emission potentials are presented in this chapter. Also, CH_4 from animal waste (slurry, urine, dung) is an important source: methods pertaining to measuring GHG potential from these sources are included.

Keywords GHGs · Animals · Direct and indirect emission · SF_6 · CH_4

6.1 Introduction

Agriculture contributes more than 25% to the total greenhouse gases (GHGs) globally, and ruminant livestock animals are one of the largest contributors to these emissions, responsible for an estimated 14% of GHGs worldwide (Tubiello et al. 2014). In non-industrialised countries, emissions from livestock (methane and nitrous oxide

K. Dawar
Department of Soil and environmental Sciences, University of Agriculture, Peshawar, Pakistan

W. X. Ding
Institute of Soil Science, Chinese Academy of Sciences, Nanjing, China

P. Dörsch · L. Molstad
Faculty of Environmental Sciences and Natural Resource Management, Norwegian University of Life Sciences (NMBU), Aas, Norway

T. Frosch
Leibniz Institute of Photonic Technology, Technical University Darmstadt, Darmstadt, Germany

J. Goopy
International Livestock Research Institute (ILRI), Nairobi, Kenya

C.-M. Görres
Department of Soil Science and Plant Nutrition/Department of Applied Ecology, Hochschule Geisenheim University, Geisenheim, Germany

A. Gupta
Independent Consultant India, Mumbai, India

S. Henjes · M. A. Horn
Institute of Microbiology, Leibniz University Hannover, Hannover, Germany

M. E. G. Hofmann
Picarro B.V., 's-Hertogenbosch, The Netherlands

M. M. R. Jahangir
Department of Soil Science, Bangladesh Agricultural University, Mymensingh, Bangladesh

combined) may be the most important source of GHGs. However, measurement–though critical for meeting international obligations and for assessing mitigation options–is not commonly undertaken in most countries.

Many methods are available for estimating enteric methane (CH_4) production in ruminant animals. They can be broadly classified into two groups–direct and indirect. The important distinction is that direct methods measure CH_4 produced by the ruminant animals in some manner, whereas indirect methods infer CH_4 production from parameters that are associated with CH_4 production in the ruminant animals. In all cases, the methods have strengths and weaknesses and need to be selected with care for the particular objective in mind. The choice of method will depend on available financial and technical resources and the purpose of the measurement, including whether interactions between ruminant animals and environment are important to the research question. If it is impractical to use any of the direct methods, then less precise indirect methods can be used. This chapter will include examples of both direct and indirect measurement methods. Direct methods include enclosure techniques, artificial (e.g. SF_6) or tracer techniques based on herd-level experiments in natural ventilated cattle housings (e.g. using CO_2 as internal tracer), and micro-meteorological methods using open-path lasers. Under the indirect methods, emission mechanisms are understood, where the CH_4 emission potential is estimated based on the substrate characteristics and the digestibility (i.e. from volatile fatty acids). These approximate

D. Lewicka-Szczebak
Laboratory of Isotope Geology and Geoecology, Institute of Geological Sciences, University of Wrocław, Wrocław, Poland

G. Lucic
Picarro Inc., Santa Clara, CA, USA

L. Merbold
Mazingira Centre, International Livestock Research Institute (ILRI), Nairobi, Kenya

J. Mohn
Laboratory for Air Pollution & Environmental Technology, Empa Dübendorf, Dübendorf, Switzerland

P. Murphy
Environment & Sustainable Resource Management Section, School of Agriculture & Food Science, and UCD Earth Institute, University College, Dublin, Ireland

A. Sanz-Cobena
Research Center for the Management of Environmental and Agricultural Risks (CEIGRAM), ETSIAAB, Universidad Politécnica de Madrid, Madrid, Spain

M. Šimek
Institute of Soil Biology, Biology Centre of the Czech Academy of Sciences, and Faculty of Science, University of South Bohemia, České Budějovice, Czech Republic

M. dos Reis Martins · S. Urquiaga
EMBRAPA Agrobiologia Seropédica, Brazilian Agricultural Research Corporation, Seropedica, RJ, Brazil

R. Well
Thünen Institute of Climate-Smart Agriculture, Braunschweig, Germany

methods are useful if no direct measurement is possible. This chapter provides an overview of the different systems used to quantify these emission potentials are presented. Also animal wastes are an important source of CH_4 from animal waste (slurry, urine, dung) is an important source; thus, methods pertaining to measuring GHG potential from these sources are also included.

6.2 Direct Measurements

6.2.1 *Enclosure Techniques*

All enclosure methods rely on the principle of measuring either continuously or intermittently, the concentration of CH_4 in and the total flow of air from around the animal. Methods vary in technical complexity, ease of operation, and precision.

6.2.1.1 Total Enclosure of Animal

The technique of open-circuit indirect respiration calorimetry has been routinely used with many species of ruminant animals to determine partition of dietary energy. This involves the measurement of oxygen (O_2) consumption and carbon dioxide (CO_2) production. For ruminants, the emphasis is quite different–determination of total CH_4 production, arising largely from rumen fermentation to provide more precise estimates of relationships between dietary energy intake and CH_4 production.

Models to estimate national and global CH_4 emission from sheep and cattle at the farm level are mostly based on data of indirect calorimetric measurements, which most precisely measure the relationship between enteric CH_4 and feed intake (Johnson and Ward 1996). Respiration chambers are used to measure CH_4 at an individual animal level–their use is technically demanding, and the number of animal measurements possible will be determined by the availability of physical infrastructure (number of chambers). However, these systems provide the most precise measurements on enteric CH_4 production.

Although design of the chambers varies, the basic principle remains the same. Chambers are constructed to house the subject animals, which are then sealed off and their environment is controlled (Plate 6.1). All open-circuit chambers are characterised by an inlet and exhaust, so animals breathe in a one-way stream of air passing

S. Zaman
University of Canterbury, Christchurch, New Zealand

J. Zhang
School of Geography, Nanjing Normal University, Nanjing, China

K. Lenhart
Bingen University of Applied Sciences, Berlinstr. 109, Bingen 55411, Germany

through the chamber space. Air can be pulled through each chamber by exhaust fans, generating a negative pressure within the chamber, which minimises loss of air from the chamber. However, in practice, CH_4 can still be lost from chambers (down the concentration gradient) that are imperfectly sealed, so, Gas recovery is not a maintenance task. Thresholds for chamber temperature (<27 °C), relative humidity (<90%), CO_2 concentration (<0.5%) and ventilation rate (250–260 l min^{-1}) have been described (Pinares-Patiño et al. 2011). Although these parameters may be varied, it is very important to ensure that subject animals remain in their thermo-neutral zone while being measured, or voluntary intake is likely to be compromised. In practice, chambers in tropical and sub-tropical regions will need to be fitted with air conditioning units, which will also provide a degree of dehumidification and a ventilation system. This ensures that chambers can be maintained at a constant temperature or at near ambient temperatures so as to capture the normal diurnal pattern of CH_4 production (Tomkins et al. 2011). Feed bins and automatic water systems may also be fitted with electronic scales and meters, respectively, to monitor feed and water intake during CH_4 measurement periods.

CO_2 and CH_4 concentrations are measured by sampling incoming and outgoing air, using infrared laser gas analysers, infrared photoacoustic monitors or gas chromatography systems. The other essential measurement is airflow over the measurement period, which is generally either a 24 or 48 h period. The accuracy and thus the validity of measurements are dependent on the sensitivity of the gas analysers used and routine calibration of these devices. Direct calibration of chambers is performed by releasing a known volume and concentration of standard gas to estimate recovery values (Klein and Wright 2006). Alternatively, a gas is released at a known rate until the concentration inside the chamber has equilibrated. Measurements are also influenced by the environmental conditions including temperature, humidity, atmospheric pressure and incoming air composition, and these parameters need to be measured and recorded as part of the measurement process. The calibration of the gas analysers must be accurate and reproducible for long-term usage.

Apart from technical complexity, an important limitation of the technique is that normal animal behaviour and movement is restricted when animals are kept in the respiration chambers (Table 6.1). Animals to be measured will benefit from acclimation in chambers prior to confinement and measurement to minimise alterations in their behaviour: such as decreased feed intake. Even so, there is clear evidence that this will happen in a small proportion of animals regardless of training, which should be borne in mind when interpreting data (Robinson et al. 2014). Using transparent construction material in chamber design allows animals to have visual contact with the other housed animals. The chamber should be sufficiently rigid to tolerate normal animal behaviour and if possible, a metabolism stall should be provided within the chamber to restrain the animal. Provisions should of course be made for feeding and watering the animal.

Substantial costs will be incurred with construction and maintenance of the open-circuit respiration chambers; the requirement for high-performance and sensitive gas analysers and flow meters need consideration in design and construction. Only a few

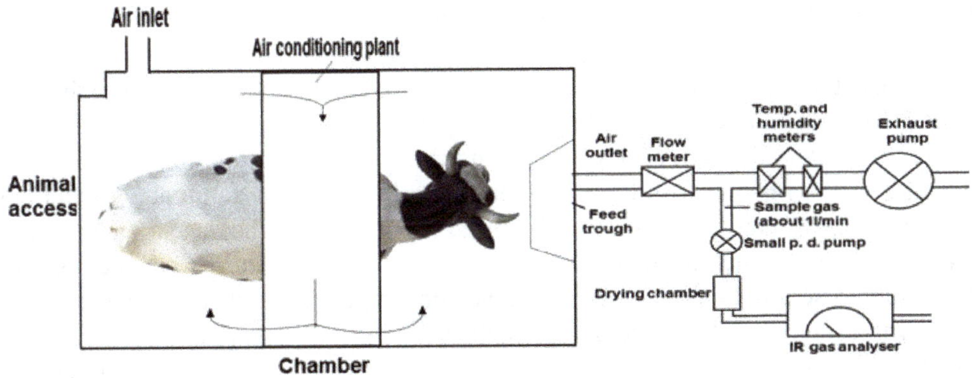

Plate 6.1 Schematic diagram of open-circuit calorimeter to measure CH_4 production in animals

Table 6.1 Recommended dimensions (m) of chambers for different livestock classes

	Height (m)	Width (m)	Length (m)	Approx. Volume (m^3)
Dairy cows	2.0	2.0	3.6	14.4
Growing beef cattle	1.8	1.8	3.0	9.7
Sheep and goats	1.6	1.0	2.0	3.2

animals can be used for measurements within chambers at any one time. Nevertheless, respiration chambers are well suited to study the differences between treatments in response to mitigation strategies and continue to be regarded as the "gold standard" for measuring individual animal emissions.

Air conditioners comprising of a cooling coil and associated refrigerator are essential in hot regions; air filter and coil heater are optional.

- Air ducting.
- Metabolism stalls with suitable feeding and watering facilities.
- Vacuum pump with suitable valve system to adjust air removal rate.

Measurement of Air Volume

Several methods are available to determine the volume of air passing through the chamber. The simplest device (which requires no electrical supply) is a commercial dry gas meter, which gives a direct measure of total gas flow. Other alternatives include electronic turbine flowmeters and venturi apparatus. These give instantaneous rates of gas flow and are highly reliable. However, the need to integrate the individual airflow rates and the relative sophistication of such apparatus precludes their use in laboratories without adequate and reliable electrical supply. It is necessary to correct the total flow of gas to Standard Temperature and Pressure (STP). This requires a determination of atmospheric pressure and temperature.

Measurement of CH_4 from Chamber Exhaust

The exhaust air needs to be analysed for CH_4 concentration. Several possibilities exist for CH_4 measurement. Most commonly, gas samples will be taken at regular (4–12 min) intervals from each chamber in turn, for ~2 min. (to allow for adequate flushing of the residual from the previous sample) using an automated switching device. The actual measurement time required will depend on the measurement apparatus itself (as outlined in the previous section). A possible alternative and simpler method is to acquire a small, representative subsample of the chamber gas over the entire run using a low flow rate pump (such as a peristaltic pump) and stored in gas-impervious (e.g. polyvinyl fluoride (PVF); Tedlar®) bags. It is important that the flow rate of the aliquot collection is constant, thereby providing an integrated representation of the entire collection. After the run has been completed, CH_4 concentration may be measured by the methods outlined in Sect. 6.8.

6.2.1.2 Head Box

Without the use of a tracer that moves with CH_4, it is essential to collect all the expired and eructated (belching) gas. Drawing a stream of air past the muzzle of the animal by enclosing the animal's head in a hood can do this. The method requires considerable training of the animals. The hood can be made sufficiently large so that it is much less restrictive–thus more easily accepted. Additionally, the animal can be fed and have access to water during the collection of gas samples. In both methods, the principle concern is to have a sufficient outflow of gas to ensure there is lower gas pressure in the hood and gas lines and that all the "leaks" are inward. However, there is still considerable scope for part of the sample to leak, up the concentration gradient and be "lost". Hence, performance of gas recoveries is essential in properly calibrating the method. With the accuracy of available analytical equipment, the dilution of CH_4 by air drawn past the animal's head does not present a problem.

The ventilated hood system is a simplification of a whole animal respiration chamber, as it measures the gas exchange only from the head rather than the whole body. Modern ventilated hood systems, applied for CH_4 measurements, have been used in Japan, Thailand (Suzuki et al. 2007, 2008), the USA (Place et al. 2011), Canada (Odongo et al. 2007), and Australia (Takahashi et al. 1999). Recently, Fernández et al. (2012) described a mobile open-circuit respiration system. The ventilated hood system used by Suzuki et al. (2007, 2008) consists of a head cage, the digestion trial pen, gas sampling and analysis, behaviour monitoring, and data acquisition system. Like the whole animal chambers, it is equipped with a digestion pen for feed intake and excreta output measurements. An airtight head cage is located in front of the digestion pen and is provided with a loose-fitting sleeve to position the animals' head. Head boxes are provided with blowers, to move the main air stream from the inlet to the exhaust. Flow meters correct the air volume for temperature, pressure and humidity. Air filters are set up to remove moisture and particles out of the gas samples, which are sent to the gas analysers (Suzuki et al. 2007). The mobile system of Fernández et al. (2012) contains a mask or a head hood connected to an open-circuit respiration system, which is placed on a mobile cart.

The ventilated hood system is a suitable method under some circumstances: especially where open-circuit chambers are not viable. A critical limitation of the hood system is that extensive training to allow the test animals to become accustomed to the hood apparatus is absolutely essential–thus, while it can be used to assess potential of feeds, it is not suitable for screening large numbers of animals, primarily because of the high degree of training required. A further consideration is that hoods capture only measurements of enteric methanogenesis, excluding the proportion in flatus.

Design of Hoods

The hood should be designed to provide sufficient feeding space and enough room for the animal to move its head in an unrestricted way. A wide variety of materials may be used to build a box that is reasonably airtight. The most common materials used are wood and metal. While they can be custom-built, it is also possible to use plastic or metal drums or pre-constructed packing crates. It is a major advantage to have a clear removable panel to provide access for feeding and for checking the animal. This clear panel helps to maintain normal animal behaviour, particularly if other animals are visible to the experimental animal during the period it is in the hood. The animal should be restrained while its head is in the hood and the design of the hood depends on the facility available for restraint. For example, if animals are held in metabolism cages where they cannot turn around, a canvas sleeve can be fitted around the neck and connected to the hood as shown in Plate 6.2. This allows the animal to stand, eat and lie down during the measurement period. It may also be necessary to restrain the animal within the hood by means of a halter or collar. Hoods may also be built around yokes or even head bales at the end of a working race. It is desirable to minimise the amount of air leakage around the neck and head: a sleeve is an effective means of achieving this. This can be tied around the neck using a drawstring. The sleeve can be constructed from any material, but canvas or heavy cotton is most suitable. The length of the sleeve should be enough to allow the animal to stand up, lie down and have unrestricted access to feed and water.

An example for sheep (arrangement similar to that shown in Plate 6.2) uses a box made of 9 mm plywood that has solid sides 0.9 m × 0.4 m top and bottom (0.4 m × 0.6 m). The front and back panels (0.9 m × 0.6 m) have windows 0.25 m × 0.25 m for the animal's head and 0.5 m × 0.5 m for feeding and observation. A removable clear plastic or Perspex panel 0.6 m × 0.6 m is required for the feeding/observation window. The dimensions of the box can be varied to accommodate standard feeders, a water trough and the layout of the animal housing. The length of the sleeve should be around 0.35 m and will taper from a diameter of 0.25 m at the point where it is attached to the hood to 0.15 m so as to fit over the animal's head and be secured around the neck. The dimensions of the hood should be increased for cattle (approximately 3 times larger than for sheep). This will vary considerably according to the size

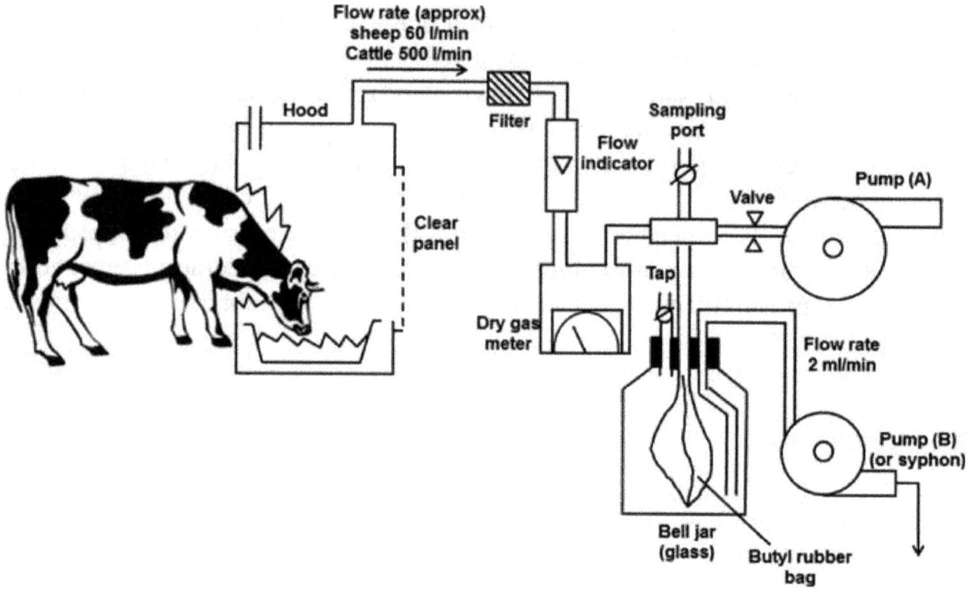

Plate 6.2 Gas sampling from hood: subsampling over 24 h for subsequent analysis

of the cattle and the type of feed (diet). The basic principle for the hood is that it should be sufficiently large for the animal's comfort while maintaining the ability to feed concentrate and roughage diets. Attention should also be given to minimise the places where gas can leak from the system

Gas sampling: The methods for collecting the gas sample, including gas lines, pumps and meters, are similar irrespective of whether hoods or chambers are used.

Sampling and airflow: Airflow of 50–70 l min^{-1} is suggested for measuring CH_4 production in sheep. This gives concentrations of between 100 and 500 ppm (v/v) of methane in the airflow from the hood, but this may need to be modified, depending on the range over which the measuring device exhibits a linear response. For example, for cattle a flow rate of anywhere from 9 to 21 l sec^{-1} the actual flow rate will be determined by the concentration range deemed most suitable for the gas measurement apparatus used, which may be indicated, depending on the measurement device, live weight of the animal and level of feeding. These levels of gas flow will also be sufficient to provide the animal with fresh air and maintain CO_2 levels below 1%. Gas flow rates are easily controlled by altering the speed of the pump or even by an in-line flow restrictor but must not be altered during a run. A single pump can be used to sample a number of animals providing each sampling line has independent control of flow rate and a separate airflow meter. A schematic diagram is shown in Plate 6.2.

Sampling lines: The gas sampling lines can be secured overhead. The diameter of the tubing should be between 1.5 and 2.0 mm ID for sheep and 2–5 mm ID for cattle and can be constructed of materials such as copper, PVC or flexible rubber or plastic hoses, although CH_4 will leak from plastic or PVC tubing, and the use of these materials in a sampling line is better to avoid if possible.

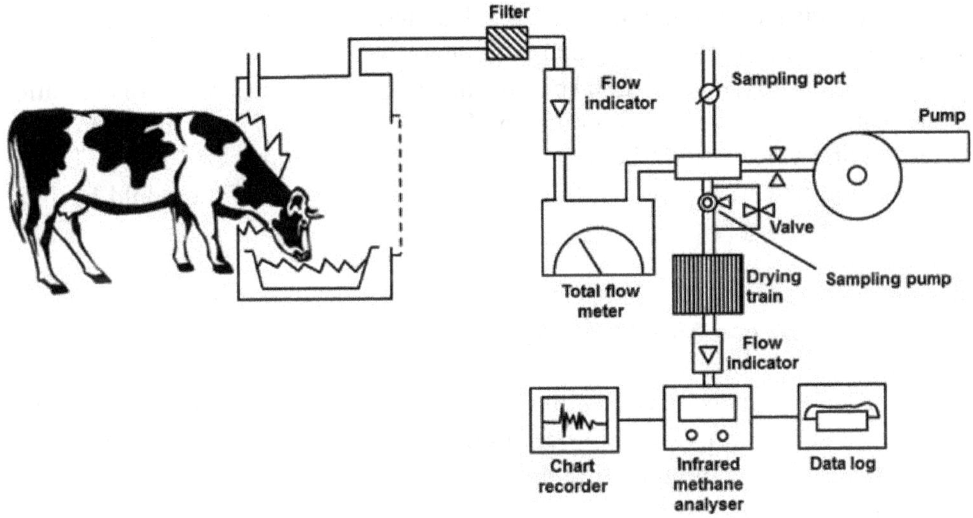

Plate 6.3 Gas sampling and continuous analysis of a subsampled air stream

Filter: It is advisable to have a simple filter system in the main sampling line in order to remove insects and feed particles that may enter the line. A jar containing plastic scourers or glass wool is adequate.

Gas flow measurement: The total volume of gas drawn past the animal must be measured accurately using a commercial dry gas meter.

Analysis and Subsampling Systems

Continuous gas analysis: It is possible to analyse the CH_4 content of the gas from the animal by incorporating an infrared gas analyser in the main line, or in a stream of gas taken from the main line (see Plate 6.3). With this system, the output from the meter needs to be recorded continuously and integrated over time.

Calibration of this equipment requires relatively large quantities of gas, and this can present a potential source of error and an on-going cost. In addition, it requires relatively sophisticated recording and integration equipment or a data logger. Placing the infrared gas analyser in the main line is not recommended unless there is a specific need to measure changes over time.

Time scale for measuring CH_4 production: The extrapolation of short-term measurements of CH_4 production by subsampling over only part of the day can be very misleading. There is considerable variation in the rate of CH_4 production during the day, and it can change considerably after feeding. Unless the measurement is for specific screening purposes, and the time of sampling is standardised with respect to feeding and other animal husbandry practices, measuring for periods of less than 24 h is not recommended.

Calculation: The production of CH_4 is calculated by multiplying the concentration of CH_4 in the subsample by the total volume of gas passing through the hood. For example, a daily airflow of 86,000 l past the head of a sheep with a CH_4 concentration of 200 ppm (v/v) gives a daily production of 17.2 l day^{-1} (0.77 mol day^{-1}). In cattle, corresponding values might be a daily airflow past the animal of 720 000 l and a concentration of 250 ppm (v/v). In this instance, CH_4 production would be 180 l day^{-1} (8.04 mol day^{-1}).

Greenfeed® Emission Monitoring Apparatus

Greenfeed® is a patented device (Zimmerman and Zimmerman, 2012) that measures and records short-term (3–6 min duration) CH_4 emissions from individual cattle repeatedly over 24 h by attracting animals to the unit using a "bait" of pelleted concentrate. By being available 24 h/day, potential sampling bias is reduced, and the technique has been shown to provide comparable estimates to that produced by both respiratory chamber and SF_6 techniques (Hammond et al. 2013). However, a significant limitation of the technique is the requirement to supply an "attractant" to lure the animal to use the facility, consisting of up to 1 kg of concentrate pellets per day. This will certainly affect DMP and may also alter VFA profiles or the overall digestibility of the diet consumed. Attempts to use energy-neutral attractants, such as water, have proven equivocal (J Velazco pers. comm.).

6.2.1.3 Polytunnel

The polytunnel is an alternative to respiration chambers although operation and measurements are somewhat simpler. Methane emissions from individuals or small groups of animals can be acquired under some degree of grazing. This allows test animals to express normal grazing behaviour, including diet selection over the forages within the polytunnel space. It has been used in the UK to measure CH_4 emissions from ruminants under semi-normal grazing conditions. Murray et al. (2001) report CH_4 emissions from sheep grazing of two ryegrass pastures and a clover/perennial ryegrass mix pasture using this methodology. Essentially, polytunnels consist of one large inflatable or tent-type tunnel made of heavy-duty polyethylene fitted with end walls and large diameter ports. Air is drawn through the internal space up to 1.0 m^3s^{-1} (Lockyer and Jarvis 1995). In general, polytunnels may be used where emissions from fresh forages are of interest because animals can be allowed to graze a confined area of known quality and quantity. When the amount of available forages is depleted, the tunnel is moved to a new patch.

Airflow rate can be measured at the same interval as the CH_4 levels are assessed or can be continuously sampled at the exhaust port (Lockyer 1997). Micropumps may be used to pass the exhausted air to a dedicated gas analyser or a GC (Murray et al. 2001). Data from all sensors can be sent to a data logger, which records flow rate, humidity and temperature within the tunnel, as well as gas production from the livestock. Samples of the incoming and exhaust air can be taken as frequently as necessary, depending on the required accuracy. The samples can be either taken manually or by an automatic sampling and injection system.

The polytunnel system requires frequent calibration to assure a good recovery rate, which is performed by the same principle as the chamber technique. Methane measurements can be collected over extended periods of time. Fluctuations can be expected to occur due to changes in animal behaviour, position relative to the exhaust port, internal temperature and relative humidity and grazing pattern of the animal: eating, ruminating or resting (Lockyer and Jarvis 1995; Lockyer and Champion 2001). The polytunnel is suitable for measuring CH_4 emissions under semi-normal grazing conditions. It has been reported that the tunnel method gives lower readings of CH_4 concentration (15%), compared to the respiration chamber method, suggesting that animals actually consume less–this requires further investigation. Recovery rate is high in both systems, 95.5–97.9% versus 89.2–96.7% for tunnel and chambers, respectively (Murray et al. 1999). With an automated system, measurements can be performed with high repeatability. The polytunnel system is portable and can be used on a number of available pastures or to also browse shrubs. Its utility is limited by the inability to capture and record feed intake.

6.2.1.4 Portable Accumulation Chambers (PAC)

PACs consist of a clear polycarbonate box of approximately 0.8 m^3, open at the bottom and sealed by achieving close contact with flexible rubber matting. Methane production is measured by the increase in concentration, which occurs over approximately a 1 h period. PACs were designed to screen large numbers of sheep to identify potentially low and high methane-emitting individuals and thus develop estimates of the genetic parameter in sheep populations. This technique initially showed good agreement with respiratory chamber measurements (Goopy et al. 2009, 2011) and subsequent investigations demonstrated such measurements to be moderately repeatable in the field and to have potential for genetic screening of the animals (Goopy et al. 2014). Longer term comparisons of PAC measurements and respiratory chamber data, however, suggest that these two methods may be measuring quite different traits and further investigations are required before committing significant resources to pursue measurements using PACs (Robinson et al. 2015a, b).

6.3 Tracer Techniques

6.3.1 Use of SF_6 Bolus

The sulphur hexafluoride (SF_6) technique is a direct measurement of the CH_4 emission of individual animals. This technique can be performed on an individual animal under normal grazing conditions and can also be employed under more controlled conditions where intake is measured and/or regulated.

The SF_6 principle relies on the insertion of a permeation tube with a predetermined release ratio of SF_6 into the rumen, via the mouth (per os). Air from around the animal's muzzle and mouth is drawn continuously into an evacuated canister connected to a halter fitted with capillary tube around the neck. Johnson and Ward (1996) provide a detailed description of the methodology.

The regulation of the duration of collection of each sample is achieved by altering the length and/or diameter of the capillary tube (Johnson and Ward 1996). Several modifications have since been reported with specific applications (Goopy and Hegarty (2004); Grainger et al. 2010; Ramirez-Restrepo et al. 2010). Most recently, Deighton et al. (2014) have described the use of an orifice plate flow restrictor which considerably reduces the error associated with sample collection and thus should be considered in preference to the traditional capillary tube flow restrictors. At completion of sample collection, the canisters are pressurised with N_2 prior to compositional analysis by gas chromatography. Enteric CH_4 production is estimated by multiplying the CH_4/SF_6 ratio by the known permeation tube release rate corrected for actual duration of sample collection and background CH_4 concentrations (Williams et al. 2011), which are determined by sampling upwind ambient air concentrations. Williams et al. (2011) emphasised the importance of a correct measurement and the reporting of the background concentrations, especially when the method is applied indoors. Methane is lighter (16 g mol^{-1}) compared to SF_6 (146 g mol^{-1}); therefore, it will disperse and accumulate differently depending on ventilation, location of the animals and other building characteristics.

This method enables sampling of gas concentrations in exhaled air of individual animals, and it also takes into account the dilution factor related to air or head movement. A significant limitation of this method, however, is the high within- and between-animal variation. Grainger et al. (2007) reported a variation within animals between days of 6.1% and a variation among animals of 19.7%. Pinares-Patiño et al. (2011), monitoring sheep in respiration chambers simultaneously with the SF_6 technique, reported a higher within (×2.5) and between (×2.9) animal variance compared to the chamber technique combined with a lower recovery rate (0.8 ± 0.15 vs. 0.9 ± 0.10). These sources of variation need to be taken into account in determining the number of repeated measures needed to obtain accurate results. Moate et al. (2015) have recently described the use of Michaelis–Menten kinetics to better predict the discharge rate of the SF_6 capsules, which should potentially reduce the error associated with estimating discharge rates and also prolong the useful life of experimental subjects through the improved predictability of discharge rates over much longer intervals.

The SF_6 technique gives animals the ability to move and graze normally on test pastures. This makes the method suitable for examining grazing management effects on CH_4 emission (Pinares-Patiño et al. 2007), but it does so at a cost. The SF_6 method is less precise, less physically robust (high equipment failures) and more labour-intensive than respiration chamber method (Plate 6.4).

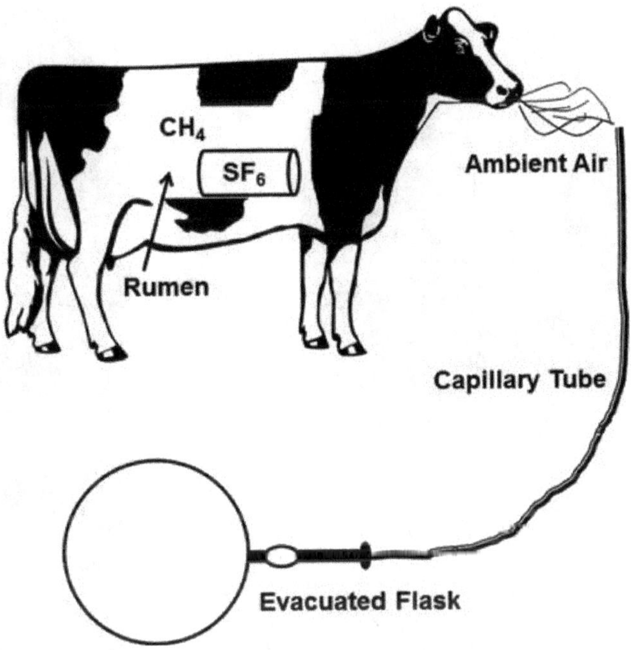

Plate 6.4 Structure of the slow permeation bolus for releasing control amounts of SF_6

The permeation tubes consist of a closed stainless-steel tube capped at one end with a Teflon disc held in place with a standard Swagelok^{Tm} nut. The tube is filled with SF_6 while being cooled with liquid N_2 causing it to reach solid state on contact. It is then allowed to equilibrate to a fixed temperature in a water bath. Permeation rates are determined gravimetrically through repeated weighing on a precision balance over a period of several weeks. Typical permeation rates are of the order of 0.5–2.5 mg day^{-1}, although they may be considerably higher if circumstances dictate. Permeation tubes should be inserted into the rumen a minimum of 3 days prior to the first scheduled collection to allow steady-state conditions to be reached. Insertion can be accomplished with a balling/drenching gun or other similar methods for inserting boluses into the rumen.

A leather pad attached to the noseband of a halter serves as an anchor point for the sample line near the animal's nose and mouth. A small piece of plastic tubing is attached to a filter and oriented such that it is placed over one of the nostrils (see Plate 6.5). A filter (10 μm) connected to the upstream end of the sample line protects the flow restrictor from becoming plugged. Fastening the tube to the sides of the halter helps to protect the capillary tubing and reduces animal irritation. The capillary tube-collection flask connection should be via a quick-connect fitting to simplify flask exchange. A soft rope fastened around the neck with a clasp that can be attached to the collection flask helps stabilise the flask and takes pressure off the capillary tube.

The collection vessel should be large enough to accommodate the size of the desired sample (i.e. ~5 l for cattle), should be able to withstand a vacuum and should have a valve for sealing the flask. Immediately prior to use, all air should be removed

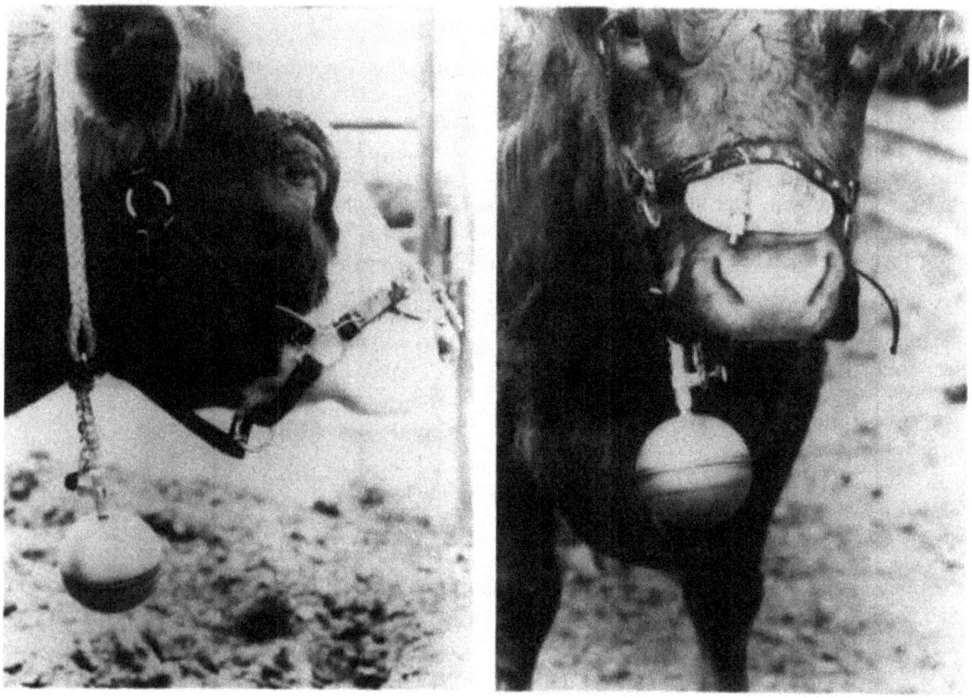

Plate 6.5 Position of the evacuated vessels on the animal

from the flask and the valve closed. After fastening the flask to the supporting neck rope and attaching the capillary tube, the time of day should be noted, and the valve opened. When the sampling time is complete, the flask is removed for analysis. If repeated collections are desired, another flask should be added after the first one is filled. It is recommended that many measurements are made on each animal, and that total 24 h emissions are also reported. The diameter and length of the capillary tubes needed depend on the rate at which sampling is desired. The size of the capillary tube bore should be such that the evacuated sample flask fills to about 1/2 atmospheric pressure over the desired sampling time. The flask pressure should be measured, and the flask then filled with N_2 to bring it to positive pressure (around 1.5 atmospheres). Both pressures need to be recorded to enable the extent of dilution by N_2 to be calculated.

The vessel samples can be quickly and accurately analysed for the tracer using an electron capture gas chromatograph. Methane can be analysed using a gas chromatograph with an FID detector (see Sect. 6.8).

The emission rate of CH_4 (L hr^{-1}) is calculated from

$$QCH_4 = QSF_6 \frac{[CH_4]}{[SF_6]} \tag{6.1}$$

where QSF_6 is the release rate of SF_6 in litre hr^{-1}, $[CH_4]$ is the concentration of CH_4 in the gas sample and $[SF_6]$ is the concentration of SF_6 in ppm (v/v).

6.3.2 *Tracer Ratio Method for Emission Measurements in Naturally Ventilated Housing*

An established approach to determine emissions at herd level in naturally ventilated housings is the use of artificial tracers with known source strength (Mohn et al. 2018; Ogink et al. 2013; Schrade et al. 2012). The most frequently used tracer gas for emission measurements in livestock husbandry is SF_6, as it is chemically extremely stable and has a concentration in ambient air of only a few ppt (10^{-12} mol mol^{-1}). The most widely used dosing principle is constant SF_6 dosing, which can be implemented for continuous or semi-continuous measurements. The quantification of emissions with the tracer ratio method is based on the assumption that the tracer gas (i.e. SF_6) release sufficiently mimics the release of the target substances from surfaces (e.g. NH_3) or animals (CH_4, CO_2). The mass flow of the target gas (m_{target}, e.g. m_{CH4}) is calculated from the ratio of the background-corrected target (c_{target}) and tracer gas concentration (c_{tracer}) and the mass flow of the tracer gas (m_{tracer}) according to $m_{target} = m_{tracer} \times c_{target}/c_{tracer}$. In contrast to CO_2 and heat balance methods, tracer ratio methods are also applicable to naturally ventilated housings with large openings and areal sources like outdoor exercise areas and open slurry storages (Schrade et al. 2012), because the mass flow of the tracer can be adapted to the dilution ratio.

The uncertainty level of tracer gas methods is highly dependent on the accuracy and distribution of the tracer gas dosing and the number and positioning of the sampling points (Calvet et al. 2013). Homogenous dispersion can best be achieved by dosing a diluted tracer gas, with similar density to ambient air, in a dosing grid next to the emitting areas or objects. Representative, preferably homogenously mixed, air samples can be collected using an air sampling grid with critical orifices. Validation experiments of the tracer ratio method have demonstrated that the technique is suitable for both areal and point emission sources and can achieve an uncertainty of less than 10% for target gas mass emissions, which is superior to other currently available methods (Mohn et al. 2018; Schrade et al. 2018). The sensitivity of the technique is mainly given by the capability for background correction, i.e. to separate and subtract emissions in the close vicinity of the housing. This is especially important for calm nights where, depending on the topographic conditions, the target gas accumulates in the nocturnal boundary layer.

A specific variant of the tracer ratio method, using two (or more) different tracer gases (e.g. SF_6, SF_5CF_3), enables the independent assessment of housing areas (Mohn et al. 2018; Schrade et al. 2018). In an experimental housing with identical, but spatially separated housing areas, this approach can be applied to quantify the reduction potential of abatement measures (e.g. flooring, feeding) under comparative measurement conditions (e.g. climate). Additionally, using two different tracers, cross-contaminations between experimental areas are readily apparent by enhanced concentrations of the tracer gas in the respective other section.

6.3.3 Application of CH_4: CO_2 Ratio

Madsen et al. (2010) proposed using the ratio of CH_4: CO_2 in exhaled breath to assess enteric CH_4 production in ruminants. The principle relies on knowledge of intake, energy content and heat increment of the feed ration consumed. Haque et al. (2014) applied this method using a fixed heat increment factor. Hellwing et al. (2013) has regressed open-circuit chamber measurements of Daily Methane Production (DMP) in cattle against estimates calculated using CH_4: CO_2 ratios and found them to be only moderately correlated ($R^2 = 0.4$), suggesting that this method is unsuitable for precision measurements.

6.4 Micrometeorological Estimates

6.4.1 Open-Path Lasers

The use of open-path lasers combined with a micrometeorological dispersion method can now be used to measure enteric CH_4 emissions from herds of animals and facilitates whole-farm CH_4 measurements across a number of pastures.

The open-path laser method for whole-farm CH_4 measurements is already in use in Canada (McGinn 2006; Flesch et al. 2005, 2007), Australia (Loh et al. 2008; McGinn et al. 2008; Denmead 2008; Tomkins et al. 2011), New Zealand (Laubach and Kelliher 2005) and China (Gao et al. 2010). Methane concentration measurements are performed using one or more tuneable infrared diode lasers mounted on a programmable and motorised scanning unit (Tomkins et al. 2011). The tuneable infrared laser diode beams to a retroreflector along a direct path, which reflects the beam back to a detector. The intensity of the received light is an indicator of the CH_4 concentration (ppm) along the path. In an optimal situation, there should be at least one path for each predominant wind direction: one path upwind (background CH_4) and multiple paths downwind (CH_4 emission) of the herd. This method assumes that the herd acts as a surface source or, when individual animals can be fitted with GPS collars, individual animals are treated as point sources. Regardless of application, the CH_4 concentration is calculated as the absorption ratio of the external absorption to internal reference-cell absorption, of the infrared laser beam as it travels along the path (Flesch et al. 2004, 2005). Continuous measurements are required for the CH_4 concentration and the environmental measurements, the latter being recorded by a weather station, i.e. atmosphere temperature, pressure, and wind direction and speed (Loh et al. 2008, 2009). Data management can be achieved statistically by merging all data, including GPS coordinates of the field (paddock) or individual animals, from a number of averaging time periods using statistical software. After integrating, WindTrax software (Thunder Beach Scientific, Nanaimo,

Canada) uses a backward Lagrangian Stochastic (bLS) model to simulate CH_4 emission (grams day^{-1} animal^{-1}), by computing the line average CH_4 concentrations with atmospheric dispersion conditions.

The data integrity of the open-path laser method will be highly dependent on environmental factors and the location of test animals. Flesch et al. (2007) described several criteria to determine data integrity using an open-path laser method. These criteria are based on wind turbulence statistics, laser light intensity, R^2 of a linear regression between received and reference waveforms, surface roughness, atmospheric stability, and the source location (surface or point source). Invalid data can be generated by misalignment of the laser, unfavourable wind directions, surface roughness or periods in which the atmospheric conditions (rain, fog, heat waves, etc.) are unsuitable for applying the model (Freibauer 2000; Laubach and Kelliher 2005; Loh et al. 2008). To optimise the positioning of the equipment, all of these meteorological and physical aspects of the experimental site must be taken into account (Flesch et al. 2007; Loh et al. 2008, 2009). Moreover, the measurement area is restricted by the length of laser paths when using a surface source approach. It is especially important to define the herd location, as uneven distribution of the herd results in miscalculations of the CH_4 concentration. Tomkins et al. (2011), comparing open-circuit respiration chambers with the open-path laser technique, reported estimated CH_4 emissions using the bLS dispersion model of 29.7 ± 3.70 g kg^{-1} dry matter intake (DMI) relative to 30.1 ± 2.19 g kg^{-1} DMI measured using open-circuit respiration chambers.

The open-path laser method does not interfere with the normal grazing behaviour of the cattle and is non-invasive. Spatial variability is taken into account in these measurements, as the method can simulate gas fluxes over a large grazing area. Moreover, the tuneable diode laser is highly sensitive and has a fast response to changes in CH_4 concentration with detection limits at a scale of parts per trillion (ppt) (McGinn 2006). The labour intensity is low, although the equipment requires continuous monitoring. This method is expensive, which reflects not only the requirement of sensitive and rapid-response instruments to analyse CH_4 concentration, but also the requirement to capture micrometeorology data. Diurnal variations due to grazing and rumination pattern, pasture composition, and individual variation need to be considered in planning experimental protocols and are dependent on the purpose of the study to prevent over- or under-estimation of the total CH_4 emission. Furthermore, DMI determination is not very accurate as these are based on predictive models using the relationship between live weight (LW) and LW gain, following assumptions of the ARC (1980).

6.5 Short-Term Measurements

While most assessments of enteric methane emissions are focused on Daily Methane Production (DMP), or the derivative: Methane Yield (MY), there is an increasing impetus to estimate the emissions of large numbers of animals in their productive environment. This is driven both by the demand for data to establish genetic parameters for DMP and/or to verify mitigation strategies or GHG inventories. Here, this area will be discussed only briefly, as there is at present limited scope for the application of these technologies in the developing countries. For the interested reader, the area has been ably reviewed by Hegarty (2013).

6.5.1 *Spot Sampling Using Lasers*

Spot measurements of CH_4 in the air around cattle's mouths have been made using laser devices to provide short-term estimates of enteric CH_4 flux (Chagunda et al. 2009; Garnsworthy et al. 2012). These estimates are then extrapolated to represent DMP. Chagunda and Yan (2011) have claimed correlations of 0.7 between laser and respiratory chamber measurements, but this claim is based on the laser apparatus measuring CH_4 concentrations in the outflow of respiratory chambers, rather than from the animals themselves; thus, this "method" remains unproven under realistic field conditions.

6.6 Indirect Methods

The procedures described below are "indirect" methods for approximating CH_4 production and do not measure levels of CH_4 per se that are produced by ruminant animals. Therefore, they should be used only if more precise measurements are logistically unachievable (e.g. for regional or country-wide surveys), or there is a need to fill gaps in a much larger survey, or obtain preliminary data before embarking on more extensive studies using more accurate direct techniques.

6.6.1 *Methane Emissions from Feed and Feed Characteristics*

Enteric methane production (EMP) can be estimated from intake and diet quality (digestibility), and there are a number of extant algorithms that attempt to quantify this, although it has been demonstrated that estimates of emissions can vary by 35% or more for a particular diet (Tomkins et al. 2011). Diet quality (i.e. digestibility) can be inferred from the analysis of representative samples of the rations or pasture consumed, but where intake is not measured, estimation of EMP faces considerable challenges. Models which estimate intake based on diet quality or particular feed fractions assume ad libitum access, and in situations where animals are corralled without access to feed overnight (as is frequently the case in developing countries), the validity of this assumption is likely violated (Hendricksen and Minson 1980; Jamieson and Hodgson 1979). In such cases, intake can be inferred from energy requirements (Energy for LW flux; maintenance+lactation and pregnancy+locomotion) using published estimates such as the Nutrient Requirements of domestic ruminants (CSIRO 2007) to convert physical values to energy values and so infer intake of the estimated diet. However, the inherent variability in feed composition, intake patterns of fermentation and feed partitioning, makes precise prediction of CH_4 production from feed characteristics problematic, though there are many examples in the literature of attempting to do so.

Blaxter and Clapperton's (1965) landmark review of experimental results from measuring CH_4 generation in cattle and sheep indicates that both intake and digestibility determine the amounts of methane generated and do so in an interrelated manner. However, based on several hundred animal measurements over a range of diets, they proposed the following equation:

$$CH_4(\text{kcal}/100\,\text{kcal GE}) = 1.30 + 0.112D + L(2.37 - 0.05D) \tag{6.2}$$

where

D = digestibility of gross energy, and L = level of intake relative to maintenance.

Moe and Tyrell (1979) and Holter and Young (1992) both developed more complex equations using carbohydrate fractions to predict CH_4 generation in dairy cattle:

$$\begin{aligned} CH_4\left(\text{MJ d}^{-1}\right) = {} & 3.406 + 0.510\,\text{soluble residue (kg fed)} + 1.736\,\text{hemicellulose (kg fed)} \\ & + 2.648\,\text{cellulose (kg fed)} \end{aligned} \tag{6.3}$$

The ability of each of these equations to accurately predict methane generation has been assessed by Wilkerson et al. (1995). Using data from 602 cattle, they concluded that all equations showed moderate to large errors of prediction, with Blaxter and Clapperton (1965) and Moe and Tyrell's (1979) equations having the least error. Contrary to the conclusions of Wilkerson et al. (1995), Benchaar et al. (1998), in reanalysing data from published studies, concluded that the equations of both Blaxter and Clapperton (1965) and Moe and Tyrell (1979) had coefficients of determination of less than 0.6, with high errors of prediction.

Recently, an alternative algorithm for Methane Production Rate (MPR-CH_4 g d^{-1}) for cattle consuming tropical forages has been proposed by Charmley et al. (2016) to address the lack of data for estimating EMP for ruminants grazing in tropical systems.

Estimations of CH_4 production from feed characteristics alone should at best be considered as being preliminary calculations, to only be used prior to the start of proper measurements; however, a methodology based on IPCC Tier II enteric CH_4 estimation promises more accurate estimates of DMP of ruminants under field conditions. The general approach, which was developed specifically for smallholder farming systems in Africa (Goopy et al. 2018), uses total metabolic energy requirements (MERtotal) of individual cattle on a seasonal basis calculated by summing the estimated MER for maintenance (MERM), LW gain or loss (MERG/L), lactation (MERL) and locomotion/traction(MERT). Intake can be inferred as a function of MERtotal and the weighted mean DM digestibility (DMD) of the seasonal feed baskets in the study area. DMI was used as the basis to estimate daily CH_4 production rate (MPR).

6.6.2 Emissions from Volatile Fatty Acids (VFAs)

Increasing interest in reducing emission of the important GHG methane has created a need for reliable indicators of daily methane production by ruminants without resorting to lengthy and intensive calorimetry studies. Feed intake explains the majority of differences seen in DMP. However, difficulty in assessing feed intake by grazing animals limits the application of this predictor and thus, a predictor of daily CH_4 production, which could be measured in the field, was sought. Enteric methane production is a direct function of the net amount of H_2 liberated through fermentation of feed, and there is an extensive literature which relates in vitro total fermentation to total gas production (e.g. Menke et al. 1979), and to CH_4 production in particular (i.e. Demeyer and Van Nevel 1975). Stoichiometric relationships developed in vitro typically explain more than 95% of observed H_2 present in VFAs. Thus, it has been considered that measurement of VFA levels in one or more rumen fluid samples may give a useful prediction of CH_4 production on a given day. Isotopic studies (Leng 1970; Sharp et al. 1982; Sutton et al. 2003) relating VFA concentration to VFA production indicated that a moderately strong relationship exists, suggesting that spot measures of VFA concentration (as a proxy for daily VFAs production) may provide a useful way of estimating daily CH_4 production in the field.

However, one study comparing multiple samples of VFA taken over the day, with simultaneous enteric methane production measurements, found that the average concentration of individual or total VFAs explained less than 25% of the variance in daily CH_4 production (Robinson et al. 2010) and is, thus, a poor predictor of methane production of sheep, which have typical morning and afternoon feeding periods. Sharp et al. (1982) also found poor-to-moderate correlations between CH_4 production rate and concentrations of acetate ($r = -0.1$), propionate ($r = 0.63$), and butyrate ($r = 0.92$). Sutton et al. (2003) found similar relationships to Sharp et al. (1982) between VFA production rate and VFA concentration for cattle fed a mixed diet twice daily.

The disparate response of methane production and VFA concentration to feeding may result from changes in VFA absorbance and rumen volume with feeding. Variable absorption of VFAs due to differences in factors such as pH and osmolarity that change with feeding is known (Dijkstra et al. 1993). Additionally, rumen volume would itself directly affect VFA absorption by means of changing absorptive surface area available. Increases in rumen volume with feeding level are also known (Purser and Moir 1966) and could also explain how fermentation and methane production could increase without causing a proportional increase in VFA concentration, by simply increasing the volume of rumen water in which the VFAs were dissolved. Rumen volume has been observed to increase by 20–34% in the first hour after feeding (Stewart et al. 1958). Such an increase would serve, not only to proportionally dilute VFA concentrations, but also to increase rumen surface area by an estimated 13–21%, thereby considerably enhancing the opportunity for diffusion of VFA across the rumen wall. Thus, using spot sampled concentrations of VFA in the rumen cannot be recommended as a method for estimating DMP in ruminants.

6.6.3 *In Vitro Incubations*

The amount of gas released from the fermentation process and from the buffering of VFAs is related to the kinetics of the fermentation of a known amount of feedstuff (Dijkstra et al. 2005). Several systems for measuring in vitro gas production have been developed which vary considerably in complexity and sophistication. Menke et al. (1979) described a manual method using gastight syringes, which involves constant registering of the gas volume produced. More recently others have described a system using pressure transducers (Pell and Schofield 1993; Theodorou et al. 1994; Cone et al. 1996). Variants now available as proprietary systems (RF, ANKOM Technology) use radio frequency pressure sensor modules that communicate with a computer interface and dedicated software which records gas pressure values.

The basic principle of the in vitro technique relies on the incubation of rumen inoculum with a feed substrate under an anaerobic environment in gastight culture bottles. Gas accumulates throughout the fermentation process, and a cumulative volume is recorded with gas volume curves being generated over time. To estimate kinetic parameters of total gas production, gas production values are corrected for the amount of gas produced in a blank incubation, and these values can be fitted with time using a non-linear curve fitting procedure in GenStat (Payne et al. 2011) or other suitable software. Headspace gas samples are taken to analyse the gas compositions and determine actual CH_4 concentrations, typically by gas chromatography, although a "quick and dirty" alternative has been developed–specifically a strongly basic solution such as NaOH is introduced into the vessel, which will subsequently cause the CO_2 to enter solution. The remaining gas is assumed to be CH_4.

Gas production is only one of the outputs of microbial fermentation, and the quality of the information derived can be improved by also considering substrate disappearance with concomitant production of VFAs (Blümmel et al. 2005).

There are two main forms of artificial rumen. In one form, rumen contents, freshly removed from a donor animal, are incubated in vitro in batch culture, while in the other form continuous culture is established. The former type of culture is valid for a period of hours, whereas the latter culture, which represents an anaerobic system not identical to the rumen, can be sustained for days or weeks. A procedure for batch-type culture is described below, while the rumen simulation technique (RUSITEC) (Czerkawski and Breckenridge 1977) is a proven system used widely throughout the world.

Neither type of in vitro system is very good for predicting methane production in vivo. The batch culture system generally lacks pH control, and the stoichiometry of product formation cannot be guaranteed to be the same as that occurring in the donor animal. For example, lactic acid is often detected in short-term in vitro incubations but not in the animal. In the RUSITEC system, the pattern of VFAs produced is usually similar, though not identical, to the in vivo situation.

Nevertheless, in vitro systems have some value for comparative purposes, e.g. for measuring the effects of additives on factors that control fermentation. The continuous culture is more valuable than the batch system, because it can take into account any adaptation the microbial population makes in response to the additive. This is particularly important for additives that affect CH_4 production. Some types of material, such as chlorinated hydrocarbons, are highly effective in the short term, but the rumen microbial population adapts eventually to become insensitive to the inhibitor. Microbial additives, on the other hand, require adaptation of the population to detect an effect, a situation that does not occur in the batch system.

6.6.4 Batch Systems

A variety of types of apparatus can be used. These differ in their cost and ease of use and range from a simple conical flask to a pH and Eh-controlled fermenter. The following criteria are essential to all of the systems:

Rumen contents must be removed from the rumen just before use; anaerobic conditions must be maintained during both transfer of the sample to the incubation vessel and during the incubation. The fermentation liquid must be incubated at 39 °C and agitated sufficiently to maintain some of the ciliate protozoa in suspension. The donor animals should be fed the feed to be used during the incubations.

6.6.4.1 Incubation of Rumen Contents in Glass Syringes: A Simple Artificial Rumen

The simplest type of short-term artificial rumen consists of test tubes or flasks with rumen contents incubated at 39 °C. Anaerobic conditions are maintained by bubbling CO_2 during incubation or by providing the vessels with Bunsen valves. The latter arrangement is difficult to manipulate (addition of substances, withdrawal of samples); an important component of fermentation products (gas) cannot be quantified, and the free venting of gas can result in contamination when radioisotopes are used. The procedure described below, which was developed by Czerkawski and Breckenridge (1977), is simple, and the apparatus is inexpensive; it does not suffer from the disadvantages discussed above, and it can be readily adapted to different requirements.

Plate 6.6 Simple apparatus for incubating small samples of rumen contents in a water bath

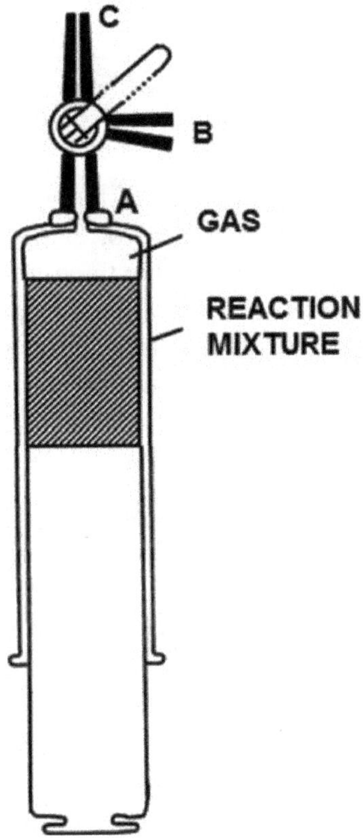

Procedure

(1) Connect a 3-way tap to a 50-ml glass syringe as shown in Plate 6.6. When the stopcock is turned to any particular opening, that opening is closed.

(2) Fill the syringe with water and then empty it in order to wet the plunger and the barrel. Turn the tap to C and inject 20 ml rumen contents through the opening B, using a 20 ml syringe. Turn the tap to A and remove the 20 ml syringe.

(3) Fill a 10 ml syringe with an inert gas (e.g. N_2), connect it to B and press the plunger, allowing the gas to escape through C until exactly 5 ml of gas is left in the syringe. Turn the tap to C and transfer the 5 ml of gas to the bigger syringe. Turn the tap to A and remove the small syringe. The syringe now contains 20 ml rumen contents and 5 ml of gas which makes it possible to agitate the liquid when the syringe is lying on its side.

(4) Place the syringe in a water bath and incubate. If a shaking water bath is available, fix the syringes to a frame with clips. If no such bath is available, allow the syringes to float (they will be 70–80% submerged) and remove them periodically in order to agitate the contents by inversion at regular intervals.

(5) Take each syringe out at regular intervals, keep it vertical as shown in Plate 6.6 (making sure that the plunger is free to move) and read the volume of gas on the scale. The difference between this reading and the amount of gas added gives the amount of gas produced. This is not a very accurate parameter, but it is very

useful since, with care, a curve relating gas production with time can be drawn. The shape of this curve can provide useful information about the extent and magnitude of the reaction in the incubated sample.

(6) At the end of incubation, take the syringe out of the bath, connect an empty syringe (20 ml) to the opening B, turn the tap to C and transfer the gas to this syringe. If the volume of gas produced is small, use a 10 ml syringe; the measurement of gas produced will be more accurate. Turn the tap to B and disconnect the three-way tap (plus the small syringe) from the big syringe at A. Read the volume of gas and set the sealed gas sample aside for further analysis.

(7) Small amounts of liquid (e.g. substrates, labelled compounds, inhibitors) can be introduced into the reaction mixture in the same way as the gas (see step (3) above), though it is best to use a small (1 ml) plastic syringe. Similarly, samples of liquid or gas can be taken (the former, with the syringe upside down).

6.6.4.2 Continuous Fermenters

Three main types of continuous systems are used. The first is actually a sequential culture, where a small amount of solid feed is added to a sample of rumen fluid which is then sub-cultured daily (Merry et al. 1987). Whether methane production is sustained and mimics accurately, the real situation is unclear. The second type involves the continuous flow of both solid and liquid phases (Hoover et al. 1976). The third type is the RUSITEC scheme where solids are received once daily, but in the liquid phase by continuous flow. Use of the RUSITEC scheme requires a considerable input of time and expertise and is not recommended unless the specific objective is particularly suited to the apparatus, i.e. only where small quantities of the potential modifying agents are available, and the small-scale and between-vessel reproducibility offers a major advantage over animal experiments.

6.7 Methane from Animal Wastes

Animal waste is a significant source of CH_4. The amount produced depends on the diet of the animal. If the waste is stored anaerobically in lagoons or in liquid slurry tanks, CH_4 production can be closely related to the disappearance of organic matter. Temperature and relative humidity affect the rate at which the faecal matter dries out under grazing and feed conditions and determines how long fermentation continues. Estimates of CH_4 production from these systems can be obtained by adapting techniques developed for field use. The closed chamber method described for groups of animals (Chap. 2) and in rice paddies is likely to be the best method to use in measuring CH_4 production from lagoons and tanks. The emission rate of CH_4 that is likely to arise from animal waste can be very large. It may require a shorter sampling time and may also require mechanical mixing of the vented gas to prevent concentration gradients from forming in the chamber or space.

For waste lagoons, floating chambers of the types used in water studies can be employed (Chap. 2). Alternatively, modifications of the hood and facemask technique may be used, whereby air is drawn over the surface of the animal waste and subsampled for analysis. Non-isotopic tracers such as SF_6 can be used for measuring CH_4 production in barns and other enclosed areas where animal waste may accumulate.

6.8 Storage and Analysis of Samples

6.8.1 Storage of Samples

Unlike solid or liquid biological samples, which can often be frozen, dried, inactivated or preserved chemically, gas samples require very specialised handling during all stages of their analysis. Furthermore, the gases dealt with here are colourless and they have no odour. Thus, whatever the procedure, whether during sampling, storage and analysis, exceptional care should be taken to avoid losses and to eliminate any chance of such losses going undetected.

Methane passes through plastic materials, and care is required when storing gas samples prior to analysis. Glass and metal provide the most reliable materials for storing gas-containing CH_4. Methane is also relatively insoluble in water, and storage under water can provide a useful means of containing the gas.

Metal and glass syringes greased with Vaseline provide safe storage over periods of 2–3 days. These should be sealed with metal taps or hypodermic needles inserted into a rubber stopper. An important rule is that gas samples should be analysed as soon as possible after sampling; if it is necessary to store gas samples, this storage should be reduced to a minimum. Metal taps (not plastic) should be used to seal syringes. Alternatively, steel needles can be used and pushed into a rubber stopper to seal them. Bags made from PVF are satisfactory for temporary storage of gas prior to analysis, as are metal-coated liners for wine and fruit juice casks. If the aim is to store the samples for long time before analysis is carried out, then Exetainers shall be used (Plate 6.7).

6.8.2 Analysis of Samples

The methods shown in the above section require concentration measurements of CH_4, SF_6, volatile fatty acids, O_2, N_2, H_2 or CO_2. For concentration analysis, the appropriate method must be chosen under consideration of prevailing conditions, such as available apparatus, required sensitivity and the need for information about other associated gases (e.g. CO_2 or H_2). Essentially three methods are available. Whichever method is chosen, all the precautions listed above must be taken.

Plate 6.7 Exetainers for gas storage

6.8.2.1 Flammable Gases (CH_4 and VFA)

Flammable gases can be measured with a GC equipped with a flame ionisation detector (FID). A detailed description of this method is given in Chap. 2.

The main principle of a GC is that the gas sample (with is a mixture of several gases, e.g. O_2, CO_2, CH_4 and others like VFA or SF_6) is split into its components in a separation column. The separation columns differ in their composition, length and diameter. Make sure the column of your analytical system is appropriate for the respective gases to be measured. The manufacturer is the best source of information on appropriate columns for your measurements.

When a sample of CH_4-containing gas is injected into the GC, the gas mix is separated into its components. Depending on the retention time in the column, one compound after the next reaches the FID (CH_4 will pass through very fast). Flammable gases will trigger a signal in the detector, which is registered as a sharp peak. The area of this peak will be proportional to the concentration of CH_4.

To determine the gas concentration in your sample, it is necessary to use reference gases (i.e. gases with a known gas concentration) during all GC measurements. A detailed description of the subsequent calculations and interpretation of your samples is given in Chap. 2.

6.8.2.2 Thermal Conductivity Detector (TCD)

Gas chromatographs fitted with this type of detector come with comprehensive descriptions for use supplied by the maker, who will also supply suitable columns and other auxiliary equipment. The use of the standard gas samples and calibration of the instrument are also dealt with by the manufacturer.

The value of this technique is that it is possible to separate all the permanent gases (O_2, N_2, H_2, CO_2, and, of course, CH_4) and to determine their relative concentrations. It is customary to use two columns, a Porapak Q column which will separate CO_2 from a composite peak containing all the other gases, and a Molecular Sieve 5A column which separates H_2, O_2, N_2 and CH_4, with CO_2 becoming part of the baseline. This means that samples have to be analysed twice. However, it is also possible to put the two columns in series on the two sides of the detector (Czerkawski and Clapperton 1968). In this method, the gas (argon) passing through one side of the detector acts as reference, and then the polarity is reversed and the other side of the detector acts as reference. It is then only necessary to inject one sample and simply to reverse the polarity of the detector during the analysis.

6.8.2.3 Infrared Detector (IRD)

An infrared gas analyser is capable of measuring CH_4 within the required range of concentrations using,

1. A flow meter.
2. Copper tubing and associated fittings.
3. A small air pump (rates as recommended by the supplier of the IR apparatus).
4. Gas driers.
5. Reference gas (N_2 if possible).
6. Standard gas of known concentration.

6.8.2.4 Laser Techniques

As described in Sect. 2.7, open-path lasers and also so-called cavity ring-down spectroscopy (CRDS) are nowadays available to determine the concentration of all relevant gases. Details on the use of these techniques can be found in Chap. 2, Sect. 2.7.

Choice of analyser: In choosing the most appropriate CH_4 analyser, it is necessary to have a suitable measurement range. In most situations, 0–500 ppm CH_4 is quite satisfactory, but by variations in analytical tube length, this range can easily be attenuated or extended. Most manufacturers can supply such options.

Of equal importance is the choice of analyser, i.e. whether it is a single channel or dual channel. In the former type, the single analytical tube is used to analyse both zero and CH_4-containing samples, but it is not possible to accommodate changes in background CH_4 concentrations or atmospheric pressure changes. In a dual-channel analyser, the use of two optically balanced analytical tubes permits the sample (unknown) gas to be measured against background air at all times in a different mode. By this procedure, changes in background CH_4 concentrations and atmospheric pressure can be fully considered, and the reliability of the resultant data increases.

Analysis: Infrared gas analysers measure CH_4 concentration in a steady stream of sample gas. Calibration procedures generally use a reference gas, i.e. N_2 or outside air if N_2 is unavailable, and a higher standard gas of known concentration. The reference and standard gases should be on either side of the range of the expected concentration of the sample gas to ensure accurate measurement of sample gas concentration. Care must be taken to ensure that the flow of sample gas to the analyser is within the recommended guidelines for the analyser and is constant throughout the analysis period. The sample gas should be dried before use. This may be accomplished by sending the gas stream through a short piece of pipe or tubing that has been loosely filled with a drying compound such as Aquasorb^{Tm} or Drierite^{Tm} or even anhydrous Na_2SO_4.

After the system is equilibrated and the reading is steady, the analyser reading is recorded and inserted into the calculation equation specific for the analyser in use.

6.8.2.5 Non-flammable Gases (CO_2 and SF6)

Concentration measurements of CO_2 and SF_6 require a GC equipped with an electron capture detector (ECD) and TCD. Note that some GCs comprise both detectors in a row so that the gas sample passes both detectors. In this case, you get gas concentration of flammable gases and non-flammable gases.

Standards should be analysed daily. A 30 ml sample of gas from a collection flask should be removed with a syringe (plastic or glass), and part of this gas is passed through the loop and the loop full of gas is injected onto the column. SF_6 will elute at approximately 20 s, and a sharp peak will be produced. O_2 has a retention time of 50 s and will be the next peak produced. It will be well resolved from the SF_6. Samples may be analysed at approximately four-minute intervals. The detection limit is approximately one part per trillion (ppt).

6.8.2.6 Standards and Calibration

In open-circuit calorimetry for measurements of CH_4 concentration and ultimately its production, it is necessary to conduct regular calibration of the CH_4 analyser and of the whole system.

(i) *Analyser calibration*

With respect to the analyser, where an infrared apparatus is available, this should be calibrated before and after use in order to establish both zero and set point deflection and to accommodate any analyser drift. The zero-point measurement can be achieved by appropriate use of a CH_4-free air sample obtained from a fresh air source at a suitable distance from any livestock or livestock waste source. It is advisable that the gas sample is filtered and dried (e.g. through a calcium chloride tower) before injection into the analyser. A primary standard containing CH_4 can be acquired in the first instance through a suitable commercial company or an international organisation. This standard ought to contain between 250 and 500 ppm CH_4 and should be used initially to calibrate the analyser. Once this has been achieved, gas from a larger cylinder containing a similar but unspecified concentration of CH_4 should be introduced into the analyser, and through repeated measurement its concentration should be established (secondary standard). Thereafter, this standard sample should be used regularly to standardise the analyser, and the primary standard should only be used to recalibrate new cylinders of the CH_4 standard as required.

Experience shows that often successful research work involves some improvisation. This is particularly true when one develops new methods. An example of the type of problems that can be encountered relates to efforts made by scientists in a particular institute to develop a simple method for measuring the concentrations of CH_4 in rumen gases. The apparatus required frequent calibration with standard gas mixtures. These gases were contained in two small aerosol cans and were considered to be so valuable that they were kept in a safe. During the development work, all the gas in one can was used up and it became imperative to do something before the remaining standard gas in the other can was used up since the delivery of another can would take 6–8 months. This was done by partly evacuating an old N_2 cylinder in the laboratory and carrying it to a nearby anaerobic digester on the institute farm where it was filled with CH_4-containing gas. The cylinder was then taken back to the laboratory and "topped up" with nitrogen to about 10 atmospheres, resulting in some 7% CH_4 in nitrogen. The accurate concentration of CH_4 in this secondary standard was determined using the precious primary standard. The secondary standard was then used during the remainder of the study and for many months thereafter.

Local conditions are very important. For instance, anaerobic digesters are very common in some countries, and it is easy to obtain CH_4 for the secondary standard. If there are no digesters, some other source of CH_4 has to be found; even a simple chemical method (i.e. methane derived from aluminium carbide) may be suitable.

(ii) *Whole systems*

With respect to calibration of whole systems such as respiration chambers, etc., to estimate the recovery of CH_4 and check both the analyser and non-analyser components of the system, the preferred procedure is to introduce a gas of known CH_4 concentration into the chamber over a minimum period of 6–8 h, and to determine its recovery downstream. Methane emission from the cylinder can be determined by a recording of cylinder weight loss over the test period and the CH_4 concentration of the cylinder gas, whilst quantitative recovery of CH_4 has to cover the period of time required for the concentration to return to baseline. Ideally, the recovery of CH_4 should be within $\pm 3\%$ of added methane. If the results are outside this range, the test must be repeated, and if still unsatisfactory, individual components of the system should be isolated and checked for satisfactory function.

A direct method of calibration involves weighing the CH_4 source and adding CH_4 to the chamber. CH_4 emission and analysis of outflow gas are continued until a measurable quantity of CH_4 has been removed from the source. The source is then closed off, and analysis of outflow gas for CH_4 content continued until outflow concentration returns to original background level.

An alternative to the burning of alcohol to produce CO_2 can be simulated quite easily by acidifying a $BaCO_3$ solution with dilute hydrochloric acid within the chamber. The generation of CO_2 should be controlled by drip addition of the acid or by slow pumping of the acid to avoid massive fluctuations in CO_2 content of the outflowing air.

References

Agricultural Research Council (ARC) (1980) The nutrient requirements of ruminant livestock. Technical Review by an Agricultural Research Council Working Party, Commonwealth Agricultural Bureau, Farnham Royal, UK

Benchaar C, Rivest J, Pomar C, Chiquette J (1998) Prediction of methane production from dairy cows using existing mechanistic models and regression equations. J Anim Sci 76(2):617–627

Blaxter KL, Clapperton JL (1965) Prediction of the amount of methane produced by ruminants. Br J Nutr 19:511–522

Blümmel M, Givens DI, Moss AR (2005) Comparison of methane produced by straw fed sheep in open-circuit respiration with methane predicted by fermentation characteristics measured by an in vitro gas procedure. Anim Feed Sci Tech 123–124(Part 1):379–390

Calvet S, Gates RS, Zhang G, Estellés F, Ogink NWM, Pedersen S, Berckmans D (2013) Measuring gas emissions from livestock buildings: a review on uncertainty analysis and error sources. Biosyst Eng 116:221–231

Chagunda MGG, Yan T (2011) Do methane measurements from a laser detector and an indirect open-circuit respiration calorimetric chamber agree sufficiently closely? Anim Feed Sci Tech 165(1–2):8–14

Chagunda MGG, Ross D, Roberts DJ (2009) On the use of a laser methane detector in dairy cows. Comput Electron Agric 68(2):157–160

Charmley E, Williams SRO, Moate PJ, Hegarty RS, Herd RM, Oddy VH, Reyenga P, Staunton KM, Anderson A, Hannah MC (2016) A universal equation to predict methane production of forage-fed cattle in Australia. Anim Prod Sci 56(3):169–180

Cone JW, van Gelder AH, Visscher GJW, Oudshoorn L (1996) Influence of rumen fluid substrate concentration on fermentation kinetics measured with a fully automated time related gas production apparatus. Anim Feed Sci Tech 61:113–128

CSIRO (2007) Nutrient requirements of domesticated ruminants. CSIRO publishing

Czerkawski JW, Breckenridge (1977) Design and development of a long-term rumen simulation technique (Rusitec). Br J Nutr 38:371–384

Czerkawski JW, Clapperton JL (1968) Analysis of gases produced by metabolism of microorganisms. Lab Pract 17:994

Deighton MH, Williams SRO, Hanna MC, Eckard RJ, Boland TM, Wales WJ, Moate PJ (2014) A modified sulphur hexafluoride tracer technique enables accurate determination of enteric methane emissions from ruminants. Anim Feed Sci Tech 197:47–63

Demeyer DI, Van Nevel CJ (1975) Methanogenesis, an integrated part of carbohydrate fermentation and its control. In: Donald IWMc, Warner ACI (eds) Digestion and metabolism in the ruminant. The University of New England Publishing Unit, Armidale, Australia, pp 366–382

Denmead OT (2008) Approaches to measuring fluxes of methane and nitrous oxide between landscapes and the atmosphere. Plant Soil 309:5–24

Dijkstra J, Boer H, Van Bruchem J, Bruining M, Tamminga S (1993) Absorption of volatile fatty acids from the rumen of lactating dairy cows as influenced by volatile fatty acid concentration, pH and rumen liquid volume. Br J Nutr 69:385–396

Dijkstra J, Kebreab E, Bannink A, France J, López S (2005) Application of the gas production technique to feed evaluation systems for ruminants. Anim Feed Sci Tech 123–124:561–578

Fernández C, López MC, Lachica M (2012) Description and function of a mobile open-circuit respirometry system to measure gas exchange in small ruminants. Anim Feed Sci Tech 172:242–246

Flesch TK, Wilson JD, Harper LA, Crenna BP, Sharpe RR (2004) Deducing ground-to-air emissions from observed trace gas concentrations: a field trial. J Appl Meteorol 43:487–502

Flesch TK, Wilson JD, Harper LA, Crenna BP (2005) Estimating gas emission from a farm using an inverse dispersion technique. Atmos Environ 39:4863–4874

Flesch TK, Wilson JD, Harper LA, Todd RW, Cole NA (2007) Determining ammonia emissions from a cattle feedlot with an inverse dispersion technique. Agr Forest Meteorol 144:139–155

Freibauer A (2000) New approach to an inventory of N_2O and CH_4 emissions from agriculture in Western Europe. Kluwer Academic, Dordrecht

Gao Z, Desjardins RL, Flesch TK (2010) Assessment of the uncertainty of using an inverse dispersion technique to measure methane emissions from animals in a barn and in a small pen. Atm Environ 44:3128–3134

Garnsworthy PC, Craigon J, Hernandez-Medrano JH, Saunders N (2012) On-farm methane measurements during milking correlate with total methane production by individual dairy cows. J Dairy Sci 95(6):3166–3180

Goopy J, Hegarty R (2004) Repeatability of methane production in cattle fed concentrate and forage diets. J Anim Feed Sci 13(1):75–78

Goopy JP, Hegarty R, Robinson D (2009) Two-hour chamber measurement provides a useful estimate of daily methane production in sheep. In: Proceedings of the XIth international symposium on ruminant physiology, ruminant physiology: digestion, metabolism and effects of nutrition on reproduction and welfare; 6–9 Sept 2009. Wageningen Academic Pub, Clermont-Ferrand, France

Goopy JP, Woodgate R, Donaldson A, Robinson DL, Hegarty RS (2011) Validation of a short-term methane measurement using portable static chambers to estimate daily methane production in sheep. Anim Feed Sci Tech 166–167:219–226

Goopy JP, Robinson D, Woodgate RT, Donaldson AJ, Oddy VH, Vercoe PE, Hegarty RS (2014) Estimates of repeatability and heritability of methane production in sheep using portable accumulation chambers. Anim Prod Sci 56(1):116–122

Goopy JP, Onyango AA, Dickhoefer U, Butterbach-Bahl K (2018) A new approach for improving emission factors for enteric methane emissions of cattle in smallholder systems of East Africa—Results for Nyando, Western Kenya. Agric Syst 161:72–80

Grainger C, Clarke T, McGinn SM, Auldist MJ, Beauchemin KA, Hannah MC, Waghorn GC, Clark H, Eckard RJ (2007) Methane emissions from dairy cows measured using the sulfurhexafluoride (SF_6) tracer and chamber techniques. J Dairy Sci 90:2755–2766

Grainger C, Williams R, Clarke T, Wright ADG, Eckard RJ (2010) Supplementation with whole cottonseed causes long-term reduction of methane emissions from lactating dairy cows offered a forage and cereal grain diet. J Dairy Sci 93:2612–2619

Hammond K, Humphries D, Crompton LA, Kirton P, Green C, Reynolds CK (2013) Methane emissions from growing dairy heifers estimated using an automated head chamber (GreenFeed) compared to respiration chambers or SF_6 techniques. Adv. Anim Biosci 4(2):391

Haque MN, Cornou C, Madsen J (2014) Estimation of methane emission using the CO_2 method from dairy cows fed concentrate with different carbohydrate compositions in automatic milking system. Livest Sci 164:57–66

Hegarty RS (2013) Applicability of short-term emission measurements for on-farm quantification of enteric methane. Animal 7(2):401–408

Hellwing A, Lund P, Madsen J, Weisberg MR (2013) Comparison of enteric methane production from the CH_4/CO_2 ratio and measured in respiration chambers. Adv Anim Biosci 4(2):557

Hendricksen R, Minson DJ (1980) The feed intake and grazing behaviour of cattle grazing a crop of Lablab purpureus cv.Rongai. J Agric Sci 95(3):547–554

Holter JB, Young AJ (1992) Methane prediction in dry and lactating holstein cows. J Dairy Sci 75(8):2165–2175

Hoover WH, Knowlton PM, Stem MD, Sniffen CJ (1976) Effect of differential solid-liquid removal rates on fermentation parameters in continuous cultivation of rumen contents. J Anim Sci 43:535–542

Jamieson WS, Hodgson J (1979) The effects of variation in sward characteristics upon the ingestive behaviour and herbage intake of calves and lambs under a continuous stocking management. Grass Forage Sci 34(4):273–282

Johnson D, Ward G (1996) Estimates of animal methane emissions. Environ Monit Assess 42(1–2):133–141

Klein L, Wright ADG (2006) Construction and operation of open-circuit methane chambers for small ruminants. Aust J Exp Agric 46(10):1257–1262

Laubach J, Kelliher JM (2005) Methane emissions from dairy cows: comparing open-path laser measurements to profile-based techniques. Agr Forest Meteorol 135:340–345

Leng RA (1970) Formation and production of volatile fatty acids in the rumen. In: Phillipson AT (ed) Physiology of digestion and metabolism in the ruminant. Oriel Press, Newcastle upon Tyne, pp 407–421

Lockyer DR (1997) Methane emission from grazing sheep and calves. Agr Ecosyst Environ 66:11–18

Lockyer DR, Champion RA (2001) Methane production in sheep in relation to temporal changes in grazing behaviour. Agr Ecosyst Environ 86:237–246

Lockyer DR, Jarvis SC (1995) The measurement of methane losses from grazing animals. Environ Pollut 90:383–390

Loh Z, Chen D, Bai M, Naylor T, Griffith D, Hill J, Denmead T, McGinn S, Edis R (2008) Measurement of greenhouse gas emissions from Australian feedlot beef production using open-path spectroscopy and atmospheric dispersion modelling. Aust J Exp Agric 48:244–247

Loh Z, Leuning R, Zegelin S, Etheridge D, Bai M, Naylor T, Griffith D (2009) Testing Lagrangian atmospheric dispersion modeling to monitor CO_2 and CH_4 leakage from geosequestration. Atmos Environ 43:2602–2611

Madsen J, Bjerg BS et al (2010) Methane and carbon dioxide ratio in excreted air for quantification of the methane production from ruminants. Livest Sci 129(1–3):223–227

McGinn SM (2006) Measuring greenhouse gas emissions from point sources in agriculture. Can J Soil Sci 86:355–371

McGinn SM, Chen D, Loh Z, Hill J, Beauchemin KA, Denmead OT (2008) Methane emissions from feedlot cattle in Australia and Canada. Aust J Exp Agric 48:183–185

Menke KH, Raab L, Salewski A, Steingass H, Fritz D, Schneider W (1979) The estimation of the digestibility and metabolizable energy content of ruminant feeding stuffs from the gas production when they are incubated with rumen liquor in vitro. J Agric Sci 93:217–222

Merry RJ, Smith RH, McAllan AB (1987) Studies of rumen function in an in vitro continuous culture system. Archiv fuer Tierernaehrung 37(6):475–488

Moate PJ, Deighton MH et al (2015) Michaelis-Menten kinetics predict the rate of SF6 release from permeation tubes used to estimate methane emissions from ruminants. Anim Feed Sci Technol 200:47–56

Moe PW, Tyrrell HF (1979) Methane production in dairy cows. J Dairy Sci 62:1583–1586

Mohn J, Zeyer K, Keck M, Keller M, Zähner M, Poteko J, Emmenegger L, Schrader S (2018) A dual tracer ratio method for comparative emission measurements in an experimental dairy housing. Atmos Environ 179:12–22

Murray PJ, Moss A, Lockyer DR, Jarvis SC (1999) A comparison of systems for measuring methane emissions from sheep. J Agric Sci 133:439–444

Murray PJ, Gill E, Balsdon SL, Jarvis SC (2001) A comparison of methane emissions from sheep grazing pastures with different management intensities. Nutr Cycl Agroeco 60:93–97

Odongo NE, Alzahal O, Las JE, Kramer A, Kerrigan B, Kebreab E, France J, McBride BW (2007) Development of a mobile open-circuit ventilated hood system for measuring real-time gaseous emissions in cattle. CABI Publishing, Wallingford

Ogink NWM, Mosquera J, Calvet S, Zhang G (2013) Methods for measuring gas emissions from naturally ventilated livestock buildings: developments over the last decade and perspectives for improvement. Biosyst Eng 116:297–308

Payne RW, Murray DA, Harding SA (2011) An introduction to the GenStat command language, 14th edn. VSN International, Hemel Hempstead

Pell AN, Schofield P (1993) Computerized monitoring of gas production to measure forage digestion in vitro. J Dairy Sci 76:1063–1073

Pinares-Patiño CS, Hour PD, Jouany JP, Martin C (2007) Effects of stocking rate on methane and carbon dioxide emissions from grazing cattle. Agr Ecosyst Environ 121:30–46

Pinares-Patiño CS, Lassey KR, Martin RJ, Molano G, Fernandez M, MacLean S, Sandoval E, Luo D, Clark H (2011) Assessment of the sulphur hexafluoride (SF_6) tracer technique using respiration chambers for estimation of methane emissions from sheep. Anim Feed Sci Tech 166(C):201–209

Place SE, Pan Y, Zhao Y, Mitloehner FM (2011) Construction and operation of a ventilated hood system for measuring greenhouse gas and volatile organic compound emissions from cattle. Animal 1:433–446

Purser DB, Moir RJ (1966) Rumen volume as a factor involved in individual sheep differences. J Anim Sci 25:509–515

Ramirez-Restrepo CA, Barr TN, Marriner A, López-Villalobos N, McWilliam EL, Lassey KR, Clark H (2010) Effects of grazing willow fodder blocks upon methane production and blood composition in young sheep. Anim Feed Sci Tech 155:33–43

Robinson D, Goopy J, Hegarty R (2010) Can rumen methane production be predicted from volatile fatty acid concentrations? Anim Prod Sci 50:630–636

Robinson DL, Goopy JP, Donaldson AJ, Woodgate RT, Oddy VH, Hegarty RS (2014) Sire and liveweight affect feed intake and methane emissions of sheep confined in respiration chambers. Anim Int J Anim Biosci (Cambridge University Press)8(12):1935–1944

Robinson DL, Goopy JP, Hegarty RS, Oddy VH (2015a) Comparison of repeated measurements of CH4 production in sheep over 5 years and a range of measurement protocols. J Anim Sci 93(10):1–15

Robinson DL, Hegarty RS, Oddy VH (2015b) Comparison of repeated measurements of CH_4 production in sheep over 5 years and a range of measurement protocols. J Anim Sci 93(10):4637–4650

Schrade S, Zeyer K, Gygax L, Emmenegger L, Hartung E, Keck M (2012) Ammonia emissions and emission factors of naturally ventilated dairy housing with solid floors and an outdoor exercise area in Switzerland. Atmos Environ 47:183–194

Schrade S, Zeyer K, Keck M, Keller M, Zähner M, Mohn J (2018) Validation of the tracer ratio method for emission measurements in naturally ventilated housing. Agrarforschung Schweiz 9:340–347

Sharp JH, Culberson CH, Church TM (1982) The chemistry of the Delaware estuary. Limnol Oceanogr John Wiley & Sons, Ltd 27(6):1015–1028

Stewart WE, Stewart DG, Schultz LH (1958) Rates of volatile fatty acid production in the bovine rumen. J Anim Sci 17:723–736

Sutton JD, Dhanoa MS, Morant SV, France J, Napper DJ, Schuller E (2003) Rates of production of acetate, propionate, and butyrate in the rumen of lactating dairy cows given normal and low-roughage diets. J Dairy Sci 86:3620–3633

Suzuki T, McCrabb GJ, Nishida T, Indramanee S, Kurihara M (2007) Construction and operation of ventilated hood-type respiration calorimeters for in vivo measurement of methane production and energy partition in ruminants. Springer, Dordrecht

Suzuki T, Phaowphaisal I, Pholson P, Narmsilee R, Indramanee S, Nitipot T, Haokaur A, Sommar K, Khotprom N, Panichpol V, Nishida T (2008) In vivo nutritive value of pangola grass (Digitaria eriantha) hay by a novel indirect calorimeter with a ventilated hood in Thailand. Jarq-Jpn Agric Res Q 42:123–129

Takahashi J, Chaudhry AS, Beneke RG, Young BA (1999) An open-circuit hood system for gaseous exchange measurements in small ruminants. Small Rumin Res 32:31–36

Theodorou MK, Williams BA, Dhanoa MS, McAllan AB, France J (1994) A simple gas production method using a pressure transducer to determine the fermentation kinetics of ruminant feeds. Anim Feed Sci Technol 48:185–197

Tomkins NW, McGinn SM, Turner DA, Charmley E (2011) Comparison of open-circuit respiration chambers with a micrometeorological method for determining methane emissions from beef cattle grazing a tropical pasture. Anim Feed Sci Technol 166–167:240–247

Tubiello F, Salvatore M, Cóndor Golec RD, Ferrara A, Rossi S, Biancalani R, Federici S, Jacobs H, Flammini A (2014) Agriculture, forestry and other land use emissions by sources and removals by sinks. Statistics Division, Food and Agriculture Organization, Rome

Wilkerson VA, Casper DP, Mertens DR (1995) The prediction of methane production of holstein cows by several equations. J Dairy Sci 78(11):2402–2414

Williams SRO, Moate PJ, Hannah MC, Ribaux BE, Wales WJ, Eckard RJ (2011) Background matters with the SF_6 tracer method for estimating enteric methane emissions from dairy cows: a critical evaluation of the SF_6 procedure. Anim Feed Sci Technol 170:265–276

Zimmerman PR, Zimmerman RS (2012) Method and system for monitoring and reducing ruminant methane production. Google Patents

7

Stable Isotope Methods for Measurement of N_2O and N_2

M. Zaman, K. Kleineidam, L. Bakken, J. Berendt, C. Bracken, K. Butterbach-Bahl, Z. Cai, S. X. Chang, T. Clough, K. Dawar, W. X. Ding, P. Dörsch, M. dos Reis Martins, C. Eckhardt, S. Fiedler, T. Frosch, J. Goopy, C.-M. Görres, A. Gupta, S. Henjes, M. E. G. Hofmann, M. A. Horn, M. M. R. Jahangir, A. Jansen-Willems, K. Lenhart, L. Heng, D. Lewicka-Szczebak, G. Lucic, L. Merbold, J. Mohn, L. Molstad, G. Moser, P. Murphy, A. Sanz-Cobena, M. Šimek, S. Urquiaga, R. Well, N. Wrage-Mönnig, S. Zaman, J. Zhang, and C. Müller

Abstract GHG emissions are usually the result of several simultaneous processes. Furthermore, some gases such as N_2 are very difficult to quantify and require special techniques. Therefore, in this chapter, the focus is on stable isotope methods. Both natural abundance techniques and enrichment techniques are used. Especially in

---

M. Zaman (✉) · L. Heng
Soil and Water Management & Crop Nutrition (SWMCN) Section, Joint FAO/IAEA Division of Nuclear Techniques in Food and Agriculture, International Atomic Energy Agency (IAEA), Vienna, Austria
e-mail: m.zaman@iaea.org; zamanm_99@yahoo.com

K. Kleineidam · C. Eckhardt · A. Jansen-Willems · G. Moser · C. Müller
Institute of Plant Ecology, Justus Liebig University Giessen, Giessen, Germany

L. Bakken
Norwegian University of Life Sciences (NMBU), Aas, Norway

J. Berendt · S. Fiedler · N. Wrage-Mönnig
University of Rostock, Rostock, Germany

C. Bracken
School of Agriculture and Food Science and Earth Institute, University College Dublin, Dublin, Ireland

K. Butterbach-Bahl
Institute of Meteorology and Climate Research, Atmospheric Environmental Research (IMK-IFU), Karlsruhe Institute of Technology, Karlsruhe, Germany

Z. Cai
School of Geography Sciences, Nanjing Normal University, Jiangsu, China

S. X. Chang
Department of Renewable Resources, University of Alberta, Edmonton, AB T6G 2E3, Canada

T. Clough
Department of Soil & Physical Sciences, Faculty of Agriculture & Life Sciences, Lincoln University, Lincoln, New Zealand

the last decade, a number of methodological advances have been made. Thus, this chapter provides an overview and description of a number of current state-of-the-art techniques, especially techniques using the stable isotope ^{15}N. Basic principles and recent advances of the ^{15}N gas flux method are presented to quantify N_2 fluxes, but also the latest isotopologue and isotopomer methods to identify pathways for N_2O production. The second part of the chapter is devoted to ^{15}N tracing techniques, the theoretical background and recent methodological advances. A range of different methods is presented from analytical to numerical tools to identify and quantify pathway-specific N_2O emissions. While this chapter is chiefly concerned with gaseous N emissions, a lot of the techniques can also be applied to other gases such as methane (CH_4), as outlined in Sect. 5.3.

Keywords $^{15}N_2O$ · $^{15}N_2$ · ^{15}N tracer technique

7.1 Introduction

In this chapter, we are presenting techniques utilising the stable isotope ^{15}N to better understand the N cycle but more importantly to determine GHG gas fluxes that cannot be quantified or are difficult to quantify with any non-isotopic technique. The

K. Dawar
Department of Soil and environmental Sciences, University of Agriculture, Peshawar, Pakistan

W. X. Ding
Institute of Soil Science, Chinese Academy of Sciences, Nanjing, China

P. Dörsch · L. Molstad
Faculty of Environmental Sciences and Natural Resource Management, Norwegian University of Life Sciences (NMBU), Aas, Norway

T. Frosch
Leibniz Institute of Photonic Technology, Technical University Darmstadt, Darmstadt, Germany

J. Goopy
International Livestock Research Institute (ILRI), Nairobi, Kenya

C.-M. Görres
Department of Soil Science and Plant Nutrition/Department of Applied Ecology, Hochschule Geisenheim University, Geisenheim, Germany

A. Gupta
Independent Consultant India, Mumbai, India

S. Henjes · M. A. Horn
Institute of Microbiology, Leibniz University Hannover, Hannover, Germany

M. E. G. Hofmann
Picarro B.V., 's-Hertogenbosch, The Netherlands

M. M. R. Jahangir
Department of Soil Science, Bangladesh Agricultural University, Mymensingh, Bangladesh

stable isotope ^{15}N was discovered in the 1920s (Naudé 1929a, b) and the advantage of using this isotope in agriculture, for the determination of the N use efficiency has been recognised and applied since 1943 (Norman and Werkman 1943). Also, microbiologists have utilised the new possibilities that ^{15}N can offer, to quantify the turnover rates of individual processes in the N cycle (Hiltbold et al. 1951) based on dilution principles (Kirkham and Bartholomew 1954). Moreover, ^{15}N allowed for the first time the development of techniques to quantify the loss of N_2 against a huge atmospheric N_2 background (Hauck et al. 1958). Also, the identification which of the processes contributing to total N_2O emissions (Butterbach-Bahl et al. 2013) is unthinkable without the use of advanced ^{15}N tracing techniques (Müller et al. 2014). With the development of new and advanced analytical techniques, it is now possible to also use information on the position of the ^{15}N (i.e. central, alpha and terminal, beta position) in N_2O, i.e. the isotopomers (of one isotopologue), providing information

D. Lewicka-Szczebak
Laboratory of Isotope Geology and Geoecology, Institute of Geological Sciences, University of Wrocław, Wrocław, Poland

G. Lucic
Picarro Inc., Santa Clara, CA, USA

L. Merbold
Mazingira Centre, International Livestock Research Institute (ILRI), Nairobi, Kenya

J. Mohn
Laboratory for Air Pollution & Environmental Technology, Empa Dübendorf, Dübendorf, Switzerland

P. Murphy
Environment & Sustainable Resource Management Section, School of Agriculture & Food Science, and UCD Earth Institute, University College, Dublin, Ireland

A. Sanz-Cobena
Research Center for the Management of Environmental and Agricultural Risks (CEIGRAM), ETSIAAB, Universidad Politécnica de Madrid, Madrid, Spain

M. Šimek
Institute of Soil Biology, Biology Centre of the Czech Academy of Sciences, and Faculty of Science, University of South Bohemia, České Budějovice, Czech Republic

M. dos Reis Martins · S. Urquiaga
EMBRAPA Agrobiologia Seropédica, Brazilian Agricultural Research Corporation, Seropedica, RJ, Brazil

R. Well
Thünen Institute of Climate-Smart Agriculture, Braunschweig, Germany

S. Zaman
University of Canterbury, Christchurch, New Zealand

J. Zhang
School of Geography, Nanjing Normal University, Nanjing, China

K. Lenhart
Bingen University of Applied Sciences, Berlinstr. 109, Bingen 55411, Germany

on the origin without the addition of ^{15}N labelled fertiliser. Note, isotopologues are molecules that differ in their isotopic composition, isotopomers are molecules with the same isotopic atoms but differing in their position, and isotopocules is the generic term for both isotopologues and isotopomers. There is a wealth of information that we can obtain from using diverse isotopic approaches based on ^{15}N or ^{18}O labelling but also on natural abundance techniques that take advantage of the different metabolism with which for instance N_2O is produced. Thus, ^{15}N provides us with a toolbox to identify emission pathways and in turn provides information on effective mitigation techniques.

7.2 ^{15}N Gas Flux Method (^{15}N GFM) to Identify N_2O and N_2 Fluxes from Denitrification

7.2.1 Background

N_2O reduction to N_2 is the last step of microbial denitrification, i.e. anoxic reduction of nitrate (NO_3^-) to dinitrogen (N_2) with the intermediates NO_2^-, NO and N_2O (Firestone and Davidson 1989; Knowles 1982). Commonly applied non-isotopic techniques enable us to quantitatively analyse only the intermediate product of this process including NO and N_2O, but not the final product, N_2, a non-greenhouse gas. The challenge to quantify denitrification rates is largely related to the difficulty in measuring N_2 production due to its spatial and temporal heterogeneity and the high N_2-background of the atmosphere (Groffman et al. 2006). There are three principal ways to overcome this problem: (i) adding NO_3^- with high ^{15}N enrichment and monitoring ^{15}N labelled denitrification products (^{15}N gas flux method, ^{15}N GFM) (e.g. (Siegel et al. 1982)); (ii) adding acetylene to block N_2O reductase quantitatively and estimating total denitrification from N_2O production (acetylene inhibition technique, AIT) (Felber et al. 2012); (iii) measuring denitrification gases during incubation of soils in absence of atmospheric N_2 using gas-tight containers and an artificial helium/oxygen atmosphere (HeO_2 method; (Butterbach-Bahl et al. 2002; Scholefield et al. 1997; Senbayram et al. 2018)). Each of the methods to quantify denitrification rates in soils has various limitations with respect to potential analytical bias, applicability at different experimental scales and the necessity of expensive instrumentation that is not available for routine studies. Today the AIT is considered unsuitable to quantify N_2 fluxes under natural atmosphere, since its main limitation among several others is the catalytic decomposition of NO in presence of O_2 (Bollmann and Conrad 1997), resulting in unpredictable underestimation of gross N_2O production (Nadeem et al. 2012). The ^{15}N gas flux method requires homogenous ^{15}N-labelling of the soil (Mulvaney and Vandenheuvel 1988) and under natural atmosphere, it is not sensitive enough to detect small N_2 fluxes (Lewicka-Szczebak et al. 2013). Direct measurement of N_2 fluxes using the HeO_2 method is not subject to the problems associated with ^{15}N-based methods (Butterbach-Bahl et al. 2013) but the need for sophisticated gas-tight incubation systems limits its use. When applying ^{15}N GFM in the laboratory, sensitivity can be augmented by incubation under an N_2-depleted atmosphere (Lewicka-Szczebak et al. 2017; Meyer et al. 2010; Spott et al. 2006). In the following, the basic principle, limitations, bias and application examples are presented and discussed.

7.2.2 *Principles of the ^{15}N Gas Flux Method*

The ^{15}N gas flux method consists of quantifying N_2 and or N_2O emitted from ^{15}N-labelled NO_3^- applied to soil in order to quantify fluxes from canonical denitrification (Mulvaney and Vandenheuvel 1988; Stevens et al. 1993), where N_2 and N_2O are formed from the combination of two NO precursor molecules. Under certain preconditions, it is also possible to identify the production of hybrid N_2 or N_2O (i.e. molecules formed from the combination of N atoms from one source of oxidised N, e.g. NO_2^-), and another source of reduced N (e.g. NH_3 or NH_2OH) via anaerobic ammonia oxidation (annamox) or co-denitrification (Laughlin and Stevens 2002; Spott and Stange 2007; Spott et al. 2011). To quantify canonical denitrification, experimental soil is amended with NO_3^- highly enriched with ^{15}N. The ^{15}N gases evolved are collected in closed chambers and ^{15}N emission is calculated from the abundance of N_2 and N_2O isotopologues in the chamber gas. ^{15}N enrichment of N_2 in the gas samples are typically close to natural abundance because the amount of N emitted from the ^{15}N-labelled soil is small compared to the atmospheric background. Precise techniques of isotope analysis are, therefore, necessary.

7.2.2.1 The Non-random Distribution of Atoms in the N_2 Molecule

The ^{15}N gas flux method is based on the assumption that within N_2 or N_2O from a single source of a given ^{15}N abundance, the N_2O isotopologues of a distinct number of ^{15}N substitutions follow a random (binomial) distribution, as given by the terms in (Eq. 7.1):

$$(p+q)^2 = p^2 + 2pq + q^2 \tag{7.1}$$

where p is the atom fraction of ^{14}N, q the atom fraction of ^{15}N and p + q is equal to unity (Hauck et al. 1958).

If N_2O or N_2 from two different N pools, one background pool of natural ^{15}N abundance (0.3663 atom %) and the second enriched in ^{15}N are mixed, the distribution deviates from the binomial pattern. Given the distribution of N_2 or N_2O isotopologues emitted from the first (background) N pool (a_{bg}) including non-labelled N_2 and N_2O (derived from the atmosphere and possibly non-labelled N_2O from non-labelled N sources in soil) and the resulting mixture (a_m), the ^{15}N abundance in the ^{15}N-labelled second pool (a_p) and the fraction of N_2O or N_2 originating from that labelled pool (f_p) can be determined (e.g. Bergsma et al. 2001; Spott et al. 2006). To calculate f_p values, the nitrogen isotope ratios $^{29}R(^{29}N_2/^{28}N_2)$ and $^{30}R(^{30}N_2/^{28}N_2)$ are used. In case of N_2, the three isotopologues $^{14}N^{14}N$ and $^{14}N^{15}N$ and $^{15}N^{15}N$ are detected. For N_2O, one option is to directly analyse intact N_2O molecules, consisting of N and oxygen (O) and analysing molecular masses 44, 45 and 46. It has to be taken into account that these molecular masses include not only N- but also O-substituted isotopocules and thus the following 6 species: $^{14}N^{14}N^{16}O$ with mass-to-charge (m/z) 44, $^{14}N^{14}N^{18}O$ (m/z 46), the isotopomers $^{14}N^{15}N^{16}O$ and $^{15}N^{14}N^{16}O$ (both m/z 45), $^{14}N^{14}N^{17}O$ (m/z 45) and $^{15}N^{15}N^{16}O$) (m/z 46). To calculate ^{15}N pool-derived N_2O, ^{29}R and ^{30}R of the N_2O–N is calculated taking into account the natural abundance of ^{17}O- or ^{18}O-substituted isotopocules ($^{14}N^{14}N^{18}O$ and $^{14}N^{14}N^{17}O$) due to their mass overlap with the ^{15}N-substituted isotopocules (Bergsma et al. 2001). Alternatively, N_2O can be reduced to N_2 prior to IRMS analysis (Lewicka-Szczebak et al. 2013), thereby allowing direct determination of ^{29}R and ^{30}R of N_2O–N.

There are various calculation procedures that have evolved over time (Hauck et al. 1958; Mulvaney 1984; Arah 1992; Nielsen 1992, Well et al. 1998; Spott et al. 2006). In Eqs. 7.2 and 7.3 we show one example (Spott et al. 2006), where the fraction of N_2 or N_2O evolved from the ^{15}N-labelled NO_3^- pool (f_p) is calculated:

$$f_p = \frac{a_m - a_{bg}}{a_p - a_{bgd}} \tag{7.2}$$

where a_m is the ^{15}N abundance of the total gas mixture

$$a_m = \frac{^{29}R + 2{*}^{30}R}{2\left(1 + {}^{29}R + {}^{30}R\right)} \tag{7.3}$$

and a_{bg} is the ^{15}N abundance of atmospheric background N_2.

The ^{15}N abundance of the ^{15}N-labelled nitrate pool undergoing denitrification is

$$a_p = \frac{^{30}x_m - a_{bgd} * a_m}{a_m - a_{bgd}} \tag{7.4}$$

where $^{30}x_m$ is the measured fraction of m/z 30 in the total gas mixture:

$$^{30}x_m = \frac{^{30}R}{1 + {}^{29}R + {}^{30}R} \tag{7.5}$$

The same calculations can be used for N_2 and N_2O, resulting in respective values for fractions of pool-derived N (f_{p_N2};f_{p_N2O}) and for the respective ^{15}N abundances of the active N pools (a_{p_N2}; a_{p_N2O}).

If only m/z = 28 and m/z = 29 are determined during isotope analysis of N_2, then emission of $^{15}N_2$ is underestimated (Hauck et al. 1958). The extent of underestimation is related to the ^{15}N atom fraction of the NO_3^- pool from which N_2 is emitted (Well et al. 1998) and f_p can thus be calculated if the ^{15}N enrichment of the denitrified N pool is known (Mulvaney 1984):

$$f_p = \left({}^{29}R_{sa} - {}^{29}R_{bg}\right)/\left(2\,a_p\left(1 - a_p\right)\right) \tag{7.6}$$

where lower case sa and bg denote sample and background (typically ambient air), respectively. An alternative equation yielding f_p from ^{29}R that is more complex, but also more precise, is given by Spott et al. (2006).

In many studies, a ^{15}N atom fraction of 0.99 was selected for the ^{15}N enrichment of applied NO_3^- ($^{15}a_{NO3}$) in order to maximise ^{30}R (see Fig. 7.1), thus yielding better ^{30}R signals. However, there are also reasons to keep $^{15}a_{NO3}$ between about 0.6 and 0.4, since ^{30}R is only detectable with high fluxes due to a typical high IRMS background signal at m/z 30 (see next section), so that f_p has to be calculated from ^{29}R only using Eq. 7.6. But f_p calculated from Eq. 7.6 with a given ^{29}R is relatively insensitive to changes in a_p between 0.4 and 0.6 since the nominator yields, e.g. for a_p between 0.4 and 0.6, values between 0.48 and 0.5. Hence, uncertainty in the estimation of a_p within that range causes minor uncertainty in calculated f_p (Well and Myrold 1999).

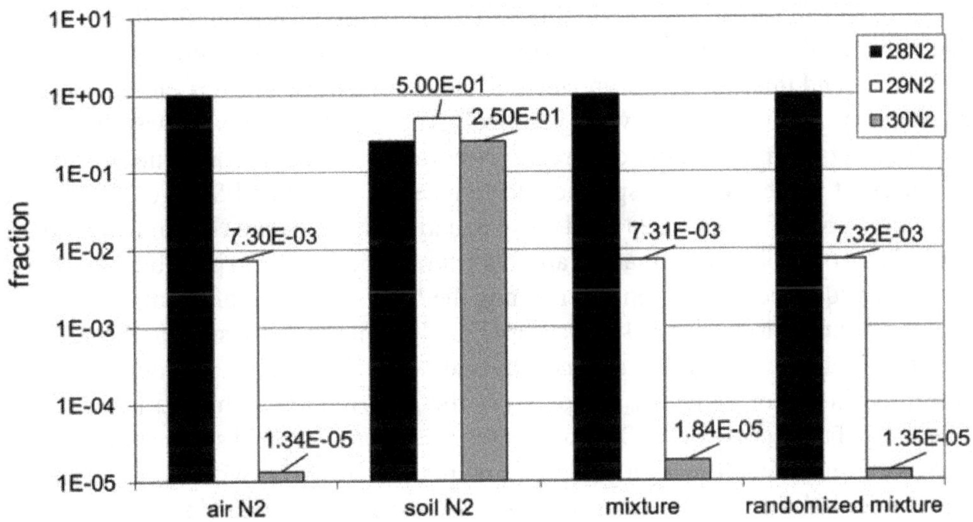

Fig. 7.1 Abundance of $^{28}N_2$, $^{29}N_2$ und $^{30}N_2$ in air, in soil-emitted N_2 evolved from NO_3^- with 50 atom% ^{15}N, and in a 1:1000-mixture without and with randomisation of isotopologues by N_2 dissociation, respectively

To illustrate how the combination of denitrification rates (i.e. f_p) and homogenous or non-homogenous ^{15}N enrichment of the soil NO_3^- pool affect instrumental raw data as well as calculated f_p and a_p values, some theoretical data are shown (Table 7.1). Three cases are represented, (1) the soil is homogenously labelled with ^{15}N, (2) non-labelled soil-derived NO_3^- dilutes the labelled pool to a different extent in the 0 to 10 and 10 to 20 cm layers, but N_2 and N_2O production rates in both layers are equal and (3) like case (2) except that production rates of both layers differ. It can be seen that only case (1) calculated using Eq. 7.4 yields results identical to ideal a_p and f_p. Equation 7.6 gives deviating results when used with $^{15}a_{NO3}$ as this value differs from a_p. In the case of (2) and (3), all calculations lead to some deviation due to the non-homogeneity in label distribution. Moreover, isotope ratios show that even at the high denitrification rate assumed (case 2, 542 g N ha^{-1} 20 cm^{-1} d^{-1}), the increase in ^{29}R (^{29}Rm–^{29}Ra) and ^{30}R (^{30}Rm–^{30}Ra) was 9.2×10^{-6} and 3×10^{-6}, respectively, and thus only about one order of magnitude above typical instrumental precision (see Table 7.2).

7.2.3 *Identifying the Formation of Hybrid N_2 and/or N_2O*

When N_2 and N_2O are formed from denitrification, both N atoms are derived from the ^{15}N labelled pool, and in hybrid N_2 or N_2O only one N atom comes from the labelled pool (N oxides, i.e. NO_2^-) and the other one comes from non-labelled reduced N (e.g. NH_3, NH_2OH or organic N). Hence, the contribution of hybrid processes is reflected by an increase in ^{29}R only, while denitrification increases both ^{29}R and ^{30}R (Clough et al. 2001). Laughlin and Stevens (2002) derived equations to calculate the fraction of hybrid and non-hybrid N_2, assuming that the measured ^{15}N atom fraction of NO_3^- also reflected the enrichment of the NO_2^- that contributed one N atom to the hybrid molecules, and that the ^{15}N abundance of the non-labelled sources (atmospheric N and non-labelled reduced N) were identical. An extended approach was developed allowing to take into account different ^{15}N enrichment for all contributing sources, i.e. different values for atmospheric and reduced N (Spott and Stange 2007; Spott et al. 2011). Spott et al. (2011) used those equations to calculate co-denitrification in a soil slurry but pointed out that the approach would be subject to possible bias due to difficulty and inaccuracy when determining the ^{15}N enrichment of the nitrite (NO_2^-) pool contributing to the hybrid formation. For N_2O mixtures consisting of N_2O from only two sources, i.e. hybrid and non-hybrid N_2O, the authors, therefore, suggest to use the indicator value R_{binom} to assess the contribution of hybrid N_2O. R_{binom} reflects the fact that N_2 or N_2O isotopocules of each non-hybrid source contributing to a gas mixture are following a random (binomial) distribution, whereas this is not the case for the hybrid N_2O. R_{binom} values >1 indicate a significant hybrid contribution. While fluxes excluding hybrid N_2O would always yield $R_{binom} \leq 1$, respective R_{binom} values would not exclude the possibility of some hybrid contribution. Hence, R_{binom} can only prove the existence (but not the absence) of hybrid fluxes. The limitation of this approach is that it does not work in the presence of additional sources, e.g. if

Table 7.1 Mole fractions (X) of $^{28}N_2$, $^{29}N_2$ und $^{30}N_2$ in air (subscript a), in soil-emitted N_2, N_2O or $N_2 + N_2O$ evolved from ^{15}N-labelled NO_3^- pool (subscript p), and of their mixture (subscript m), resulting isotope ratios (^{29}R, ^{30}R) and calculated (Spott et al. 2006) values of a_p and f_p. Values are computed for individual fluxes from two soil layers (0–10 and 10–20 cm) and the mixed flux from both layers. The ratio of soil NO_3^--N to added $^{15}NO_3^-$-N (at 60 at % ^{15}N) is varied to obtain differing $^{15}a_{NO3}$ and a_p values. Denitrification rates are calculated from f_p assuming micro-plots size of 30 cm diameter, chamber height of 10 cm and 1 h closing time

	Case 1	Case 1	Case 2			Case 3			Case 3
	$N_2 + N_2O$ 0–10 cm	$N_2 + N_2O$ 10–20 cm	$N_2 + N_2O$ 0–20 cm	N_2 0–10 cm	N_2 10–20 cm	N_2 0–20 cm	N_2O 0–10 cm	N_2O 10–20 cm	N_2O 0–20 cm
f_p	1.00E-05	1.00E-05	2.00E-05	3.00E-06	7.00E-06	1.00E-05	7.00E-06	3.00E-06	1.00E-05
Production (g N ha^{-1} d^{-1})	3.71E + 02	3.71E + 02	5.42E + 02	1.11E + 02	2.60E + 02	3.71E + 02	2.60E + 02	1.11E + 02	3.71E + 02
Soil NO_3^--N (mg N kg^{-1})	5.00E + 00	3.00E + 01		5.00E + 00	3.00E + 01		5.00E + 00	3.00E + 01	
Added $^{15}NO_3^-$-N (60 at.%, mg N kg^{-1})	3.00E + 01	3.00E + 01		3.00E + 01	3.00E + 01		3.00E + 01	3.00E + 01	
$^{15}a_{NO3}$	5.15E-01	3.02E-01	3.80E-01	5.15E-01	3.02E-01	3.80E-01	5.15E-01	3.02E-01	3.80E-01
a_p	5.15E-01	3.02E-01	4.08E-01	5.15E-01	3.02E-01	3.66E-01	5.15E-01	3.02E-01	4.51E-01
^{28}Xa	9.93E-01	9.93E-01	9.93E-01	9.93E-01	9.93E-01	9.93E-01	9.93E-01	9.93E-01	9.93E-01
^{29}Xa	7.30E-03	7.30E-03	7.30E-03	7.30E-03	7.30E-03	7.30E-03	7.30E-03	7.30E-03	7.30E-03
^{30}Xa	1.34E-05	1.34E-05	1.34E-05	1.34E-05	1.34E-05	1.34E-05	1.34E-05	1.34E-05	1.34E-05
^{28}Xp	2.35E-01	4.87E-01	3.61E-01	2.35E-01	4.87E-01	4.12E-01	2.35E-01	4.87E-01	3.11E-01
^{29}Xp	5.00E-01	4.21E-01	4.61E-01	5.00E-01	4.21E-01	4.45E-01	5.00E-01	4.21E-01	4.76E-01
^{30}Xp	2.65E-01	9.11E-02	1.78E-01	2.65E-01	9.11E-02	1.43E-01	2.65E-01	9.11E-02	2.13E-01
^{28}Xm	9.93E-01	9.93E-01	9.93E-01	9.93E-01	9.93E-01	9.93E-01	9.93E-01	9.93E-01	9.93E-01

(continued)

Table 7.1 (continued)

	Case 1	Case 1	Case 2			Case 3			Case 3
	$N_2 + N_2O$ 0–10 cm	$N_2 + N_2O$ 10–20 cm	$N_2 + N_2O$ 0–20 cm	N_2 0–10 cm	N_2 10–20 cm	N_2 0–20 cm	N_2O 0–10 cm	N_2O 10–20 cm	N_2O 0–20 cm
^{29}Xm	7.30E-03	7.30E-03	7.31E-03	7.30E-03	7.30E-03	7.30E-03	7.30E-03	7.30E-03	7.30E-03
^{30}Xm	1.61E-05	1.43E-05	1.70E-05	1.42E-05	1.41E-05	1.49E-05	1.53E-05	1.37E-05	1.55E-05
^{29}Rm	7.36E-03	7.36E-03	7.36E-03	7.35E-03	7.36E-03	7.36E-03	7.36E-03	7.35E-03	7.36E-03
^{30}Rm	1.62E-05	1.44E-05	1.71E-05	1.43E-05	1.42E-05	1.50E-05	1.54E-05	1.38E-05	1.57E-05
$^{29}Rm - {}^{29}Ra$	5.02E-06	4.21E-06	9.22E-06			4.45E-06			4.77E-06
$^{30}Rm - {}^{30}Ra$	2.67E-06	9.18E-07	3.59E-06			1.44E-06			2.14E-06
a_p, calculated	5.15E-01	3.02E-01	4.36E-01	5.15E-01	3.02E-01	3.92E-01	5.15E-01	3.02E-01	4.72E-01
f_p, calculated, Eq. 7.2	1.00E-05	1.00E-05	1.87E-05	3.00E-06	7.00E-06	9.32E-06	7.00E-06	3.00E-06	9.55E-06
f_p, calculated, Eq. 7.6, using $^{15}a_{NO3}$	1.06E-05	8.92E-06	1.95E-05			9.43E-06			1.01E-05
f_p calculated, Eq. 7.6, using a_p of N_2O	1.00E-05	1.00E-05	1.87E-05			8.90E-06			9.55E-06

Table 7.2 Typical precision (standard deviation, SD) for the nitrogen isotope ratios ^{29}R ($^{29}N_2/^{28}N_2$) and ^{30}R ($^{30}N_2/^{28}N_2$) by IRMS

Instrument	SD of ^{29}R	SD for ^{30}R
Lewicka-Szczebak et al. (2013)	5.88E-08	3.06E-07
Siegel et al. (1982)	9.00E-07	2.30E-07
Stevens et al. (1993)	5.30E-06	5.30E-07

there is N_2O from unlabelled sources including atmospheric N_2O. Thus, R_{binom} does not work for N_2 because there is always a high background of atmospheric N_2. To our knowledge, systematic and quantitative studies on hybrid fluxes from soils, including quantification of average pool enrichment and its homogeneity or non-homogeneity, and estimation of resulting uncertainties, have not yet been accomplished.

7.2.4 Analysis of N_2 and N_2O Isotopologues

Precise quantification of N_2 and N_2O isotopocules requires isotope ratio mass spectrometry (IRMS) where ^{29}R and ^{30}R are obtained from ion current ratios detected at Faraday collectors tuned for m/z 28, 29 and 30 (e.g. Lewicka-Szczebak et al. 2013). A double collector IRMS was used before multi-collector IRMS became available. Double collector IRMS required two measurements with the IRMS so that either $^{29}N_2$ or $^{30}N_2$ is positioned on the first collector (Siegel et al. 1982). Emission spectroscopy has also been used in the past to detect $^{28}N_2$, $^{29}N_2$ and $^{30}N_2$ (Kjeldby et al. 1987), but its relatively low precision enabled only detection of large N_2 fluxes. While dual inlet IRMS had been used with manual measurement of samples in glass containers that were sealed (Well et al. 1993) or isolated by stopcocks (Siegel et al. 1982), continuous flow IRMS enables automated injection of samples from septum capped vials since the 1990s (Stevens et al. 1993). Recently, further progress was obtained by automated analysis of N_2, N_2+N_2O and N_2O in one run, including N_2O reduction to N_2 (Lewicka-Szczebak et al. 2013). The latter enables the analysis of N_2O-N at m/z 28, 29 and 30, thus excluding the need to conduct ^{17}O and ^{18}O corrections, yielding better precision, since O corrections are biased to some extent by natural variation of ^{17}O and ^{18}O (Deppe et al. 2017).

While quantification of ^{29}R is quite robust, ^{30}R is affected by the mass overlap of $^{30}N_2$ with the most abundant isotopocule of NO ($^{14}N^{16}O$), since NO^+ is formed at the hot filament in the ion source of the IRMS (Brand et al. 2009, Siegel et al. 1982) due to the omnipresence of oxygen traces. NO^+ formation can be quantified by the ratio between ideal and measured ^{30}R of standard gases, giving values of 0.15 to 0.06 for atmospheric N_2 analysed in the instrumentation proposed by Lewicka-Szczebak et al. (2013). NO^+ formation can be minimised by removal of all O sources (O_2, H_2O) from the samples and also from the carrier and reference gases. In some types of IRMS the NO^+ background is too high and associated with extreme tailing of the m/z

30 peak. This makes it impossible to quantify ^{30}R (Well et al. 1993). To overcome this limitation, a procedure to quantify ^{30}R indirectly from ^{29}R was developed where ^{29}R had to be analysed twice, (i) in samples where the non-random distribution of N_2 isotopocules was randomised by the temporary splitting up of N_2 molecules during a gas discharge (see change in ^{29}R due to randomisation in Fig. 7.1). Discharge was actuated using a microwaves source, initially offline in sealed glass tubes, later with online continuous flow IRMS, where the discharge occurred in the gas circuit connecting and IMRS (Well and Meyer 1998). An overview of the IRMS precision for ^{29}R and ^{30}R in N_2 standard gases is given in Table 7.2, showing that repeatability for ^{29}R varied significantly between instruments, but ^{30}R is comparable. However, it is also evident that during the last 35 years (Siegel et al. 1982) there has been no substantial improvement in the measurement precision.

7.2.5 *Detection Limit for a_p and f_p*

Because f_p is calculated from two quantities, ^{29}R and ^{30}R, and the relationship between them depends on the ^{15}N enrichment of the active N pool (a_p, see Fig. 7.1), the limit of detection (LOD) for f_p at given repeatability of ^{29}R and ^{30}R is variable. LOD for f_p was thus determined for varying conditions using equations from Spott et al. (2006) using Monte Carlo modelling assuming a normal distribution of ^{29}R and ^{30}R errors (Standard deviation of repeated analysis of standard gas samples). The MS-Excel function *norm.inv* was used to create the normal distribution of values but allowing only a maximum deviation of 3 standard deviations, otherwise unrealistic outlier of ^{29}R or ^{30}R yield unrealistically high uncertainty. Different scenarios were tested ($f_p = 1$ to 100 ppm; $a_p = 0.055$ to 0.75 using repeatability for ^{29}R and ^{30}R of the first IRMS listed in Table 7.1). LOD is obtained for two cases: 1. Both ^{29}R and ^{30}R are taken into account to calculate both a_p and f_p; 2. f_p is calculated using only ^{29}R (using Eq. 7.4 in (Spott et al. 2006)) and a_p is estimated either from soil extract analysis or from a_p of N_2O (e.g. Stevens and Laughlin 2002). Note that a_p of N_2O is usually much more reliable than a_p of N_2 since f_p of N_2O is typically large (often between 0.1 and 1) due to the fact that, in contrast to N_2, N_2O is an atmospheric trace gas. Conversely, f_p of N_2 is typically very small (usually $<10^{-5}$ in ambient atmosphere).

The first calculation is preferable because a_p of N_2 and N_2O can be different (see Fig. 7.3) and a_p of N_2O can only be obtained if N_2O can be directly measured by IRMS, which is only the case if concentrations are high enough (about 0.3 to 3 ppm necessary, depending on ^{15}N enrichment of N_2O). Since incubation under N_2-depleted atmosphere improves f_p sensitivity, LOD is also given for an artificial gas mixture containing 2% N_2.

LOD results are as follows (Table 7.3): with high f_p (i.e. $\geq$10 ppm) and high a_p (i.e. $\geq$0.5) and ideal IRMS performance (Table 7.2) both calculations yield precise results. Under N_2-depleted atmosphere, LOD is excellent (2 to 7 ppb N_2, last columns). With lowering of a_p, LOD gets worse if a_p has to be calculated using ^{30}R. But without using

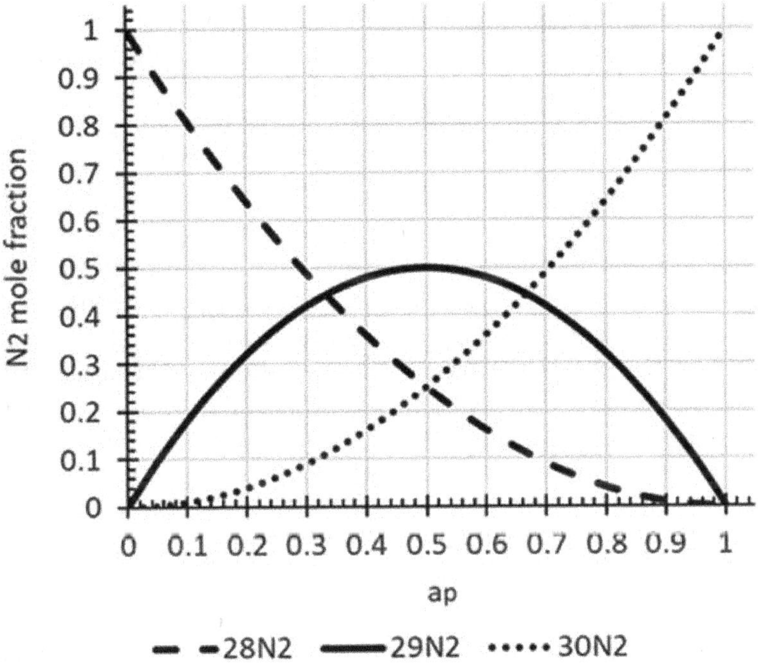

Fig. 7.2 Abundance of $^{28}N_2$, $^{29}N_2$ and $^{30}N_2$ in N_2 evolved from the ^{15}N–labelled NO_3^- depending on a_p (Siegel et al. 1982)

^{30}R and assuming an ideal a_p value or estimating a_p of N_2 from direct determination of a_p of N_2O, LOD of f_p is still excellent. This is because with decreasing a_p, abundance of $^{15}N^{15}N$ ($^{30}N_2$), and thus ^{30}R, decreases exponentially whereas the decrease of ^{29}R ($^{29}N_2$) is much slower (see Fig. 7.2).

7.2.6 Limitations of the ^{15}N Gas Flux Method (^{15}N GFM)

The following factors limit the applicability of the ^{15}N GFM

7.2.6.1 Inaccurate Definition of the Soil Volume Represented by Denitrification Measurements and Incomplete Recovery of Denitrification Gases

The denitrifying soil volume is clearly defined if soil cores are entirely labelled with ^{15}N and are incubated in closed systems. However, in situ measurement of denitrification in surface soils or subsoils with approaches other than the core methods do not include complete enclosure of the investigated soil. It is not possible to control the application of $^{15}NO_3^-$ accurately. Consequently, the soil volume represented by the detected denitrification gases is not exactly defined, and calculated denitrification rates are associated with uncertainty. Partial enclosure of the investigated soil is typically achieved by driving cylinders into surface soils. This option reduces the problem to a certain extent. Moreover, measuring the spatial distribution of the ^{15}N label at the end of experiments (Well and Myrold 2002) helps to constrain the soil volume contributing to soil N_2 fluxes that can be "seen" by ^{15}N analysis of headspace gases.

An additional problem of open systems is the difficulty to determine the direction and strength of diffusional gas transport. When chamber methods are used to determine denitrification of surface soils, a significant fraction of the denitrification gases produced in the ^{15}N-labelled soil diffuses into the subsoil and is thus not recovered in the chambers. Principally, this can be solved by modelling diffusion of ^{15}N labelled gaseous denitrification products (see Sect. 7.2.7).

7.2.6.2 The Problem of Non-homogenous ^{15}N Enrichment of the NO_3^- Pool

An overview of techniques to supply ^{15}N-labelled NO_3^- to the soil is given in Table 7.4. The ^{15}N GFM is based on the assumption of an isotopically homogenous NO_3^- pool. Because this condition is rarely achieved in soils, underestimation of denitrification rates up to 30% can result (Arah 1992; Mulvaney and VandenHeuvel 1988). An initial homogeneity can be obtained by intensive mixing of the soil, but this is a massive disturbance with huge potential effects on N processes including denitrification dynamics and is only adequate to simulate soil tillage with similar disturbance. But even with initially ideal tracer distribution, non-homogeneity inevitably develops over time, since N transformations including nitrification, denitrification and immobilisation are never homogenous in structured soil where aerobic and anaerobic domains coexist and organic matter fractions of varying reactivity are unevenly distributed. Injection of ^{15}N tracer solution (Wu et al. 2012) increases moisture and inevitably produces non-homogeneity with maximum label concentration at the injection spots. Saturation and drainage (Nõmmik 1956) or soil water displacement by irrigation of lysimeters (Well et al. 1993) leads to an interim increase in moisture and causes loss of DOC. Labelling with gaseous NO_2 was not a suitable way to achieve high and homogenous enrichment of soil NO_3^- (Stark and Hart 1996). Consequently, non-homogeneity of the label distribution is probably the main source of bias of the ^{15}N GFM. Often ^{15}N tracer has been applied to the surface similar to conventional fertilisation (Baily et al. 2012). However, in this case, only fertiliser derived fluxes are detected initially, while during ongoing diffusion and leaching of NO_3^-, the ^{15}N labelled NO_3^- pool rapidly changes its dimensions and thus non-homogeneity complicates the interpretation of results.

Possible causes and consequences of non-homogenous distribution of the ^{15}N-label and denitrification /nitrification dynamics is illustrated using two conceptual models (Well et al. 2015). The first model shows how a_p of N_2 and N_2O can differ due to non-homogeneity in ^{15}N enrichment and also non-homogeneity in N_2 and

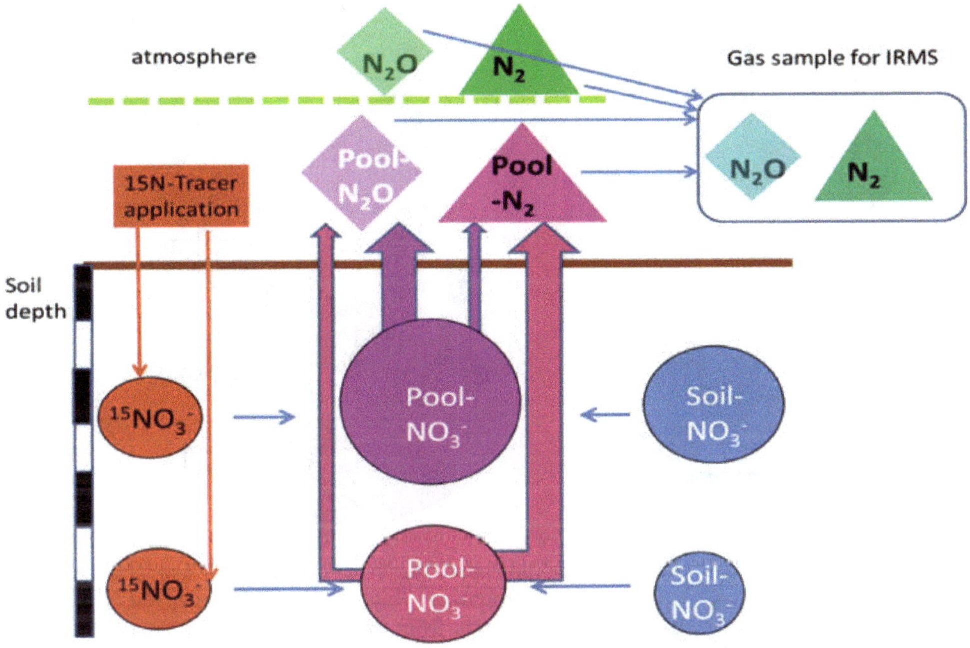

Fig. 7.3 Model 1 to explain why N_2 and N_2O from denitrification can originate from different effective ^{15}N pools: In the lower pool with a higher ^{15}N enrichment, N_2 fluxes dominate over N_2O, whereas the opposite is the case for the shallow pool with lower enrichment. Hence, emitted N_2 is more enriched compared to emitted N_2O

N_2O production rates (Fig. 7.3). Even if equal amounts of ^{15}N tracer solution could be applied to each soil layer, ^{15}N enrichment of NO_3^- would be variable due to the different dilution of the label via soil-derived NO_3^-. Additionally, production rates of N_2 and N_2O and their ratio are typically spatially variable, which results in differing a_p values for N_2 and N_2O (Fig. 7.4). The development of spatial heterogeneity in ^{15}N enrichment and the consequences arising from the fact that nitrification and denitrification typically occur in different soil niches is shown with the second conceptual model (Fig. 7.4) that had been used to explain observations (Deppe et al. 2017). In that study, the soil had been mixed with ^{15}N labelled NO_3^- and non-labelled NH_4^+ and isotopic values of initial NO_3^- and final NO_3^- and N_2O had been compared. Results showed that a_p of N_2O was similar to initial enrichment of soil NO_3^- (13 atom% ^{15}N), but final NO_3^- enrichment of the bulk soil was much lower (about 3 atom% ^{15}N) whereas a_p of N_2O did not change significantly. This was postulated to result from the dilution of the label only in aerobic domains where nitrification occurred, whereas in anaerobic microsites there was no nitrification, and hence no dilution of the label. But the undiluted microsites produced all or most of the N_2O whereas there was negligible N_2O flux from aerobic domains. While this discrepancy between ^{15}N enrichment of NO_3^- in the bulk soil and a_p of N_2O was certainly extreme in that study, similar process dynamics can be expected in many cases. Such

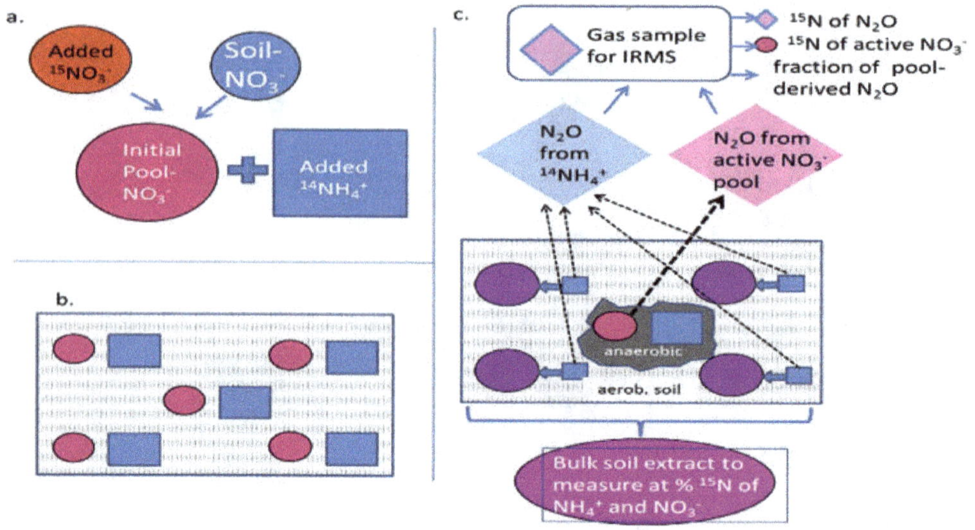

Fig. 7.4 Model 2 to explain possible non-homogeneity in ^{15}N-labelling of NO_3^- in NH_4^+-fertilised soil (Deppe et al. 2017). Colours represent enrichment (blue = nat abundance, red = max. ^{15}N enrichment). a. Initial enrichment of NO_3^- results from mixing of soil NO_3^- and added ^{15}N-NO_3^-. b. Initial homogenous distribution of labelled NO_3^- and non-labelled NH_4^+ in the soil matrix. c. In anaerobic microsites, nitrification is inhibited and the NO_3^- pool of initial ^{15}N enrichment is denitrified and produces N_2O of identical enrichment. In aerobic domains, nitrification of non-labelled NH_4^+ produces non-labelled NO_3^-, thus diluting the initial labelled NO_3^- pool and emitting unlabelled N_2O. Note that the ^{15}N enrichment of NO_3^- undergoing denitrification is larger than the average ^{15}N enrichment of extracted NO_3^- and of emitted N_2O

non-homogeneity in label distribution and its dilution as well as N_2 and N_2O production leads to uncertainties in calculation of f_p (see Table 7.1). But these examples also show that comparison of a_p of N_2 and a_p of N_2O can be used to identify heterogeneity in labelling and thus stress the importance of using analytical methods including ^{29}R and ^{30}R of N_2 and N_2O-N (Lewicka-Szczebak et al. 2013). Moreover, it shows that calculating f_p based on ^{15}N enrichment of bulk NO_3^- from soil extraction (Eq. 7.6) can lead to severe bias, since the ^{15}N enrichment of the active pool can strongly deviate from the bulk pool. Moreover, an advantage of the non-random distribution approach with N_2 and N_2O is that non-homogeneity is indicated by discrepancies between a_p of N_2, a_p of N_2O and $^{15}a_{NO3}$, which is quite useful (Lewicka-Szczebak and Well 2020). But it also shows that hybrid fluxes are difficult to identify if label distribution is non-homogenous.

Further limitations of the ^{15}N GFM have been reviewed previously (Aulakh et al. 1991; Groffman et al. 2006; Sgouridis et al. 2016). They include enhancement of denitrification by NO_3^- application in unfertilised systems, gas entrapment in very wet or fully water-saturated soils or sediment, and limited residence time of applied $^{15}NO_3^-$-N due to plant uptake and leaching.

Table 7.3 Detection limit of the ^{15}N GFM determined by Monte-Carlo modelling. Detection limit for the fraction of pool derived N_2 (f_p of N_2) is given as 1 standard deviation (SD) in dependence of ^{15}N enrichment of active labelled NO_3-pool (ap) and magnitude of f_p in atmospheres with 100% or 2% N_2 and assuming IRMS precision of ^{29}R and ^{30}R according to the first instrument in Table 2

Scenario #	Ideal f_p (ppm)	Ideal a_p	SD of f_p; modelled based on ^{29}R and ^{30}R (ppm in pure N_2)	SD of f_p; modelled based on ^{29}R and assuming ideal a_p (ppm in pure N_2)	SD of f_p; modelled based on ^{29}R and ^{30}R (ppm in 2% N_2)	SD of f_p; modelled based on ^{29}R and assuming ideal a_p (ppm in 2% N_2)
1	10	0.75	0.35	0.16	0.007	0.003
2	10	0.50	0.11	0.12	0.002	0.002
3	10	0.25	138.62	0.16	2.772	0.003
4	10	0.10	231.32	0.34	4.626	0.007
5	10	0.05	2370.94	0.67	47.419	0.013
6	1	0.50	4.16	0.12	0.083	0.002
7	1	0.25	11.16	0.16	0.223	0.003
8	1	0.10	2.13	0.34	0.043	0.007
9	100	0.50	0.11	0.12	0.002	0.002
10	100	0.25	2.32	0.16	0.046	0.003
11	100	0.10	139.23	0.34	2.785	0.007

7.2.6.3 Combining the ^{15}N GFM with Modelling of Gross N Transformation

The current model to analyse data for the ^{15}N GFM cannot be used to solve situations that include multiple labelled pools and heterogeneity of process activity and thus yield variable results in terms of flux quantification. Therefore, more complex models are needed to fill this gap. A ^{15}N tracing model had been developed to analyse N_2O dynamics in terrestrial ecosystems, which builds on previous tracing models for the quantification of the main mineral N transformations and soil nitrite (NO_2^-) dynamics (Müller et al. 2014). This model is thus a first step in taking more complex dynamics into account. Extending this approach to model heterogeneity of processes and pools might be a promising way to solve current limitations of the ^{15}N GFM. For more information on the tracing technique see Sect. 7.5 of this chapter.

7.2.7 *Evaluation of the ^{15}N GFM*

While quantification of N_2 and N_2O fluxes from distinct N pools remains a challenge after several decades of method development and improvement, this is even more the case for robust evaluation of methods, as this requires that the reference method is quantitative and is applied under the same conditions as the tested method. From that perspective, all previous tests included some uncertainties to our knowledge and were thus not fully able to evaluate the ^{15}N GFM. There have been several comparisons between ^{15}N GFM and AIT with controversial results, i.e. reporting general agreement (Aulakh et al. 1991) and severe underestimation by AIT (Arah et al. 1993; Sgouridis et al. 2016). Aulakh et al. (1991) compared ^{15}N GFM and AIT in the field and found that ^{15}N fertiliser derived $N_2 + N_2O$ fluxes were comparable to total N_2O fluxes in presence of acetylene (C_2H_2), suggesting that both methods were in general agreement. However, in all comparisons, ^{15}N fertiliser was surface applied, so only the soil volume reached by the fertiliser contributed to the surface flux, unlike the AIT, were a larger soil volume was reached by the gaseous acetylene supplied by perforated pipes or buried calcium carbide. Hence comparisons did not reflect equal parts of the soil profile. Interestingly, in most comparisons denitrification was enhanced by soil compaction or glucose amendment, to achieve detectable $^{15}N_2$ fluxes against the atmospheric N_2 background. Sgouridis et al. (2016) compared closed chamber ^{15}N GFM using needle injection to distribute $K^{15}NO_3^-$ evenly with the AIT "soil core" variant finding 3 to 5 times higher rates with ^{15}N GFM. Kulkarni et al. (2014) conducted an extensive comparison of the HeO_2 method using small cores (5 cm diameter × 5 cm height, incubated under HeO_2 in the lab) with in situ measurement using the ^{15}N GFM where KNO_3^- with 99 atom% was sprayed on the soil surface. Authors discussed difficulties to compare measurements in view of O_2 manipulation in the lab and uneven label distribution in the field as well as variable moisture and temperature conditions in the field, and also that there are N_2 fluxes from sources other than NO_3^- (Butterbach-Bahl et al. 2013). What is still needed for a quantitative evaluation of the ^{15}N GFM is to incubate ^{15}N-labelled soil in a HeO_2 setup to allow direct comparison of GC- and IRMS based N_2 fluxes.

If ^{15}N GFM is conducted under conditions maximising sensitivity and minimising bias, it can be used to evaluate other methods as for example the N_2O isotopocule approach to determine N_2O reduction (Lewicka-Szczebak et al. 2017, Buchen et al. 2018) (see Sect. 7.3).

7.2.8 *Lab and Field Experiments*

Initial application of ^{15}N GFM in lab incubations was carried out in closed vessels (Melin and Nõmmik 1983; Siegel et al. 1982). Recently, some studies used N_2-depleted atmosphere to increase sensitivity in soil incubations (Lewicka-Szczebak et al. 2017; Schorpp et al. 2016) achieving sensitivities for pool-derived N_2 of approximately 50 ppb which is thus comparable to GC sensitivity for N_2O and two order of magnitude more sensitive compared to ^{15}N GFM under ambient atmosphere. Important to note is that this also improves precision for quantifying a_p and thus yields more precise estimates for the dilution of the denitrified pool by soil-derived NO_3^-.

A key feature of ^{15}N GFM is in situ measurement of denitrification and today it must be considered the only available field method, since AIT has been found unsuitable (Felber et al. 2012; Nadeem et al. 2012; Sgouridis et al. 2016). But ^{15}N GFM has been used far less compared to the AIT probably due to its low sensitivity and high effort and expense to keep high ^{15}N labelling in the field for extended periods, and also because of the multiple sources of bias. ^{15}N GFM has thus been primarily used for soil types and/or conditions with high denitrification potential, e.g. due to abundant organic C (e.g. in organic soils or after soil compaction Arah et al. 1993). Typically, experiments covered only certain phases of the year. Maybe the most extensive study (including an extensive review of past in situ measurements) was by Sgouridis et al. (2016) who conducted ^{15}N GFM in 4 sites monthly during about 18 months. But it has recently been found that during field application of the ^{15}N GFM, denitrification is severely underestimated because a large fraction of the labelled N_2 and N_2O produced is not emitted from into the soil surface but diffuses to the subsoil or accumulates in pore space (Well et al. 2019a). This was confirmed experimentally and production–diffusion modelling showed that under typical experimental conditions, denitrification rates would be underestimated by more than 50%. It was concluded that field surface fluxes of ^{15}N-labelled N_2 and N_2O have been severely underestimated in the past, but that diffusion modelling can be used to correct data. Moreover, to overcome the poor sensitivity of in situ ^{15}N GFM, a new procedure was developed to conduct the ^{15}N gas flux method using artificial N_2-depleted atmosphere also for field application (Well et al. 2019b), giving a sensitivity for $N_2 + N_2O$ fluxes up to 80-fold better compared to the conventional ^{15}N GFM under ambient atmosphere. Consequently, recent methodical improvements are promising to yield good progress in the study of denitrification control at the field scale. ^{15}N GFM has been used extensively with water saturated cores of aquatic sediments, e.g. Enrich-Prast et al. (2015), where sensitivity is less critical due to the possibility to measure ^{15}N-labelled N_2 dissolved in pore water where atmospheric N_2 background is small.

7.2.8.1 In Situ Measurement in Subsoil and Groundwater

Some modifications of the ^{15}N GFM for subsurface applications had been proposed and applied. For water saturated subsoil of hydromorphic soils or deeper groundwater, the "push–pull" type experimental setup (Istok et al. 1997) was combined with ^{15}N tracing (Addy et al. 2002; Well et al. 2003; Well and Myrold 1999), where ^{15}N tracer solution is injected in groundwater wells and groundwater samples are subsequently extracted over time and analysed for ^{15}N labelled N_2 and N_2O. Similar to using ^{15}N GFM in water saturated sediment in the lab (see above, Enrich-Prast et al. 2015), this approach is quite sensitive since produced N_2 mixes with the small N_2 background of N_2 dissolved in groundwater. The ^{15}N push–pull approach has been compared to slurry incubations of aquifer samples in the lab (Eschenbach et al. 2015; Well et al. 2005) finding good agreement between both approaches. It has also been successfully applied for deeper groundwater up to 90 m depth (Eschenbach et al. 2015).

In the unsaturated zone, subsoil denitrification has been quantified in situ from the steady-state $^{15}N_2 + ^{15}N_2O$ concentration within a defined ^{15}N-labelled soil volume (Well and Myrold 2002). Diffusion-reaction modelling has been used to quantify rates by fitting measured and modelled f_p values, but accuracy of this approach was limited by the difficulty to quantify the volume of ^{15}N-labelled soil, its gas diffusivity and its distribution in ^{15}N enrichment.

7.2.9 Conclusions and Outlook

The ^{15}N GFM is a powerful approach to quantify soil denitrification and its $N_2O/(N_2 + N_2O)$ mole ratio, to distinguish N_2O fluxes derived from NO_3^- and other N sources and, under certain conditions, also to identify the formation of hybrid N_2 and N_2O fluxes. It is applicable in the lab as well as in the field. But it is based on a variety of assumptions and prerequisites that are not always easy or possible to validate or to fulfil. Therefore, and because of its high expense for isotope tracers, IRMS analysis and demanding experimental setups, it has until now rarely been used routinely to study denitrification. Moreover, systematic evaluation using independent methods, e.g. using the HeO_2 method, is still pending. Progress has been made in automated IRMS approaches that can be established using commercially available devices with some custom-made modifications. While sensitivity was clearly improved in the lab by incubation under N_2 depleted atmosphere, this has not yet been fully realised for field conditions. These are good reasons to intensify the use of ^{15}N GFM in future N cycle research, since despite large efforts during preceding decades, the magnitude of denitrification is still the big unknown of the N cycle (Butterbach-Bahl et al. 2013; Müller and Clough 2014).

7.3 Isotopocule Techniques to Identify Pathway-Specific N_2O Emissions

7.3.1 Introduction

N_2O isotopocules are the chemically identical N_2O molecules but differing either in the atomic mass due to a substitution of one atom with heavy isotope ^{15}N or ^{18}O (isotopologues: $^{14}N^{14}N^{16}O$; $^{15}N^{14}N^{16}O$; $^{14}N^{14}N^{18}O$), or in the location of ^{15}N substitution (isotopomers: $^{14}N^{15}N^{16}O$; $^{15}N^{14}N^{16}O$) (Toyoda et al. 2017). Thus, the asymmetric NNO molecule has in total twelve distinct isotopocules, representing all possible combinations of the N isotopes ^{14}N and ^{15}N and the oxygen isotopes ^{16}O, ^{17}O and ^{18}O and providing a wealth of interpretation perspectives. Most commonly the three isotopic characteristics ($\delta^{18}O$, $\delta^{15}N^{\alpha}$ and $\delta^{15}N^{\beta}$) are measured, reporting the relative differences of isotope ratios of the four most abundant N_2O isotopocules

Table 7.4 Overview of current ^{15}N- labelling techniques

Technique	Initial homogeneity	Impact on denitrification	Reference
Injection of $^{15}NO_3^-$solution with needles	Depending on initial NO_3^-distribution and resolution of injections	Enhancement by moisture and NO_3^- addition	Wu et al. (2012)
Irrigation with $^{15}NO_3^-$solution	Good if soil water is completely displaced	Enhancement by moisture and NO_3^- addition	Well (1993)
Saturation and drainage	Good	Enhancement by moisture and NO_3^- addition	Melin and Nommik (1983)
Mixing with fertiliser	Ideal	Soil disturbance and enhancement by NO_3^- addition	Well et al. (2006)
Surface application of fertiliser	Poor	None, but detected N_2 flux only from fertiliser N	Kulkarni et al. (2014)
Application of gaseous $^{15}NO_2$	Variable	No moisture and structure effects, enhancement only by N addition	Stark and Firestone (1995)

$^{14}N^{14}N^{18}O$/ $^{14}N^{14}N^{16}O$ ($\delta^{18}O$), $^{14}N^{15}N^{16}O$/ $^{14}N^{14}N^{16}O$ ($\delta^{15}N^{\alpha}$) and $^{15}N^{14}N^{16}O$/ $^{14}N^{14}N^{16}O$ ($\delta^{15}N^{\beta}$) in relation to a measurement standard defined on an international isotope ratio scale, Air-N_2 for $^{15}N/^{14}N$ and Vienna Standard Mean Ocean Water (VSMOW) for $^{18}O/^{16}O$. The average of $\delta^{15}N^{\alpha}$ and $\delta^{15}N^{\beta}$ is referred to as $\delta^{15}N^{bulk}$ and the difference between $\delta^{15}N^{\alpha}$ and $\delta^{15}N^{\beta}$ (i.e. $\delta^{15}N^{\alpha}$–$\delta^{15}N^{\beta}$) is called $\delta^{15}N$-site preference ($\delta^{15}N^{SP}$), or commonly as SP (Toyoda and Yoshida 1999).

Natural abundance isotopic signatures can be used as an alternative approach to ^{15}N tracing to constrain N_2O transformations in the environment. Variations in stable isotope abundances are due to the fact that for many biotic and abiotic processes, the reaction rates differ between isotopic species, e.g. reduction of $^{15}NO_2^-$ versus $^{14}NO_2^-$, leading to a so-called isotopic fractionation. As the isotopic fractionation is distinct for certain reaction pathways, isotopic signatures of particular production pathways and reduction fractionation factors determined in laboratory pure culture studies can be used to differentiate processes from each other. Distinct process information is provided by the difference in ^{15}N substitution between the central and terminal position within the N_2O molecule (SP), which is independent of the precursor's isotopic composition and characteristic of specific reaction mechanisms or enzymatic pathways. The most common interpretation strategy used to date is the dual isotope plot, also known as "mapping" approach, presenting the relationship between two isotopic parameters–commonly $\delta^{18}O$/ $\delta^{15}N^{bulk}$, $\delta^{15}N^{SP}$/ $\delta^{15}N^{bulk}$ or $\delta^{15}N^{SP}$/ $\delta^{18}O$. From such figures, estimates can be made about trends, probable dominance of particular pathways, or reduction progress (Toyoda and Yoshida 1999; Lewicka et al. 2017; Koba et al. 2009; Ibraim et al. 2019) (Fig. 7.5).

N_2O isotopocules at natural abundance levels can be analysed by isotope ratio mass spectrometry (IRMS) (Toyoda and Yoshida 1999) and more recently mid-infrared (MIR) laser spectroscopic techniques.

With N_2O isotopic analysis, the qualitative information can be added to the quantitative information gained from the concentration measurements. This is to naturally occurring differences between N_2O from various origins as a result of isotopic fractionation, which causes enrichment or depletion of the reaction product in heavy isotope. Typically, for biochemical reactions we deal with the product depleted in heavy isotopes, but different biochemical pathways show characteristic isotope fractionation, which results in larger or smaller isotope effects (ε, difference between substrate and product (Eq. 7.7)), including also possible inverse isotope effects (product enriched in heavy isotopes, negative ε).

$$\varepsilon \cong \delta_{\text{product}} - \delta_{\text{substrate}} = \Delta_{\text{product/substrate}} \tag{7.7}$$

Isotope effect is often expressed as Δ values, representing the difference between δ values of product and substrate. The values of ε should be used for a particular chemical reaction or physical transformation and describe the characteristic isotopic fractionation for this process (so-called intrinsic isotope effects), whereas Δ values may also be applied to describe an isotopic change between initial substrate and the final product, which may be due to a chain of following reactions and diffusion. This is the case e.g. for denitrification where we can mostly only determine the overall observed isotope effect between NO_3^- and N_2O (also called apparent or net isotope effect, $\Delta^{15}N^{bulk}{}_{N2O/NO3^-}$) but without insight into intermediate products (NO_2^-, NO) we cannot determine the ε values of individual reduction steps.

Due to distinct isotopic fractionation for various biochemical reactions, the N_2O isotopic studies have been often used to distinguish between different N_2O production pathways, e.g., nitrification and denitrification (Cardenas et al. 2017; Deppe et al. 2017; Köster et al. 2015; Toyoda et al. 2011; Wolf et al. 2015), or between different microorganisms involved in N_2O production, e.g. fungal and bacterial denitrification (Kato et al. 2013; Schorpp et al. 2016; Zou et al. 2014). Moreover, also N_2O reduction can be potentially monitored by N_2O isotopic data. The possible reduction of N_2O to N_2 during denitrification is associated with isotopic fractionation, which changes the isotopic signature of the residual N_2O. Therefore, isotopic analyses of residual N_2O can be used to estimate the magnitude of its reduction and thereby the N_2 production (Kato et al. 2013; Lewicka-Szczebak et al. 2017; Toyoda et al. 2011). Comprehensive reviews on the use of N_2O isotopocules to estimate N_2O dynamics are given by Ostrom and Ostrom (2011), Decock and Six (2013), Toyoda et al. (2017) and Yu et al. (2020). The main problem in the interpretation of isotopocule analysis of emitted N_2O is the parallel production, possibly from various pathways, and consumption due to reduction to N_2.

7.3.2 *Principles*

For a proper interpretation of the analysed isotopic values of emitted N_2O, both the possible production pathways and consumption due to N_2O reduction to N_2 must be taken into account.

To be able to identify potential production pathways, we need the basic data of the characteristic isotopic signatures for particular pathways, so called **endmember values**. These are obtained from the pure culture studies, where specific microorganisms are incubated separately and N_2O is collected and analysed. Numerous pure culture studies are summarised in detail in the recent review papers (Denk et al. 2017; Toyoda et al. 2017). N_2O isotopic signatures for specific pathways were also determined in controlled incubation of the whole soil by applying conditions favouring specific pathways. Such experiments were also summarised before (Denk et al. 2017; Toyoda et al. 2017). Here we present an overview of the most common pathways including results from pure culture studies and controlled soil incubations with some necessary critical selection explained below (after (Denk et al. 2017; Lewicka-Szczebak et al. 2017; Toyoda et al. 2017, Yu et al., 2020)). For each isotopic signature ($\delta^{15}N^{sp}$, $\delta^{18}O$, *and* $\delta^{15}N^{bulk}$) the rules how to properly use endmember values are explained and for each N_2O production process the range of values (minimal and maximal literature reported value), the mean (of all literature reported values) and the median (of all literature reported values) is given.

$\delta^{15}N^{bulk}$ of the produced N_2O depends on the precursor isotopic signature, i.e. on soil NO_3^- for denitrification and soil ammonium for nitrification. Therefore, to compare any results with literature endmember values we need to calculate the N isotopic signature of the N_2O in relation to the precursor, i.e. $\Delta^{15}N^{bulk}{}_{N2O/NO3-}$ for denitrification and $\Delta^{15}N^{bulk}{}_{NH4+}$ for nitrification. Some pure culture denitrification studies also reported the isotope effect between nitrite and N_2O ($\Delta^{15}N^{bulk}{}_{N2O/NO3-}$), especially for fungal denitrification, but for field studies, we usually analyse soil NO_3^-. By calculating isotope effects between N_2O and N precursors one should be aware that the reaction progress changes the isotopic signature of the precursor: the more substrate is consumed, the more ^{15}N enriched gets its residual pool. Therefore, the precursor N isotopic signature at the beginning and at the end of an experiment may differ depending on the reaction progress. Moreover, the $\delta^{15}N$ of the measurable bulk N pools (by soil extraction) may deviate from the $\delta^{15}N$ of the active N_2O producing pools if the fractionating processes are heterogeneously distributed. This is especially the case in unsaturated soils where NO_3^- in anoxic microsites is denitrified and thus progressively enriched in ^{15}N, while in aerobic domains nitrification adds NO_3^- at a lower $^{15}N^{bulk}$ NO_{3-} enrichment. Substantial deviation between bulk soil and active pool enrichment has been recently shown in tracer studies in the laboratory (Deppe et al. 2017) and in the field (Buchen et al. 2016). This indicates that the interpretation based on $\delta^{15}N^{bulk}$ values is very complex and requires a good understanding of N transformation processes in the soil (see also Sect. 7.5).

The following endmember values can be considered:

- *heterotrophic bacterial denitrification:* $\Delta^{15}N^{bulk}{}_{N2O/NO3-}$ determined in pure culture studies from −37 to −10‰, mean −25‰, median −23‰ (Barford et al. 1999; Granger et al. 2008; Sutka et al. 2006; Toyoda et al. 2005). The controlled soil incubation experiments targeted for bacterial denitrification (the sole contribution from bacterial Denitrification was confirmed by $\delta^{15}N^{SP}$ values and ^{15}N tracing) show much lower values from −52.8 to −39.2‰ (Lewicka-Szczebak et al. 2014).
- *nitrifier denitrification:* $\Delta^{15}N^{bulk}{}_{NH4+}$ from −60.7 to −53.1‰, mean −56.9‰ (Frame and Casciotti 2010);
- *nitrification*: $\Delta^{15}N^{bulk}{}_{NH4+}$ from −64 to −47‰, mean −57‰, median −57‰ (Mandernack et al. 2009; Sutka et al. 2006; Yoshida 1988);
- *fungal denitrification*: $\Delta^{15}N^{bulk}{}_{N2O/NO3-}$ from −46 to −31‰, mean −38‰, median −38‰ (Rohe et al. 2014). The study of Maeda et al. (2015) provides only data of the produced $\delta^{15}N^{bulk}$ and not the isotope effect, therefore is not summarised here.

$\delta^{15}N^{sp}$ of the produced N_2O is independent of the precursor isotopic signature. Hence, unlike $\delta^{15}N^{bulk}$, the endmember values are identical in $\delta^{15}N^{sp}$ of the produced N_2O. Therefore, the measured N_2O $\delta^{15}N^{sp}$ values can be directly compared with the following endmember values:

- *heterotrophic bacterial denitrification*: determined in pure culture studies from −7.5 to +3.7‰, mean −1.9‰, median −1.9‰ (Sutka et al. 2006; Toyoda et al. 2005). The values obtained in the controlled soil incubation experiments targeted for bacterial denitrification from −4.7 to +1.7‰ fit within the range given by pure culture studies (Lewicka-Szczebak et al. 2014);
- *nitrifier denitrification*: from −13.6 to +1.9‰, mean −5.9‰, median −5.9‰ (Frame and Casciotti 2010; Sutka et al. 2006);
- *fungal denitrification:* from 27.2 to 39.9‰, mean 33.5‰, median 33.6‰ (Maeda et al. 2015; Rohe et al. 2014, 2017; Sutka et al. 2008). A recent study indicated also a lower $\delta^{15}N^{sp}$ value for one individual fungal species, which was disregarded here due to its very low N_2O production: *C. funicola* showed $\delta^{15}N^{sp}$ of 21.9‰ but less than 100 times lower N_2O production with nitrite compared to other species, and no N_2O production with NO_3^- (Rohe et al. 2014). Similarly, from the study of Maeda et al. (2015), only the values of strains with higher N_2O production were accepted for this summary (>10 mg N_2O-N g^{-1} biomass).
- *nitrification*: from 32.0 to 38.7‰, mean 35.0‰, median 34.6‰ (Frame and Casciotti 2010; Heil et al. 2014; Sutka et al. 2006).

$\delta^{18}O$ depends on the isotopic signature of several possible precursors: NO_3^-, NO_2^-, H_2O and O_2. For oxic processes like nitrification the incorporation of O_2 is important (Snider et al. 2011) whereas for anoxic processes the O of the substrate or of soil water can be incorporated in N_2O. Theoretically, during nitrification (hydroxylamine oxidation) O in N_2O originates from O_2 and during denitrification O from NO_3^- should be transferred to N_2O. But this is additionally complicated by the exchange of O atoms between soil water and denitrification intermediates (Kool et al. 2007). The extent of this exchange differs for various bacterial and fungal species (Rohe et al. 2017), but it has been shown recently that for soil incubations it is rather high (Lewicka-Szczebak et al. 2016). Therefore, soil water isotopic signatures show the largest impact on the final $\delta^{18}O$ values of N_2O, hence it was suggested to present the results as $\Delta^{18}O_{N2O/H2O}$ if dealt with denitrification (Lewicka-Szczebak et al. 2016). However, in pure culture studies, this rule works for fungal denitrification but not very well for bacterial denitrification where NO_3^- plays an important role as a precursor for O atoms in N_2O (Rohe et al. 2017). Because of different patterns for different processes, we present a summary of the measured, uncorrected $\delta^{18}O$ values and additionally for denitrification we also show $\Delta^{18}O_{N2O/H2O}$ values.

- *heterotrophic bacterial denitrification* based on controlled soil incubations: from 4.8 to 18.4‰, mean 10.4‰, median 10.2‰ (Lewicka-Szczebak et al. 2016, 2014). For heterotrophic bacterial denitrification, it is more reasonable to use the values of the controlled soil incubations (from 4.8 to 18.4‰) because pure culture studies show a large range of possible values (from 7.3 to 46.5‰ (Rohe et al. 2017; Sutka et al. 2006; Toyoda et al. 2005)) due to variable O-exchange with ambient water depending on the bacterial strain, whereas soil incubations indicated that this exchange is high (Kool et al. 2007; Snider et al. 2013) and the isotope effect between water and formed N_2O quite stable (Lewicka-Szczebak et al. 2016). The values calculated versus soil water ($\Delta^{18}O_{N2O/H2O}$) show a much narrower range from 16.7 to 23.3‰, mean 19.2‰, median 19.0‰ (Lewicka-Szczebak et al. 2016, 2014).
- *nitrifier denitrification* was determined in two pure culture studies (Frame and Casciotti 2010; Sutka et al. 2006). Frame and Casciotti (2010) provide the value in relation to nitrite $\delta^{18}O_{N2O/NO2}$ of -8.4 ± 1.4‰. However, $\delta^{18}O$ of N_2O originating from nitrifier denitrification is mostly governed by $\delta^{18}O_{H2O}$ due to reaction stoichiometry and additional O-exchange between water and nitrification intermediates (Frame and Casciotti 2010; Kool et al. 2010), and hence it is reasonable to express the isotope effect in relation to water, similarly as for bacterial denitrification. Based on the values presented in supplementary materials of Frame and Casciotti (2010) this value can be recalculated in relation to water giving the range of $\delta^{18}O_{N2O/H2O}$ from 12.4 to 19.4‰ (Frame and Casciotti 2010). Sutka et al. (Sutka et al. 2006) provide a raw $\delta^{18}O_{N2O}$ of 10.8 ± 0.5‰. Assuming the probable $\delta^{18}O_{H2O}$ between -8 and -4‰, the calculated $\delta^{18}O_{N2O/H2O}$ between 14.3 and 19.3‰ fits well within the defined range from (Frame and Casciotti 2010).
- *fungal denitrification* from 31.2 to 45.7‰, mean 36.8‰, median 36.6‰ (Maeda et al. 2015; Rohe et al. 2014, 2017; Sutka et al. 2008). The values calculated versus soil water ($\Delta^{18}O_{N2O/H2O}$) range from 42.0 to 55.1‰, mean 47.2‰, median 46.9‰ (Rohe et al. 2014, 2017; Sutka et al. 2008). The study of Maeda et al. (2015) provide only data of the produced $\delta^{18}O$ without the O isotope signature of water, therefore the $\Delta^{18}O_{N2O/H2O}$ values cannot be given.
- *nitrification* determined in nitrifier cultures incubated with NH_3 reported the $\delta^{18}O_{N2O}$ values close to atmospheric oxygen of 23.5 ± 1.3‰ (Sutka et al. 2006). Frame and Casciotti (2010) determined a slight isotope effect resulting

in $\delta^{18}O_{N2O/O2}$ of −2.9‰. Hence, for this process, the $\delta^{18}O_{N2O}$ range of 23.5 ± 3‰ can be accepted (Frame and Casciotti 2010; Sutka et al. 2006). For the plots in Fig. 7.5, the $\delta^{18}O_{N2O}$ values are shown, which were determined in experiments utilising the air $\delta^{18}O_{O2}$ of 23.5‰. For each case study where deviations from the typical O_2 value are known (e.g. due to consumption in water column), these values should be expressed relative to the actually measured $\delta^{18}O_{O2}$.

The most common way of identifying various N_2O producing pathways is a graphical presentation of the measured values together with the literature endmember values. From the graphs, we can often identify the dominant pathway. To obtain more precise quantitative information, the contribution of a pathway (A) can be calculated based on the measured N_2O isotopic signature (δ_{N2O}) using the **isotope mass balance**:

$$\delta_{N2O} = \delta_{\text{pathway A}} \cdot a + \delta_{\text{pathway B}} \cdot (1 - a) \tag{7.8}$$

It must be noted that for this calculation (Eq. 7.8), the δ_{N2O} value may not be changed due to N_2O reduction. This is only fulfilled if reduction is inhibited, measured to be negligible or included in calculations as described below. Using one isotope signature ($\delta^{15}N^{bulk}$, $\delta^{15}N^{sp}$ or $\delta^{18}O$), we are able to determine the mixing ratios of two pathways. Applying more isotopic signatures can theoretically enable quantification of more pathways. However, the results are not very exact due to the sometimes wide ranges of possible isotopic values for different pathways and overlapping of these ranges for more pathways. For both, $\delta^{15}N^{sp}$ and $\delta^{18}O$, the ranges for heterotrophic bacterial denitrification and nitrifier denitrification. Additional interpretation of $\delta^{15}N^{bulk}$ can further help but is often problematic due to lacking information on precursor isotope values (Lewicka-Szczebak and Well 2020). To increase precision of such calculations, controlled soil incubations with the soil under study may help to determine more narrow ranges of endmember values characteristic for the particular soil (Lewicka-Szczebak et al. 2017).

But besides the mixing processes also isotopic **fractionation during N_2O reduction** changes the final isotopic value of the residual N_2O. During N_2O reduction to N_2 (the last step of bacterial denitrification) preferentially the N-O bonds between light isotopes (^{14}N and ^{16}O) are broken and as a result the residual unreduced N_2O is enriched in $^{15}N^{\alpha}$ and ^{18}O. In consequence, $\delta^{15}N^{sp}$, $\delta^{18}O$ and $\delta^{15}N^{bulk}$ values of residual N_2O increase with progressing reduction. The magnitude of the shift towards higher values depends on the amount of reduced N_2O and the isotopic fractionation factor associated with the N_2O reduction. Hence, if we know the fractionation factor and the δ value of initially produced N_2O before reduction (δ_0), we can calculate the amount of reduced N_2O and thereby determine the magnitude of N_2 flux based on the measured δ value of the residual N_2O after reduction (δ_r). This is calculated according to the following isotopic fractionation Eqs. 7.9 to 7.11 by applying Rayleigh model that is valid for closed systems, either in its exact form (Mariotti et al. 1981):

$$\frac{1+\delta_r}{1+\delta_0} = (r_{N2O})^{\varepsilon_{N2-N2O}} \tag{7.9}$$

or in simplified, approximated form:

$$\delta_r \cong \delta_0 + \varepsilon_{N2-N2O} \cdot \ln(r_{N2O}) \tag{7.10}$$

where δ_r is the residual N_2O isotopic signature, after reduction, δ_o is the initial N_2O isotopic signature, before reduction, $\varepsilon_{N2\text{-}N2O}$ is the isotopic fractionation factor associated with N_2O reduction and r_{N2O} is the residual unreduced N_2O fraction ($r_{N2O} = y_{N2O}/(y_{N2} + y_{N2O})$; ($y$: mole fraction))

The application of the closed system model has been confirmed by several studies (Köster et al. 2015; Lewicka-Szczebak et al. 2017, 2014). However, it was also suggested that an isotopic fractionation model for open systems could be suitable (Decock and Six 2013), which is associated with smaller apparent isotope effects during N_2O reduction:

$$\delta_r = \delta_0 - \varepsilon_{\text{red}}\left(1 - r_{N_2O}\right) \tag{7.11}$$

To be able to determine r_{N2O} from N_2O isotopic values of individual samples according to the above equations, isotopic fractionation factor associated with N_2O reduction to N_2 ($\varepsilon_{N2\text{-}N2O}$) must be known. They were determined in numerous studies in controlled soil incubations (Jinuntuya-Nortman et al. 2008; Lewicka-Szczebak et al. 2014; Menyailo and Hungate 2006; Ostrom et al. 2007; Well and Flessa 2009) and the following ranges were obtained:

- $\varepsilon^{15}N^{bulk}{}_{N2\text{-}N2O}$ from −11.0 to −1.8‰ with a mean of −7.1‰ and median −7.0‰
- $\varepsilon^{15}N^{sp}{}_{N2\text{-}N2O}$ from −8.2 to −2.9‰ with a mean of −5.9‰ and median −6.0‰
- $\varepsilon^{18}O_{N2\text{-}N2O}$ values from −25.1 to −5.1‰ with a mean of −15.4‰ and median −15.9‰

In the summary, we disregarded one study which provided an inverse isotope effect for $\varepsilon^{15}N^{bulk}{}_{N2\text{-}N2O}$ and $\varepsilon^{18}O_{N2\text{-}N2O}$ (Lewicka-Szczebak et al. 2014). These values might have been a result of untypical reduction conditions in the experiment or an experimental artefact (Denk et al. 2017), therefore, they are neglected here. From the study of Lewicka-Szczebak et al. (2015) only the data of moderate reduction (from Pool1) were summarised here, because it was shown that by very intensive reduction the results can be strongly affected by N_2O diffusion. This depends on the balance between diffusive and enzymatic fractionation during N_2O reduction (Lewicka-Szczebak et al. 2014). By nearly complete N_2O reduction, we observe a relatively large impact of diffusive N_2O fractionation, resulting in residual N_2O more depleted in heavy isotopes, hence the apparent isotope effects are significantly lower, i.e. −2.7‰, −1.5‰, and −2.0‰ for $\varepsilon^{15}N^{bulk}{}_{N2\text{-}N2O}$, $\varepsilon^{15}N^{sp}{}_{N2\text{-}N2O}$, and $\varepsilon^{18}O_{N2\text{-}N2O}$, respectively (Lewicka-Szczebak et al. 2015).

It is often problematic to separate the impact on the final N_2O isotopic values by the mixing endmember for the produced N_2O and by the isotopic fractionation due to N_2O reduction. The interpretations and calculations based on N_2O isotopic studies are difficult when we deal with the simultaneous variations in r_{N2O} and δ_0 values. Usually, to calculate r_{N2O} a stable δ_0 is assumed (Lewicka-Szczebak et al. 2015) and to precisely determine temporal changes in δ_0, we need independent data on r_{N2O} (Köster et al. 2015). In field studies, both r_{N2O} and δ_0 cannot be determined precisely, but the possible ranges for each parameter can be given (Zou et al. 2014).

It is often attempted to distinguish between mixing and fractionation processes by using the changes in the isotopic signatures and their relations: $\delta^{15}N^{sp}/\delta^{18}O$, $\delta^{15}N^{sp}/\delta^{15}N^{bulk}$, $\delta^{18}O/\delta^{15}N^{bulk}$. These relations differ for the N_2O reduction process and for mixing processes due to differences in the respective isotope effects. From literature data on N_2O reduction fractionation factors (Jinuntuya-Nortman et al. 2008; Lewicka-Szczebak et al. 2014; Menyailo and Hungate 2006; Ostrom et al. 2007; Well and Flessa 2009) the following ratios are determined:

- $\varepsilon^{15}N^{sp}_{N2\text{-}N2O}/\varepsilon^{18}O_{N2\text{-}N2O}$ from 0.23 to 0.98 with a mean of 0.45 and median 0.36
- $\varepsilon^{15}N^{sp}_{N2\text{-}N2O}/\varepsilon^{15}N^{bulk}_{N2\text{-}N2O}$ from 0.51 to 2.78 with a mean of 0.96 and median 0.77
- $\varepsilon^{18}O_{N2\text{-}N2O}/\varepsilon^{15}N^{bulk}_{N2\text{-}N2O}$ values from 1.02 to 3.83 with a mean of 2.21 and median 2.25.

Although the range of possible $\varepsilon_{N2\text{-}N2O}$ variations is quite large, it has been shown recently that the mean values and typical $\varepsilon^{15}N^{sp}_{N2\text{-}N2O}/\varepsilon^{18}O_{N2\text{-}N2O}$ ratios are well applicable for oxic or anoxic conditions unless N_2O reduction is almost complete, *i.e.* the ratio $N_2O/(N_2 + N_2O) < 0.1$, meaning more than 90% of N_2O was reduced (Lewicka-Szczebak et al. 2015).

For comparison, here are the relations between isotopic signatures of emitted N_2O resulting from mixing processes calculated based on literature ranges for mixing endmembers given above. Because of the overlapping endmember ranges, we cannot distinguish between all individual pathways, and we determine the slopes of mixing lines between selected endmember values (Figs. 7.5 and 7.6) as follows:

- mixing between heterotrophic bacterial denitrification and nitrification:
  - $\boldsymbol{\delta^{15}N^{SP}/\delta^{18}O}$ from −10.5 to 4.8 with a mean of 6.1;
  - $\boldsymbol{\delta^{15}N^{SP}/\delta^{15}N^{bulk}}$ from −4.6 to −0.5 with a mean of −1.2;
  - $\boldsymbol{\delta^{18}O/\delta^{15}N^{bulk}}$ from −1.0 to 0.1 with a mean of −0.1.
- mixing between heterotrophic bacterial denitrification and fungal denitrification:
  - $\boldsymbol{\delta^{15}N^{SP}/\delta^{18}O}$ from 1.1 to 1.4 with a mean of 1.3;
  - $\boldsymbol{\delta^{15}N^{SP}/\delta^{15}N^{bulk}}$ from −3.9 to 7.9 with a mean of −2.7;
  - $\boldsymbol{\delta^{18}O/\delta^{15}N^{bulk}}$ from −2.8 to 6.4 with a mean of −2.2.

Fungal denitrification cannot be distinguished from relations including $\delta^{15}N^{bulk}$ because of the overlapping range with bacterial denitrification (see Fig. 7.5). Anyway, relations including $\delta^{15}N^{bulk}$ are difficult to use due to dependence of this isotope value on the precursor, which differ for nitrification and denitrification. Here

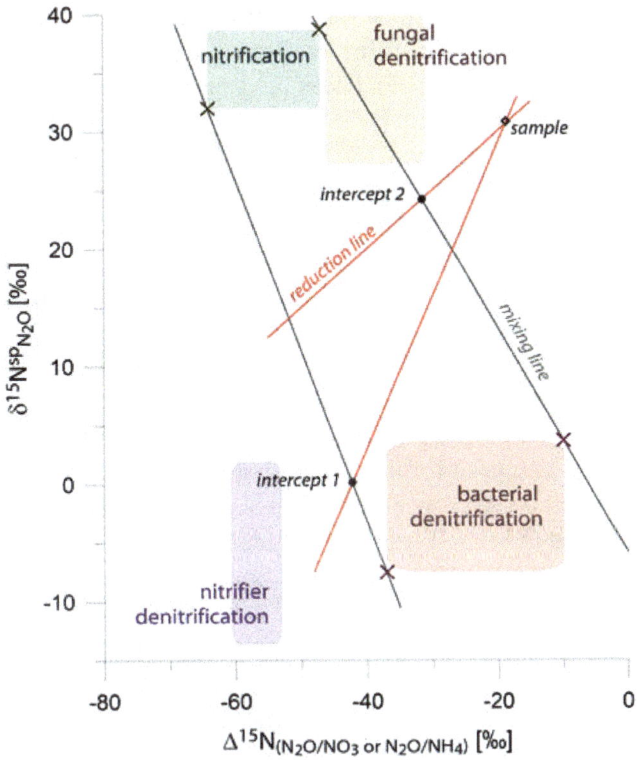

Fig. 7.5 Scheme of the $\delta^{15}N^{sp}/\delta^{18}O$ mapping approach to simultaneously estimate the possible range of N_2O reduction and the admixture of nitrification. The endmember values are shown according to the citations provided in the text. Note that δ^5N values are given in relation to N substrate, which should be determined for the particular study (here 0‰ for both NO_3^- and NH_4^+ was assumed). Here the mixing of bacterial denitrification and nitrification is considered. The method can be applied for other selected processes (Zou et al. 2014)

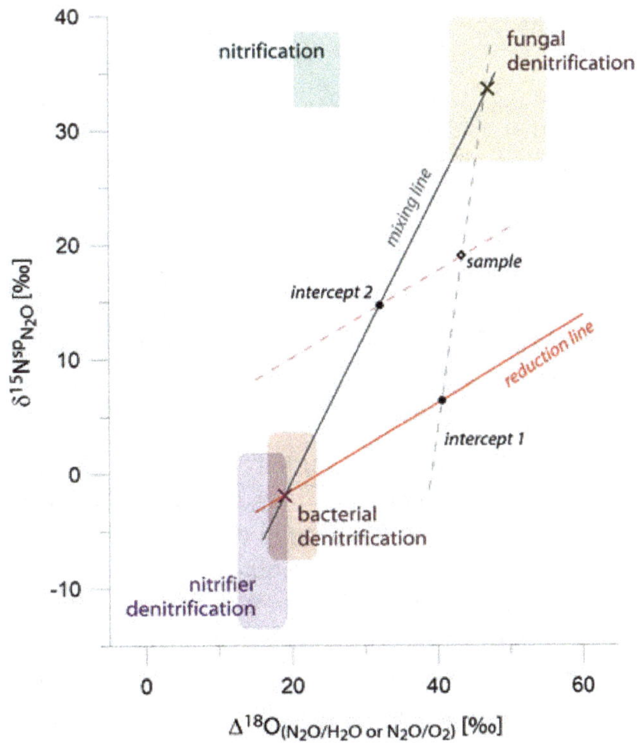

Fig. 7.6 Scheme of the $\delta^{15}N^{sp}/\delta^{18}O$ mapping approach to simultaneously estimate the magnitude of N_2O reduction and the admixture of fungal denitrification (or nitrification). The endmember values are shown according to the citations provided in the text. Note that $\delta^{18}O$ values are given in relation to water and to air oxygen (for nitrification). Here the mixing of bacterial denitrification and fungal denitrification is considered. The method can be applied for other selected processes

the relationships must be determined with isotope effect for $\delta^{15}N^{bulk}$, i.e. using $\Delta^{15}N^{bulk}(N_2O/NH_4^+)$ for nitrification and $\Delta^{15}N^{bulk}(N_2O/NO_3^-)$ for denitrification (see x-axis in Fig. 7.5). Often the isotopic signatures of the precursors are not known, which make the interpretation of $\delta^{15}N^{bulk}$ values rather ambiguous. Nevertheless, some studies apply the $\delta^{15}N^{sp}/\ \delta^{15}N^{bulk}$ isotope maps for distinction of mixing and fractionation processes, but for such isotope maps, systematic changes in $\delta^{15}N^{bulk}$ induced by systematic changes in the N isotopic composition of one of the precursors NH_4^+ or NO_3^- could be misinterpreted as reduction events (Well et al. 2012; Wolf et al. 2015). Hence, the careful monitoring of precursor isotopic signatures is needed (Zou et al. 2014).

A **$\delta^{15}N^{sp}/\ \delta^{15}N^{bulk}$ isotope mapping approach** allowing for assessment of minimal and maximal reduced N_2O fraction and nitrification and denitrification mixing ratios was proposed by Zou et al. (2014) (Fig. 7.5). Such an approach is most often used for distinguishing between nitrification and bacterial denitrification only. However, other cases have also been analysed (Zou et al. 2014). The calculation method presented (Fig. 7.5) assumes first mixing of N_2O from different endmembers and afterwards its partial reduction. Two mixing lines are defined–for the minimum and maximum values for both endmembers as well as two reduction lines–with maximal and minimal slope. From the intercept 1 the maximal denitrification contribution is determined whereas from the intercept 2 the minimal one. Based on the difference between the sample point and intercept 1 or 2 the reduction contribution, respectively, maximal and minimal, is also determined. However, it must be noted that in case of significant admixture of fungal denitrification or nitrifier denitrification the results may be biased.

The application of **$\delta^{15}N^{sp}/\ \delta^{18}O$ isotope mapping approach** may be easier since $\delta^{15}N^{sp}$ and $\delta^{18}O$ values are more stable in time (Lewicka-Szczebak et al. 2017; Wu et al. 2019), and $\delta^{18}O$ values show narrower endmember ranges when compared to $\delta^{15}N$ values. The distinction of mixing and fractionation processes is based on the different slopes of the mixing lines and the reduction line (Fig. 7.6).

Isotopic values of the samples analysed are typically located between these two, reduction and mixing, lines. Here we defined only one mixing line for the median values of bacterial and fungal denitrification and one reduction line with a mean slope. From sample's, location, we can estimate the impact of fractionation associated with N_2O reduction and admixture of N_2O originating from fungal denitrification. We can deal with two scenarios:

(i) Scenario 1: the N_2O emitted due to bacterial denitrification is first reduced (point move along reduction line up to the intercept 1 with dashed mixing line) and then mixed with the second endmember (point move along dashed mixing line to the measured sample point).

(ii) Scenario 2: the N_2O from two endmembers is first mixed (point move along mixing line up to the intercept 2 with dashed reduction line) and only afterwards the mixed N_2O is reduced (point move along dashed reduction line to the measured sample point).

While both scenarios yield identical results for the admixture of N_2O from fungal denitrification, the resulting reduction shift, and hence the calculated r_{N2O} value, is higher when using Scenario 2. It is still not clear which scenario is more realistic. The uncertainty analysis of this method has been recently presented by Wu et al. (2019) and this approach has been successfully applied in the field case studies (Buchen et al. 2018; Ibraim et al. 2019; Verhoeven et al. 2019). However, after the appearance of those publications, it has been found that other $\delta^{18}O$ values should be applied for nitrification (Yu et al., 2020). This summary reports the most current choice of endmember ranges, which differ from those presented recently (Buchen et al. 2018; Ibraim et al. 2019; Lewicka-Szczebak et al. 2017; Verhoeven et al. 2019; Wu et al. 2019).

7.3.3 *Analysis of N_2O Isotopocules by IRMS*

The most common method for N_2O isotopocule analysis is isotope ratio mass spectrometry (IRMS). In order to perform N_2O isotopic analysis the gas samples need to be purified, and N_2O must be separated and pre-concentrated. First, water and CO_2 are removed by chemical traps, and then N_2O is concentrated with liquid N traps. Afterwards, the gases are separated with gas chromatography and finally introduced in the isotope ratio mass spectrometer.

In the mass spectrometer, N_2O isotopocule values are determined by measuring *m/z* 44, 45 and 46 of the intact N_2O^+ ions as well as *m/z* 30 and 31 of NO^+ fragment ions. This allows the determination of average $\delta^{15}N$ ($\delta^{15}N^{bulk}$), $\delta^{15}N^{\alpha}$ ($\delta^{15}N$ of the central N position of the N_2O molecule), and $\delta^{18}O$ (Toyoda and Yoshida 1999). $\delta^{15}N^{\beta}$ ($\delta^{15}N$ of the peripheral N position of the N_2O molecule) is calculated from $\delta^{15}N^{bulk} = (\delta^{15}N^{\alpha} + \delta^{15}N^{\beta}) / 2$ and ^{15}N site preference ($\delta^{15}N^{sp}$) from $\delta^{15}N^{sp} = \delta^{15}N^{\alpha}–\delta^{15}N^{\beta}$. Since the IRMS approach was developed simultaneously by two groups (Brenninkmeijer and Röckmann 1999; Toyoda and Yoshida 1999), two different nomenclatures had been introduced for the two positions of N_2O-N. Hence, in some studies, the peripheral (β) position is referred to as 1- and the central (α) as 2-position (Brenninkmeijer and Röckmann 1999). The scrambling factor resulting from the exchange of ^{15}N atoms on the ion source must be taken into account. The magnitude of the scrambling factor should be determined individually for each mass spectrometer (Röckmann et al. 2003). Also, ^{17}O-correction should be taken into account, because ^{17}O substitution is indistinguishable from ^{15}N, therefore typical terrestrial ^{17}O content (0.528) is assumed (Kaiser and Röckmann 2008).

Up to now, there are still no internationally agreed gaseous N_2O reference materials for N_2O isotopocule analyses. Usually, the laboratories calibrated pure N_2O gas for isotopocule analyses in the laboratory of the Tokyo Institute of Technology according to the method of Toyoda and Yoshida (1999). Recently, the first interlaboratory comparison has been performed and now the standards from this study (REF1, REF2) are available for the laboratories and allow the performing of two-point calibration for $\delta^{15}N^{sp}$ values (Mohn et al. 2014). This intercalibration study has shown that the two-point calibration method is necessary to obtain accurate $\delta^{15}N^{sp}$ values. Recently, two N_2O standards had been tested in a further interlaboratory comparison (Ostrom et al. 2018) and is available from United States Geological Survey (USGS).

The sample volume needed for the N_2O isotopocule depends on the concentration and is about 100 ml for ambient N_2O concentration samples (about 300 ppb) and about 10 ml for N_2O concentration of above two ppm.

7.3.4 *Laser Spectroscopic Analysis of N_2O Isotopomers to Differentiate Pathways*

The invention and availability of non-cryogenic light sources in the mid-infrared (MIR) spectral range (Brewer et al. 2019) coupled with different detection schemes such as direct absorption quantum cascade laser absorption spectroscopy (QCLAS) (Mohn et al. 2010, Mohn et al. 2012, Wächter et al. 2008), cavity ring down spectroscopy (CRDS) (Erler et al. 2015) and off-axis integrated-cavity-output spectroscopy (OA-ICOS) (Wassenaar et al. 2018) has provided sensitive and field-deployable laser spectroscopic analysers for N_2O isotopocule analysis. These instruments can analyse the N_2O isotopic composition in gaseous mixtures (e.g. ambient air) in a flow-through mode, providing real-time data with minimal or no sample pre-treatment, which is highly attractive to better resolve the temporal complexity of N_2O production and consumption processes. Most importantly, MIR laser spectroscopy is selective for ^{17}O, ^{18}O and position-specific ^{15}N substitution due to the existence of characteristic rotational-vibrational spectra (Gordon et al. 2017).

Therefore, laser spectroscopy has the potential to open a new field of research in the N_2O biogeochemical cycle, but, applications remain challenging and are still scarce for the following main reasons: (1) laser spectrometers as any analytical instrument are subject to drift effects, in particular under fluctuating environmental conditions, limiting their performance (Werle et al. 1993); (2) changes in N_2O concentration affect N_2O isotope results when using the δ-calibration approach (Griffith 2018); (3) laser spectroscopic results are affected by mole fraction changes of atmospheric background gases (N_2, O_2, and Ar), called gas matrix effects, due to the difference of pressure-broadening coefficients, and potentially by spectral interferences from other atmospheric constituents (H_2O, CO_2, CH_4, and CO, etc.), called trace gas effects, depending on the wavelength region used in an instrument. Spectral interferences are particularly pronounced for N_2O due to its low atmospheric abundance in comparison to other trace gases; (4) only since recently two pure N_2O isotopocule reference materials (USGS51, USGS52) have been made available through the United States Geological Survey (USGS) (Ostrom et al. 2018), which was identified as a major reason limiting interlaboratory compatibility (Köster et al. 2013; Mohn et al. 2014, 2016).

In a recent study, the most common commercially available N_2O isotope laser spectrometers were carefully characterised for their dependence on N_2O concentration, gas matrix composition (O_2, Ar) and spectral interferences caused by H_2O, CO_2, CH_4 and CO to develop analyser-specific correction functions. In addition, the authors suggest a step-by-step workflow that should be followed (Fig. 7.7) by researchers to acquire trustworthy N_2O isotopocule results using laser spectroscopy (Harris et al. 2020).

7.3.5 *Hands-on Approach to Use a CRDS Isotopic N_2O Analyser*

Introduction

As an example, the Picarro G5101-*i* analyser can be used to determine N_2O concentration, $^{15}N^{bulk}$ isotope ratios and isotopomer values ($^{15}N^{\alpha}$ and $^{15}N^{\beta}$) by continuous or discrete sample measurement. Small volume discrete samples ($\leq$20 ml) can be measured using the SSIM (small sample isotope module) (see also Sect. 5.3.) peripheral unit in conjunction with the Picarro G5101-*i* analyser. The G5101-*i* analyzer is the predecessor of the current G5131-*i* analyzer which also measures $\delta^{18}O$ in addition to $\delta^{15}N^{bulk}$, $\delta^{15}N^{\alpha}$ and $\delta^{15}N^{\beta}$. The SSIM can also be used to dilute samples. Larger volume samples (e.g. Tedlar bags) can be measured by direct input into the G5101-*i* analyser or through the 16-port distribution manifold. The 16-port distribution manifold allows for partial automation of measurement and can be used in conjunction with the SSIM for smaller volume samples (see also Fig. 5.5 that illustrates the coupling of a 16-port manifold and a SSIM). The SSIM can also be used to dilute samples.

Principle

Samples are measured using mid-IR laser by CRDS (cavity ring down spectroscopy). Measurement precision increases with measurement time. Several options are available for delivery of N_2O samples into the analyser and how long measurements take. Sample volume and the required precision of measurements should be considered to decide which operational set up is the most appropriate.

Apparatus

- Picarro G5101-*i* isotopic N_2O analyser and pump.
- Picarro SSIM peripheral unit.
- Picarro 16-port distribution manifold.
- Gas-tight syringe.
- Pressure regulators.
- Stainless steel tubing.
- Swagelock fittings.
- Injector nut for SSIM.

Analyser specifications:

Aquire information on your laser spectrometer from manufacturer and/or own tests, e.g.:

- What is the precision / drift at different N_2O concentrations?
- Are there significant gas matrix (e.g. N_2, O_2, Ar) and trace gas (CO_2, CH_4, CO, etc.) effects on performance?
- How big are N_2O concentration effects on delta values?
- How severe are temperature effects on the performance?
- What is the required sample gas volume?

Experimental design:

Which experimental design do I need to align analyser specifications with target performance?

- Perform two-point delta calibration with reference gases with similar gas matrix and trace gas composition to sample gas.
- Regular analysis of a reference gas during experiment allows post-correction for drift.
- Are there significant changes in the gas matrix and trace gas concentrations during experiments, if yes:
  - Implement scrubbers to remove interferants (e.g. H_2O, CO, CO_2).
  - Analyse gas matrix or trace gas (e.g. O_2, CH_4) changes.
  - Install preconcentration device to normalize gas composition.
- Implement dynamic dilution if concentrations are too high and to adjust gas matrix, etc.

Data processing of delta values:

- Develop and apply analyzer-specific matrix and/or trace gas corrections at relevant N_2O concentration, if required.
- If N_2O and interferant concentrations co-vary, more complex correction functions have to be developed and applied.
- If applicable correct N_2O concentration effects on delta values.
- Apply drift correction, if required.
- Standardize delta values to international scales, based on reference gas measurements.

Fig. 7.7 Workflow to acquire trustworthy N_2O isotopocule results using laser spectroscopy (Harris et al. 2020)

Consumables

- Zero Air.
- N_2O working standards.
- Septa for injector nut on SSIM.
- Septa capped vials for discrete gas samples.
- Tedlar bags for larger volume gas samples.
- Side port needles for sample injection to SSIM.

Sampling

For discrete gas samples follow a suitable sampling procedure as outlined by De Klein and Harvey (2012). Small volume samples (≤20 ml) should be stored in septum capped vials, ensuring to overpressure when filling to prevent inward contamination by ambient air. Vials can be stored in a cool dry place. Larger volume samples in Tedlar bags should be measured ASAP as storage reliability decreases greatly after 24–48 h.

Operational Procedure

- To start the analyser, ensure the power switches are on for the pump, analyser and monitor. Turn the power switch at the rear of the analyser from O to I. **NB: the power switch on the pump should always be in the on position, the pump will power up when the analyser is turned on.** To turn on the analyser press the button on its front. Windows will load on the monitor and the analyser software will run through the system checks.
- When the analyzer is in startup mode, monitor the liquid coolant at the back of the analyzer. You should observe little to no bubbles and the fluid should be flowing. If the bubbles have not disappeared after a few minutes or the liquid is not flowing, refer to the troubleshooting section in this document.
- After the system checks are complete, the GUI (Graphical User Interface) will appear. It will begin by measuring the Cavity Pressure, DAS (Data Acquisition System, i.e. the analyser) temperature and Etalon temperature. Once the correct temperatures and pressures are reached a message will appear on the bottom of the GUI screen; e.g. "Pressure locked", "Cool Box Temperature locked", "Preparing to measure", Measuring…". The GUI will then begin to show the continuous N_2O measurements in real time. It may take up to 1 h for the analyser to begin N_2O measurements. Before measuring samples, allow the laser to stabilise for up to 24 h by measuring room air.
- Continuous samples (e.g. incubation experiments) can be measured by directly connecting a piece of tubing from the sampling container to the inlet at the back of the analyser and segmenting the data into the respective time periods.

Discrete samples (≤20 ml)

- To measure small volume discrete samples (≤20 ml) allow the laser to run continuously for 24 h to ensure that the laser has been given sufficient time to stabilise.

- Before operating the SSIM check that the "Valve Seqeuncer MPV" is turned off. To do this click Shutdown on the GUI and select "Software Only". Double click on the Picarro Utilities icon located on the desktop and double click on "Setup Tool". Under the "Port Manager" tab check that the "Valve Sequencer MPV" is turned off. If necessary change this setting to off and close the Picarro Utilities folder.
- Restart the GUI software by double clicking on the Picarro Switcher Mode icon located on the desktop and select the Isotopic N_2O option followed by clicking Launch.
- In the standard GUI mode, the H_2O parameter is not available from the Data Key drop-down menus. This is necessary to check for pressure leaks in the system. To access this, log into the service GUI mode under the settings tab of the GUI. The password is "picarro".
- In the cylinder cabinet open the zero air (ZA) cylinder followed by opening the valve to the lab (**do not open the exhaust valve–this will drain the ZA cylinder**). Record the overall pressure remaining in the cylinder before and after each use. The pressure regulator in the cabinet should be set to 3.5 bar when the cylinder and line valve are open. This will drop to around 3 bar when the pressure regulators at the lab bench are opened.
- At the lab bench, open the black valve at the first pressure regulator on the ZA line to the SSIM and adjust slowly to 1.5 bar. This will rise to 2 bar when it meets the resistance of the SSIM.
- The second pressure regulator has been set to 3 psi (following Picarro's recommendations), check to ensure this is the case and only adjust if necessary. **Never allow the final pressure into the SSIM to go above 8 psi.** The indicator on the valve may flicker during operation due to valve switching within the SSIM.
- Connect the stainless steel tubing from the SSIM to the analyser. Finger tighten and then apply a ¼ turn using the adjustable spanner.
- Connect the grey (valve switching controls) and black (pressure detector) cables from the analyser to the SSIM. This will power on the SSIM, indicated by the green light on the front of the unit.
- **NB Turn on the SSIM vacuum pump. This must be done before launching the SSIM software**.
- Launch the SSIM software by double clicking on the "SSIM Pressure Detector" icon. This locates the COM port (**COM 7**) of the analyser that the SSIM is connected to. Leave this software window open while using the SSIM.
- Measure the Zero Grade Air only for 30 min before beginning sample analysis. This is to obtain the average N_2O concentration, $^{15}N^{bulk}$, $^{15}N^{\alpha}$ and $^{15}N^{\beta}$ of the Zero Grade Air, necessary for correcting concentration dilution and isotope mixing.
- Double click on the "SSIM Coordinator" from the desktop and select G5101-*i*. Configure the settings to suit the measurement procedure required. There are nine parameters (1–9) to be set.

1. Multi-Port Valve: 1 = Use 16 Port Distribution Manifold; 2 = Don't Use 16 Port Distribution Manifold. **[Select 2 for SSIM only].**

2. If using Multi-Port Valve: Number of Sample Ports (between 1 and 8). **[Select 1 for SSIM only].**
3. Number of Repeats per Sample (between 0 and 5). **[Select 1].**
4. Number of Repeats per Standard (between 0 and 5). **[Select 0].**
5. Standard Mode: 1 = Between Each Sample; 2 = Beginning and End. **[Select 2].**
6. Measurement Mode: 1 = One Time; 2 = Continuous Loop. **[Select 2].**
7. Measurement Speed: 1 = Standard, 2 = Fast. **[Select 2] (Fast is approximately 10 min per sample/ Standard is approximately 15 min per sample).**
8. Sample Loading: 1 = Manual; 2 = Automatic; 3 = Syringe. **[Select 3].**
9. Sample Dilution: 1 = No Dilution; 2 = Dilute Sample with ZA. **[Select 2 for samples < 20 ml. Select 1 for samples >20 ml].**

   - Click OK. Select G5101-*i* for reference standard.
   - SSIM pressure measurements should be available in the GUI data key drop-down tabs. Select this parameter to monitor SSIM pressure visually on the left side of the GUI.

Note: Under vacuum, the SSIM pressure should be ~8 Torr or below. When a sample is injected the max pressure is reached upon filling the cavity with sample/ ZA. The max pressure should read between 980 and 1000 Torr. If the pressure is too high down-regulate the second pressure regulator. If the pressure is too low up-regulate the second pressure regulator being very careful not to exceed 8 psi.

- Overlay the SSIM Coordinator screen on to the bottom right corner of the GUI screen. This allows the user to monitor the parameters on the left side of the GUI while following the prompts of the SSIM Coordinator.
- Follow the steps indicated on the SSIM Coordinator screen to process each sample.
- The first prompt requires the operator to inject the sample syringe with the valve closed and to click "Resume" under "Control". The SSIM coordinator will then run through several valve sequencing steps.
- The next prompt to the operator is to open the syringe valve and to click "Resume" under "Control". The SSIM coordinator will then run through several valve sequencing steps.
- The operator will then be prompted to inject the sample. The sample will begin to draw itself in but the operator may be required to manually complete the injection depending on the sample volume. Once the sample is fully injected, close the syringe valve. Allow the SSIM pressure reading to settle and record this pressure value followed by clicking "Resume" under "Control". **NB–Always manually record the SSIM pressure as it settles after sample injection, and record the max SSIM pressure when the ZA dilution is carried out. This is used to work out the actual volume of a sample using the pressure vs volume calibration curve.**
- The SSIM coordinator will then begin the dilution process. **NB–watch the SSIM pressure readings and record the maximum pressure reached during the dilution step.**

- The SSIM will then begin the sample measurement. At this stage, the syringe can be removed from the injector nut to prepare for the next sample injection.
- Before each measurement day, complete a pressure vs volume calibration curve. Use room air injected at the following volumes: 0 ml, 5 ml, 10 ml, 15 ml and 20 ml. To complete the 0 ml point do not inject the syringe, instead allow the zero air to fill 20 ml (cavity volume) into the SSIM.

Note: The calibration curve should be almost perfectly linear with a $R^2 = 0.99$ +. Deviations from the curve or lower R^2 values may indicate a leak. Check the injector nut and septum, change septum if necessary. Check the ZA line connections from the SSIM unit to the analyser. Tighten loose connections if necessary by finger tightening + ¼ turn with an adjustable spanner. **Never over tighten as this can lead to leaks.**

- To check the instrument precision and to avoid measurement drift, it is recommended that a room air/zero blank is run after every 10 samples. A reference standard or working standard may also be used if available.
- To discontinue SSIM use and return to continuous measurement reverse the order of the SSIM setup steps. Close the SSIM Coordinator window. (**Note**: a system alarm will appear on the GUI, this is normal) Close the SSIM pressure detector window. Turn off the SSIM vacuum pump. Disconnect the grey and black cables from the SSIM. Disconnect the stainless steel tubing from the SSIM output. Close the black valve on the first pressure regulator at the lab bench and close this regulator by turning in the decrease direction. Close the valve and the ZA cylinder in the cylinder cabinet.

Note: A system alarm will probably appear on the top left of the GUI and a message stating "Pressure unlocked". This results from the SSIM being disconnected. To resolve: click Shutdown and select "Stop Analyser Software Only". Wait a couple of minutes and relaunch the analyser software by double clicking on the Picarro Switcher Mode icon on the desktop and selecting G5101-*i* Isotopic N_2O and click launch. Monitor the system as it relaunches and wait until it begins measuring N_2O parameters.

- To turn off the instrument completely click shutdown on the GUI.

NB: Never leave the analyser measuring ZA overnight, this will lead to drift.

Expression of Results

- N_2O concentration is expressed as ppb.
- $\delta\ ^{15}N^{bulk}$ is expressed as permil (‰).
- $\delta\ ^{15}N^{\alpha}$ is expressed as permil (‰).
- $\delta\ ^{15}N^{\beta}$ is expressed as permil (‰).

Quality Assurance

- Prior to taking gas samples in the field (i.e. from static chamber) ensure vials are properly sealed and that they have been flushed and evacuated three times.

- Ensure samples are injected with slight overpressure (e.g. 20 ml into 12 ml vial) to avoid inward contamination that would dilute the sample concentration.
- Ensure samples are stored in a cool dry place. Process samples as quickly as possible. Vials lose pressure over time. Avoid storing in direct sunlight.
- The laser should be given sufficient time to stabilise. 24 h is recommended prior to measuring samples.
- Before each measurement day, complete a pressure vs volume calibration curve as described above. Check for leaks based on any variation detected.
- Ensure the septum in the injector nut is replaced approximately every 100 injections.
- Use side bore needles to reduce the damage caused to the septum.
- For acceptable precision and accuracy ensure that sample concentrations are within the stated operating range of the analyser (300–1500 ppb N_2O).
- Minimise moisture (H_2O) in samples. Use drying tubes to introduce samples to the analyser if necessary.
- Use a gas-tight syringe to inject discrete samples into the SSIM. If using a plastic syringe and with a three-way valve, replace when necessary due to wear and tear.
- Never leave the analyser measuring ZA overnight. This will cause measurement drift.

Reporting of Results

Raw data files are automatically generated by the analyser and are stored on the instrument's computer as a DataLog_User file. These raw data files can be found by following the file path: C:\UserData\DataLog_User\Year\Month\Day. An example of the file naming convention is JBDS5030-20170331-140739Z-DataLog_User. JBDS5030 refers to the instrument serial number. 20170331 is the Year, Month and Date the file was started. 140739 is the Hour, Minute and Second of when the file was started. There are a number of values available for the N_2O parameters measured. The dry corrected values are the appropriate values to select for analysis.

When measuring discrete samples using the SSIM there is sufficient time between samples to record the real-time values on a separate spreadsheet that has been premade with sample reference numbers included.

Safety

- When using syringes and needles for sampling and analysis, take extra care to avoid needle stick injuries.
- Regularly check the pressure reading of the instrument and the pressure regulators on the ZA line.
- Never handle pressurised gas cylinders without the appropriate safety training and certification.
- If moving the instrument, always ensure it is shut down so that the cavity returns to ambient pressure and does not remain under vacuum.
- There are a number of valve sequences during operation of the SSIM. Ensure to follow the prompts carefully to avoid loss of sample or pressure build ups.

- Read and follow the information in the Risk Assessments for the Stable Isotope Analysis lab.

Trouble Shooting

- **Start-up:**

If the chiller line contains large air bubbles this may stop the circulation of water in the line. This can lead to the baseplate temperature being exceeded which causes the analyser to enter safe mode (error message appears in GUI). This problem should be avoided by keeping the cooling agent LIQ-702 (propylene glycol) in the buffer tank (externally mounted on the chiller cover) topped up to 90% of its full volume with deionised water. To do this unscrew the black cover and use a wash bottle to add in fresh deionised water. This can be done while the analyser is running. If the error message does appear this may require the instrument to be shut down and for the chiller line be flushed following the instructions provided in the installation manual for the installation of the water buffer tank.

7.3.6 Accuracy, Precision and Bias

The analytical precision for IRMS measurements determined as standard deviation (1σ) of the internal standards for measurements of $\delta^{15}N^{bulk}$, $\delta^{18}O$ *and* $\delta^{15}N^{sp}$ is typically 0.1, 0.1 and 0.5‰, respectively. Commercially available laser spectrometers at ambient N_2O concentrations offer a precision of 0.2 to 1 ‰ for $\delta^{15}N^{\alpha}$, $\delta^{15}N^{\beta}$ and $\delta^{18}O$, which can be reduced to 0.1 ‰ at higher concentrations, or by using a preconcentration device. However, from the inter-comparison study, we see that the bias may be much larger, up to: for $\delta^{15}N^{bulk}$ 0.8 and 2.8‰, and for $\delta^{15}N^{sp}$ 4.3 and 3.7‰ for mass spectrometry and for laser spectroscopy, respectively (Mohn et al. 2014). But these potentially large errors can be minimised by a proper data calibration using two points standardisation with the reference gases that bracket the measurement range. Care must be also given when samples with high concentrations are diluted as the dilution matrix (typically Helium or N_2) may apparently have an impact on the final result. The rule of identical treatment of standards and samples should be held, including identical dilution matrix and similar concentration range (Mohn et al. 2014).

Possible bias is also associated with calculations applied for data interpretation. Due to large ranges of literature data, the N_2O source partitioning cannot be done precisely, but rather the ranges of possible results can be given. To increase precision of such methods controlled soil incubation can be applied to determine the soil specific endmember isotopic values or fractionation factors (Lewicka-Szczebak et al. 2017).

7.3.7 Examples of Laboratory Applications

Köster et al. (2015)

This experiment applied an N_2O isotopocule approach combined with conventional N_2O and N_2 flux measurements to study microbial pathways after different organic fertiliser applications. The direct determination of emitted N_2 was used to take isotope effects during N_2O reduction to N_2 into account. The measured isotope signatures were corrected for isotope effects during N_2O reduction with Eq. 7.10 using previously determined fractionation factor ranges. Based on the corrected values the isotope mass balance equations (Eq. 7.8) for $\delta^{15}N^{sp}$ and $\delta^{18}O$ were applied. The ranges for different pathways contribution were given for $\delta^{15}N^{sp}$- and $\delta^{18}O$-based results and the common area for both was accepted as most probable. Two mixing scenarios were considered: bacterial denitrification and nitrification or bacterial and fungal denitrification. Although the range of possible results for endmembers contribution varied up to 30%, a clear increase in nitrification contribution with the incubation time has been documented.

Schorpp et al. (2016)

In this experiment, incubations with soil fauna were applied to check the impact on N_2O and N_2 emission of anecic earthworms and euedaphic collembola. Isotopocule approach was applied together with ^{15}N tracing. Interpretation of the isotopocule results based on the $\delta^{18}O$-$\delta^{15}N^{sp}$ isotope map, similar as presented in Fig. 7.6, including three possible mixing endmembers: bacterial and fungal denitrification and nitrification (hydroxylamine oxidation) and taking N_2O reduction into account. Isotope data allowed concluding that the presence of collembolans shifted the process pathways towards bacterial denitrification although no change in N_2O concentration could be noted.

Deppe et al. (2017)

In this incubation experiment high NH_4^+ concentrations in soil were established to check the supposed inhibition of nitrification. An isotopocule approach, together with ^{15}N tracing and acetylene inhibition approach, was applied to gain insight into N_2O production processes. Interpretation of the isotopocule results based on the $\delta^{18}O$-$\delta^{15}N^{sp}$ isotope map, similar as presented in Fig. 7.6, including two mixing endmembers: denitrification and nitrification (hydroxylamine oxidation) and N_2O reduction. This assumption of the mixing conditions appeared incorrect, since some data were located outside of the mixing and reduction lines. This indicated a substantial contribution of nitrifier denitrification and/or coupled nitrification-denitrification (10–40%) to total N_2O production.

Cardenas et al. (2017)

Laboratory incubation was carried out at different saturation levels for a grassland soil and emissions of N_2O and N_2 were measured as well as the N_2O isotopocules. Thanks to direct measurements of N_2 flux, the extent of N_2O reduction was known.

Hence, the measured δ values were mathematically corrected to obtain the δ values of the produced N_2O before reduction applying Eq. 7.9. An endmember mixing model (Eq. 7.8) was then used to calculate the percentage of bacterial N_2O in the total N_2O flux based on $\delta^{15}N^{sp}$ and $\delta^{18}O$. To assess the uncertainty of this approach the ranges of possible endmembers isotopic signatures and reduction fractionation factors were taken into account. The variations of the bacterial N_2O contribution due to assumed ranges of input values reached up to 40%. But still it allowed to distinguish the dominant pathways for different water saturation levels and indicated that only when the micropores become partially dry, the more aerobic soil conditions allow a higher contribution of nitrification. The dryer conditions in soil macropores did not result in significant changes in bacterial denitrification contribution.

7.3.8 Examples of Field Applications

Toyoda et al. (2011)

N_2O emitted from agricultural soils planted with rice, wheat, soybean, and vegetables, and treated with synthetic (urea or ammonium) and organic (poultry manure) fertilisers was analysed. The observed isotopic values for $\Delta^{15}N$ and $\delta^{15}N^{sp}$ were compared with literature endmembers of nitrifying and denitrifying bacteria. A characteristic relationship between $\delta^{15}N^{bulk}$ and $\delta^{15}N^{sp}$ during N_2O reduction by denitrifying bacteria was used to quantify N_2O reduction. The relative fraction of N_2O derived from nitrification and the approximate progress of N_2O reduction were calculated by a Monte Carlo method. Different scenarios for pairs of mixing endmembers were tested (nitrification and denitrification; nitrification and nitrifier–denitrification; fungal denitrification and denitrification; fungal denitrification versus nitrifier–denitrification) but due to overlapping ranges for $\delta^{15}N^{sp}$ values it was chosen to consider only the mixing between bacterial nitrification and denitrification. It was found that the contribution from nitrification was relatively high (40%–70%) in soils amended with synthetic ammonium fertiliser, while denitrification was dominant (50%–90%) in the same soils amended with poultry manure.

Kato et al. (2013)

In this study, field samples from static flux chambers located on alpine meadow, shrub and wetlands were collected and analysed. Interpretation of results based on the relationship between $\delta^{15}N^{bulk}$ and $\delta^{15}N^{sp}$ (similar as presented in Fig. 7.5). A mixing of two endmembers was assumed: bacterial and fungal denitrification and subsequent N_2O reduction. Applying literature values for endmembers and fractionation during reduction the contribution of fungal denitrification (from 23 to 41%) and degree of reduced N_2O (from 83 to 93%) was calculated. The calculations were performed with Monte Carlo simulations and the assessed uncertainty of the results ranged from 17 to 23% for contribution of mixing endmembers and from 10 to 19% for degree of reduced N_2O.

Zou et al. (2014)

Soil gas was collected from a highly fertilised tea field at 10–50 cm depths using a silicone tube. $\delta^{15}N^{sp}$ $-\Delta^{15}N^{bulk}$ isotope maps (Fig. 7.5) were applied for interpretations. The precursor isotopic signatures were determined, and the endmember ranges have been recalculated according to the measured precursor values for bacterial and fungal denitrification, nitrification and nitrifier denitrification. For the N_2O reduction two scenarios were taken into account: assuming reduction after mixing and applying closed system dynamics and assuming reduction preceding mixing and applying open system dynamics. Predictions of $\delta^{15}N^{sp}$ values for different scenarios, reduction degrees and mixing ratios were presented and compared to the measured results. The study identified the bacterial denitrification as the dominant process and allowed for indication of the particular events when the contribution of nitrification or fungal denitrification increased pronouncedly.

Wolf et al. (2015).

N_2O isotopic analyses were done directly from the atmospheric surface layer (at 2.2 m height) applying a laser spectrometer connected to an automated N_2O preconcentration unit. The isotopic signatures of soil-emitted N_2O were derived using the Keeling plot approach, where δ values measured in the atmosphere surface layer are plotted versus the inverse of N_2O mole fractions (for a background on Keeling plot analysis see Pataki et al 2003). The intercept of the linear regression line is interpreted as the isotopic composition of soil-emitted N_2O. The interpretation of the results is based on isotope maps of $\delta^{15}N^{sp}$ vs. $\delta^{15}N^{bulk}$ and $\delta^{15}N^{sp}$ vs. $\delta^{18}O$. These isotope maps allowed concluding that N_2O was predominately formed by bacterial denitrification and that variations in isotopic composition may have been caused predominately by N_2O reduction to N_2. The study did not attempt to quantify the mixing ratios or N_2O reduction. The high-frequency isotope data was combined with a biogeochemical model Landscape DNDC with a stable isotope model for nutrient cycles (SIMONE) to identify and address weaknesses in N cycling of the model (Denk et al. 2019).

7.3.9 Outlook

N_2O isotopocule analyses provide a unique possibility to get insight into processes contributing to N_2O production as well as to assess the magnitude of N_2O reduction and thereby also N_2 flux. However, the information is still rather indicative than strongly quantitative. The calculation methods presented allow estimates of ranges of possible mixing ratios and reduction contribution rather than precise numbers. However, such information is also quite precious hence often not attainable by any other methods. ^{15}N tracing, which is often a more precise tool, is much more expensive and laborious, moreover applicable only on a very limited space and time scale, hence much more constrained in application potential.

A promising perspective is to apply the N_2O isotopocule analyses in combination with other methods, like with ^{15}N tracing (Deppe et al. 2017; Schorpp et al. 2016) (see also Sect. 7.5.) or with process modelling (Bai and Houlton 2009; Denk et al. 2017) which vastly increases the interpretation potential of such studies. Moreover, more quantitative estimates can be expected if the isotopocule approach is calibrated using controlled incubations where endmember values and isotopic fractionation factors are determined for specific conditions using independent estimates of contributing processes, e.g. by direct measurement of N_2 production or ^{15}N tracing (Lewicka-Szczebak et al. 2017; Wu et al. 2019). The most recent idea for interpretation of N_2O isotope data is the application of a N_2O isotopocule model which incorporates all three measured isotopic signatures ($\delta^{15}N^{bulk}$, $\delta^{15}N^{sp}$ and $\delta^{18}O$) (Lewicka-Szczebak and Well 2020).

7.4 Dual Isotope Method for Distinguishing Among Sources of N_2O

Various microbial processes can produce N_2O (for a simplified overview, see Fig. 7.8). These may occur simultaneously in distinct soil microhabitats or take

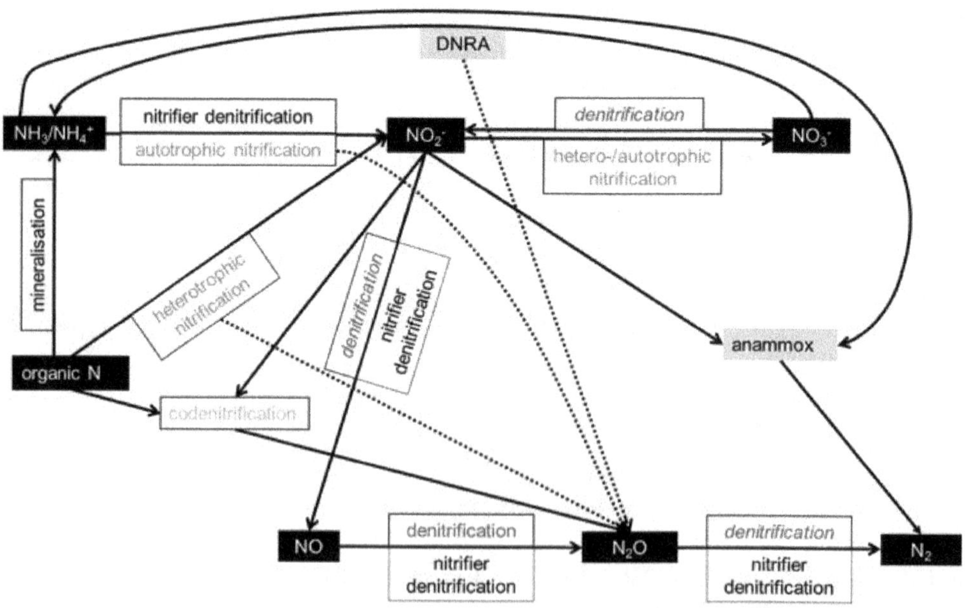

Fig. 7.8 Overview of N_2O producing processes carried out by nitrifiers and denitrifiers in soils. Shown is also the use of O_2 versus H_2O as source of oxygen. Please note that the distinction between nitrification-coupled denitrification and fertiliser denitrification is purely methodological, as the organisms and pathways are identical, but the source of NO_3^- differs. Within brackets, abbreviations of the pathways as used in the text are shown. DNRA: Dissimilatory NO_3^- reduction to NH_3 (Wrage-Mönnig et al. 2018)

Table 7.5 Treatments (TR) used for the dual isotope method

	H_2O	NO_3^-	NH_4^+
TR 1	^{18}O-enriched	Unlabelled	Unlabelled
TR 2	Unlabelled	^{18}O-enriched	Unlabelled
TR 3	Unlabelled	^{15}N-enriched	Unlabelled
TR 4	Unlabelled	Unlabelled	^{15}N-enriched

place temporally separated with fluctuating soil conditions. Often, only nitrification and denitrification are considered to be the main sources. However, the methods often applied cannot distinguish among all sources. For example, using ^{15}N tracing with labelled ammonium or NO_3^- does not allow a distinction among N_2O produced by nitrifiers either via hydroxylamine (termed here nitrification, N) or via nitrite reduction (nitrifier denitrification, ND), or by denitrifiers using NO_3^- produced by nitrifiers (nitrification-coupled denitrification, NcD). All N_2O produced by these sources is summarised as 'nitrification' by authors using this method. No method based on ^{15}N alone can so far separate the sources shown in Fig. 7.5. However, a distinction is important, as ND can under certain conditions produce all N_2O derived from NH_4^+ and has been reported to cause up to 90% of total N_2O emissions (Kool et al. 2010).

A distinction between ND and other sources of N_2O is possible if ^{18}O labelling is used in addition to ^{15}N labelling (Kool et al. 2011). As seen in Fig. 7.8, nitrifiers use distinct sources of O_2 for the oxidation of NH_3 and the subsequent oxidations of NH_2OH and NO_2^-. This is used in the dual isotope method, where ^{18}O-labelled H_2O is applied on top of ^{15}N tracers. However, care has to be taken to account for O-exchange, which can occur between H_2O and N oxides in all reactions of N oxides shown in Fig. 7.8. This is accomplished using the enrichment ratio retention (ERR) approach, where the enrichment ratio of ^{18}O:^{15}N of N_2O is compared to that of NO_3^- in incubations with either ^{15}N- or ^{18}O-labelled NO_3^-. Then, we can differentiate among N_2O produced by N, ND, NcD and fertiliser denitrification (FD).

The preparation of soil samples proceeds in a similar way as for other stable isotope methods. However, one has to keep in mind that water needs to be added as tracer, so that the water content during conditioning needs to be a bit less than intended for the incubation. So far, conditioning has been done at 40% water-filled pore space (of samples dried at 40°C), and incubation at 80%, but this is adaptable as long as the requirements for tracer additions are kept. Soil samples of 75–100 g have been incubated in glass jars of about 300 ml for 24–28 h. These ratios and times may be adapted, but care must be taken to ensure linear N_2O production over the incubation period, as well as stable concentrations of substrates (including O_2 and H_2O). Much longer incubations are difficult, as the ^{18}O enrichment of the soil H_2O might change locally due to evaporation and addition of H_2O. The occurrence of NO_3^- assimilation and DNRA needs to be checked (indicated by enrichment of NH_4^+ in incubations with ^{15}N-NO_3^-) and accounted for if necessary, as in other ^{15}N methods.

The treatments (TR) are established (Table 7.5) with proper replication (at least five times) after conditioning of the soil as needed. So far, added label has been

Table 7.6 Oxygen sources of N_2O in the different processes and pathways distinguished using the dual isotope method

	Nitrification (%)	Nitrifier denitrification (%)	Nitrification-coupled denitrification (%)	Fertiliser denitrification (%)
O_2	100	50	33	0
H_2O	0	50	67	0
NO_3^-	0	0	0	100

enriched at 1.0 atom% for ^{18}O and 40 atom% for ^{15}N, but higher enrichments may be desirable to reduce the amount of substrates added, especially concerning the N-substrates in natural systems. Usually, 100 mg N kg^{-1} soil has been applied, half each as NO_3^- and as NH_4^+. When applying less NO_3^-, one has to consider that if NO_3^- becomes limiting, the underlying assumption of the method that only NO_3^- and no NO_2^- is used in NcD and FD becomes invalid. This would result in an underestimation of NcD and an overestimation of ND. Additional incubations with ^{18}O-NO_2^- (which is currently not commercially available, though) or analysis of the ^{18}O enrichment of the NO_2^- pool may help to overcome this.

Immediately after establishing the treatments, the jars are closed, and samples are taken for N_2O content and isotopic signature as explained in Chap. 3 and above. At the end of the incubation, soil samples are taken for analysis of mineral N and its isotopic signature (the latter only in TR 3 and 4), as well as the soil moisture content to verify that this did not change during incubation. Consider that the label added with ^{18}O-H_2O is diluted when mixed with moist soil.

For quantifying the O-exchange between N oxides and ^{18}O-H_2O, the ERR approach is used. It is assumed that the O-exchange is similar for denitrifiers and nitrifiers. This need not be true, as O-exchange by nitrifiers has often been found to be less than in denitrifiers. Such a discrepancy would lead to an underestimation of the N_2O produced by ND and NcD. No O-exchange is assumed to affect N_2O derived from N. The ERR is calculated in Eqs. 7.12 to 7.17 as follows:

$$ERR[\%] = 100\frac{^{18}O(N_2O_{TR2})}{^{15}N(N_2O_{TR3})} / \frac{^{18}O\left(NO^-_{3_{TR2}}\right)}{^{15}N\left(NO^-_{3_{TR3}}\right)} \tag{7.12}$$

where $^{18}O(Y_{TRx})$ and $^{15}N(Y_{TRx})$ denote the ^{18}O or ^{15}N enrichment, respectively, of substance Y from treatment x. Without O-exchange, ERR is 100%. O-exchange (O_{ex}) is then quantified as

$$O_{ex} = 100 - ERR \tag{7.13}$$

Next, the percentage of N_2O derived from NO_3^- $\left(N_2O_{NO_3^-}\right)$ and NH_4^+ $\left(N_2O_{NH_4^+}\right)$ is calculated. $N_2O_{NO_3^-}$ is defined as N_2O from FD, whereas N_2O_{NH4+} comprises the other three sources.

$$N_2O_{NO_3^-}[\%] = FD = 100\frac{^{15}N(N_2O_{TR3})}{^{15}N(N_2O_{TR3}) +^{15} N(N_2O_{TR4})} \tag{7.14}$$

$$N_2O_{NH_4^+}[\%] = N + ND + NcD = 100\frac{^{15}N(N_2O_{TR4})}{^{15}N(N_2O_{TR3}) +^{15} N(N_2O_{TR4})} \tag{7.15}$$

If the ^{15}N enrichment of N_2O in TR4 does not exceed the ^{15}N enrichment of NO_3^- in the same treatment, all $N_2O_{NH_4^+}$ might have been derived from NcD (maximum contribution of NcD, $NcD_{max} = N_2O_{NH_4^+}$, implying that ND and N were equal to zero). If not, NcD_{max} is calculated as follows:

$$NcD_{max}[\%] = N_2O_{NH_4^+} \times \frac{^{15}N\left(NO^-_{3_{TR4}}\right)}{^{15}N\left(NO^-_{3_{TR4}}\right) + {}^{15}N\left(NH^+_{4_{TR4}}\right)}, \text{ if } ^{15}N(N_2O_{TR4}) >^{15} N\left(NO_3{}^-{}_{TR4}\right) \tag{7.16}$$

To distinguish among the other N_2O producing pathways considered, the actual O incorporation from H_2O into N_2O (AOI) is determined from TR1:

$$AOI[\%] = 100\frac{^{18}O(N_2O_{TR1})}{^{18}O(H_2O_{TR1})} \tag{7.17}$$

This AOI may come from O_{ex} quantified as shown above and the reaction stoichiometry of the different pathways as shown in Table 7.6.

A large AOI may thus be caused either by a larger contribution of pathways with a larger incorporation of ^{18}O-H_2O (ND or NcD) or by a larger O_{ex}. For further evaluation, O_{ex} is maximised, i.e. assumed to take place in the $NH_4{}^+$-derived pathways to the same extent as in FD (Scenario A) or minimised, i.e. assumed to be absent in nitrification pathways (Scenario B). Furthermore, in Scenario A, the contributions of N and NcD are maximised, while in Scenario B, ND is maximised. Under both scenarios, a theoretical O incorporation (TOI) is calculated and compared to the AOI.

Under Scenario A, the TOI (TOI_A) is calculated (Eq. 7.18) as

$$TOI_A = N_2O_{NO_3^-} \times O_{ex} + NcD_{max}\left(2/3 + 2/3O_{ex} - 1/3(O_{ex})^2\right) \tag{7.18}$$

This calculation comprises O_{ex} occurring during D ($N_2O_{NO_3^-} \times O_{ex}$) as well as from NcD stoichiometry (2/3 O from H_2O) and O_{ex} occurring during N to $NO_3{}^-$ and NcD (for further explanation, see Appendix 1 in Kool et al. 2009). If $TOI_A \geq$ AOI, no contribution by ND is necessary to explain the AOI. The minimal contribution of ND, ND_{min}, is then set to zero, and the maximum contribution of N, $N_{max} = N_2O_{NH_4^+} - NcD_{max}$. If not, ND must have contributed to N_2O production ($ND_{min} > 0$), which implies at the same time a maximum contribution of N, N_{max} $\left(N_{max} < N_2O_{NH_4^+} - NcD_{max}\right)$. In this case, we can calculate the contribution of ND_{min} (Eq. 7.19) as follows:

$$ND_{\min} = \frac{AOI - (N_2O_{NO_3^-} \times O_{ex})}{2/3 + 2/3O_{ex} - 1/3(O_{ex})^2} - NcD_{\max} \quad (7.19)$$

N_{max} is then equal to $N_2O_{NH_4^+}$−NcD_{max}−ND_{min}.

Under Scenario B, ND is maximised by assigning $N_2O_{NH_4^+}$ to ND and assuming no O_{ex} during this pathway, and in Eq. 7.20 TOI_B is calculated as

$$TOI_B = N_2O_{NO_3^-} \times O_{ex} + N_2O_{NH_4^+} \times 0.5 \quad (7.20)$$

If $TOI_B > AOI$, not all $N_2O_{NH_4^+}$ can have been derived from ND $\left(ND_{max} < N_2O_{NH_4^+}\right)$. In that case, some N_2O must have originated from N (i.e. the minimum contribution of N, $N_{min} > 0$), which will lower the TOI. However, if $TOI_B \leq AOI$, all $N_2O_{NH_4^+}$ may have come from ND $\left(ND_{max} = N_2O_{NH_4^+}\right)$ and the contribution of N_{min} was zero. A larger AOI ($TOI_B < AOI$) may either come from a contribution of NcD or O_{ex} during ND, which was assumed not to take place under this scenario. As both may equally well explain the numbers, NcD_{min} is set to zero in this case and O_{ex} assumed to have occurred during ND.

If N_{min} was found to be larger than zero, we can calculate ND_{max} from this scenario as follows (Eq. 7.21):

$$ND_{\max} = \frac{AOI - N_2O_{NO_3^-} \times O_{ex}}{0.5} \quad (7.21)$$

In that case, $N_{min} = N_2O_{NH_4^+}$−ND_{max}.

Thus, in the dual isotope method, the contribution of NcD is always maximised, and consequently minimum and maximum contributions of N and ND are estimated based on Scenario A and B. Applying this method allows insight into these three potential sources of N_2O plus fertiliser denitrification. However, in soils, further microbial processes can lead to N_2O production. In the following, we will briefly discuss potential effects of nitrification by heterotrophs and archaea, fungal denitrification, as well as DNRA and co-denitrification.

If N_2O of nitrification by heterotrophs and archaea is produced by the same sources and similar pathways as in autotrophic nitrifiers, this should not interfere with the calculations. The contribution of N would then comprise that of other nitrifiers. However, archaea have also been suggested to produce N_2O in a pathway similar to ND (Jung et al. 2014). If so, this would be included in the contribution of ND. However, the pathway of N_2O production by archaea is not clear yet and needs further study (Stieglmeier et al. 2014), the outcome of which will also affect the calculations presented here. In soils where fungal denitrification occurs, this is counted as FD using the dual isotope method, if the fungi use added NO_3^- as a source in a reaction similar to denitrification. Fungal denitrification may be quantified using the isotopomer method (Sect. 7.3), calling for a combination with the dual isotope method.

The occurrence of DNRA should be tested for as explained above. Should it lead to N_2O production (Stevens et al. 1998), this would lead to an overestimation of N_2O from FD. As DNRA may be important in soils (Rütting et al. 2011), this pathway should always be considered by checking for enriched NH_4^+ in incubations with added $^{15}N\text{-}NO_3^-$.

Co-denitrifiers combine NO_3^- or NO_2^- with other nitrogenous compounds to produce N_2O or N_2. The occurrence of such a process could be quantified using the triple labelling ^{15}N tracing model (Müller et al. 2014) in combination with non-random ^{15}N distribution (Laughlin and Stevens 2002). Incorporating this would be an improvement of the dual isotope method, as co-denitrification could interfere with the source estimations presented above.

The dual isotope method could be further developed by incorporating better rates of O_{ex} for the pathways starting from NH_4^+. Despite potential for improvements, however, this method allows an estimation of the contributions of N, ND, NcD and FD to N_2O production and should be applied to a range of soils to further our understanding of these sources of N_2O and potential mitigation strategies.

7.5 Quantification of Gross N Transformation Rates and Process Specific N_2O Pathways via ^{15}N Tracing

7.5.1 Background

The N cycle is a conceptual model that illustrates where and in which form N is present in the environment and how N is transformed and exchanged between organic, mineral and gaseous N forms. Since the N cycle is a dynamic system not only the sizes of the different N pools, e.g. NH_4^+, NO_3^- or organic N but also the rates between the pools provide an understanding of the dynamic nature of this important elemental cycle in soils and aquatic systems (Ryabenko 2013). The most common and easiest approach to understand the dynamic nature of the N pools is the determination of net process rates, such as net mineralisation rates by calculating the difference in the size of the mineral N pool between two time points. If this rate is positive, we refer to a net mineralisation, if it is negative then we call it net immobilisation. Thus, a net rate always refers to the difference between the production and consumption of the N pools in question. It can easily be shown that different pairs of production and consumption rates will lead to exactly the same net result. Thus, the analyses of net rates do not provide a measure of the individual rates that are contributing to the observed net rate. The individual rates associated with the N pool in question are called gross transformation rates. However, the quantification of these individual rates is not trivial because they cannot be measured directly. The most commonly used method to quantify the gross rates is the isotopic dilution technique (Stark 2000). The principle of this technique relies on the ^{15}N labelling of a certain N pool so that

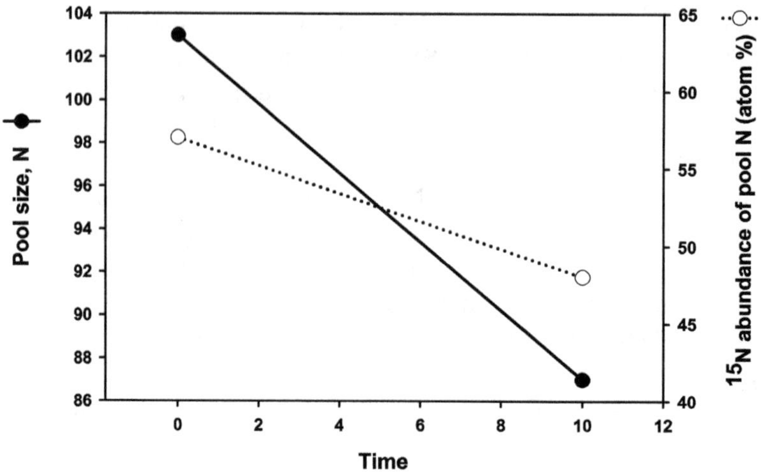

Fig. 7.9 Illustration of the dilution technique by considering pool size and ^{15}N abundance at two time points according to Kirkham and Bartholomew (1954)

the gross rate entering this pool can be quantified by taking into account the change in pool size and ^{15}N enrichment of at least two times after label addition (Fig. 7.9).

The example in Fig. 7.9 shows that the pool size is decreasing which means that a net immobilisation of N occurred. However, the decline in ^{15}N abundance of the pool N during the same period also shows that N at natural abundance or low ^{15}N abundance must have entered the pool N. Thus, via visual inspection of the data we can say that N must have entered but also left the pool and that the rate leaving the pool must have been faster than the rate entering the pool. To quantify the individual rates requires a numerical analysis via a suitable N cycle model. Based on a simple two-pool N model, Kirkham and Bartholomew (1954) were the first to derive analytical equations that allowed the calculations of the two rates between two-time points, i.e. the gross mineralisation and immobilisation rates. The underlying assumptions are (i) ^{15}N is homogeneously labelled and no preferential usage of either ^{15}N or ^{14}N occurs in the soil, (ii) immobilised N will not re-mineralise and (iii) N transformation rates follow zero-order kinetics (constant rates). The conceptual model of the Kirkham and Bartholomew approach and the equations derived for their model are illustrated in Fig. 7.10.

Since Kirkham and Bartholomew's pioneering work in the 1950s, analyses techniques have been developed which are based on more realistic conceptual N models. These include the division of the mineral N pool into NH_4^+ and NO_3^- pools with separate immobilisation rates, the consideration of more than one organic N pool and additional N loss rates such as ammonia volatilisation and denitrification (Myrold and Tiedje 1986). The dilution technique works well in simple systems where the inflow into a pool occurs via a single gross N transformation rate. However, in reality, often more than one pathway contributes to the buildup of a pool size. This can be illustrated by the NO_3^- pool in soil. Production of NO_3^- can occur via oxidation

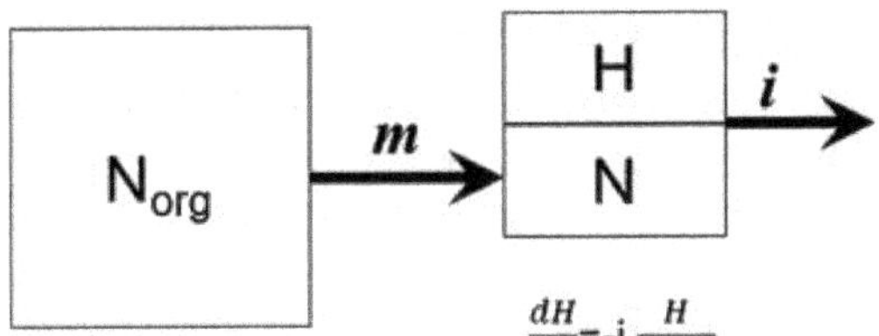

$$\frac{dN_{org}}{dt} = -m \qquad \frac{dH}{dt} = -i\,\frac{H}{H+N} \qquad \frac{dN}{dt} = m - i\,\frac{N}{H+N}$$

$$m = \frac{M_0 - M}{t}\,\frac{\ln H_0 M / H M_0}{\ln M_0 / M}$$

$$i = \frac{M_0 - M}{t}\,\frac{\ln H_0 / H}{\ln M_0 / M}$$

Fig. 7.10 The conceptual model, the differential equations of the various pools and the closed-form analytical solutions for the individual gross rates (m and i) according to Kirkham and Bartholomew (1954). Note, N_{org} (assumed to contain only ^{14}N) depicts the organic N pool which mineralises into mineral N (M) which consist of H (^{15}N) and N (^{14}N), M = H +N. The subscript 0 refers to the concentrations of the pools at time zero

of NH_4^+ to NO_3^- (usually termed autotrophic nitrification), and via oxidation of organic N to NO_3^- (usually termed heterotrophic nitrification) (Fig. 7.11).

Following the principles of the dilution technique, the total gross rate of NO_3^- production can be quantified by labelling the NO_3^- pool and following the concentrations and ^{15}N dilution of this pool over time. This total NO_3^- production rate includes both, autotrophic (oxidation of NH_4^+) and heterotrophic nitrification (oxidation of N_{org}). To separate the two processes, in addition to the $^{15}NO_3^-$ label also the NH_4^+ pool should be labelled in a separate ^{15}N labelling treatment. To keep the conditions in the two ^{15}N labels the same, it is important to also apply NH_4^+ in the soil that has received the $^{15}NO_3^-$ label while NO_3^- should be applied in the $^{15}NH_4^+$ treatment. Now, the ^{15}N enrichment in the NO_3^- when only NH_4^+ has been labelled

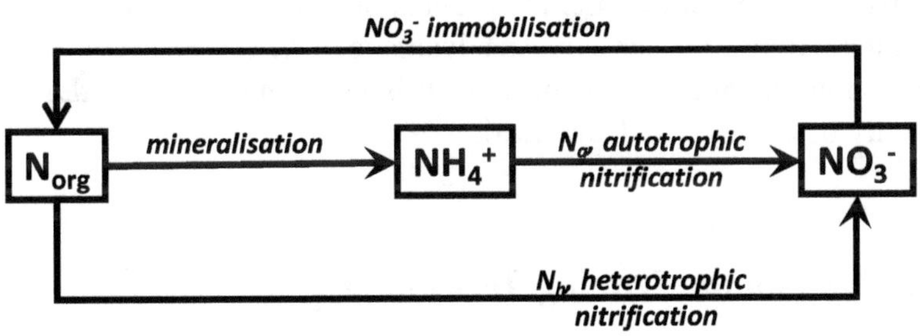

Fig. 7.11 Conceptual model for nitrification

will provide a measure of autotrophic nitrification while heterotrophic nitrification can be calculated by difference: $N_h = N_{tot} - N_a$ where N_{tot}, N_a and N_h refer to total, autotrophic and heterotrophic nitrification, respectively. In practice, to quantify N_{tot} the dilution of the ^{15}N labelled NO_3^- pool (Fig. 7.10) can be used while N_h and N_a can only be estimated via a simulation model that takes into account both nitrification rates (Barraclough and Puri 1995). A parameter optimisation technique can also be used to estimate N_a or N_h (Myrold and Tiedje 1986). Thus, in modern ^{15}N tracing applications dilution-enrichment principles will be taken into account which we refer to as tracing.

In models with several simultaneous N transformations, it is impossible to derive analytical solutions. Therefore, the development of ^{15}N tracing models which use numerical solutions has become the state-of-the-art approach (Mary et al. 1998). These models rely on a set of differential equations for example of a simple system that describes the N cycle. Transformation rates between the various pools can be constant (zero-order kinetics) or are dependent on the pool size where the rate is originating from (first-order kinetics) or follow enzyme kinetics (i.e. Michaelis–Menten kinetics). While zero and first-order kinetics are described by one parameter, rates calculated via Michaelis–Menten kinetics are dependent on two parameters, i.e. the maximum velocity of the reaction rate and the half-saturation constant (Müller 2000). The determination of the parameters in such equation systems rely on parameter optimisation tools. A whole range of parameter optimisation tools are available and different algorithms have been used in ^{15}N tracing models (Mary et al. 1998; Myrold and Tiedje 1986). More recently, parameter optimisation tools based on Bayesian probability have become more common because they allow the simultaneous optimisation of a large number of parameters (for more details see Müller et al. 2007). It should be noted that the sole purpose of tracing models is to quantify gross transformation rates and are therefore data analysis tools and should not be confused with simulation models.

In the following sections, current ^{15}N tracing techniques are illustrated. This includes the description of experimental requirements to obtain suitable data sets and the subsequent model analysis. A number of ^{15}N tracing models have been developed (e.g. FLUAZ, Mary et al. 1998), and here, the data analysis will be illustrated by the *Ntrace* model. This model is based on the tool presented by Müller et al. (2007) and has since been developed further to analyse data sets from a range of differently complex setups, including dynamics of nitrite, gaseous N emissions, soil–plant interactions, biochar, etc. An advantage of *Ntrace* is its flexibility to adapt to various conditions and models because it is programmed in MatLab with code that can easily be changed and amended.

7.5.2 Stable Isotope Tracing Technique

A stable isotope tracing study consists of two parts, an experimental study where one or more pools are isotopically labelled and a data analysis tool (e.g. *Ntrace*) to quantify individual gross transformation rates. The technique can be regarded as a calculation procedure to quantify gross rates which cannot be quantified via any other means. Thus, both the tracing experiment and the numerical tool are building an analysis unit and it is important that the experimental approach is taking into account the requirements of the numerical analysis and vice versa. What is also important is that the quality of the final results critically depends on the data quality and therefore on the careful execution of the experimental part of the tracing study. For instance, data with high uncertainties may also result in gross N rates that are associated with large errors.

7.5.3 Setup of Tracing Experiments

To be able to analyse experimental data with the *Ntrace* model, the experiment needs to be set up in a certain way. Based on the research questions both field and laboratory experiments can be carried out. The research question usually requires the setup of several treatments (e.g. effect of various soil amendments). To quantify the individual gross N transformation rates in each treatment usually a set of at least two ^{15}N labels should be employed per treatment (i.e. ^{15}N-labelled NH_4^+ and ^{15}N-labelled NO_3^-, typically applied as NH_4NO_3 to ensure the application of equal quantities of each N species). However, often multiple labels are used (e.g. very common is a triple labelling approach with NH_4NO_3 where either NH_4^+, NO_3^- or both moieties are ^{15}N labelled).

Ideally the ^{15}N label should be applied without enhancing the concentration because this will also have an impact on the N transformations. Thus, in ecosystems that are not used to receive large N concentrations often a high ^{15}N enrichment (e.g. 99 atom% ^{15}N) is applied at a very low application rate. However, in agricultural soils which receive N in the form of fertiliser, this is less of a problem. The advantage of applying a reasonable, but not unrealistically high, N concentration is that it can more homogenously be applied to the soil. In most cases, a ^{15}N enrichment of a few percent (e.g. 10 atom% ^{15}N) is sufficient to determine gross rates. However, in situations where, for instance, the nitrite or gaseous N species such as N_2O are analysed, the labelled N pool (e.g. NO_3^-) should ideally be enriched by approximately 50 atom% ^{15}N which allows most precise analysis based on the expected 29/28 iron current (Stevens et al. 1993). The ^{15}N solutions are made up according to standard calculations which are, for instance, summarised by Cabrera and Kissel (1989). To homogeneously label the soil a variety of application techniques are described in the literature ranging from application via side port needles in different depth, multiple needle applicators and automated techniques (Buchen et al. 2016) (Table 7.4). In field tracing studies often an application via a watering can is preferred, simply, because under field conditions when large plots of several m^2 have to be treated, it is critical that the solutions are applied within a short time window to ensure the same starting conditions (Plate 7.1). This is particularly important if dynamically changing N species such as N_2O should be compared among treatments (Moser et al. 2018).

Plate 7.1 Application of ^{15}N solution ($^{15}NH_4NO_3$, $NH_4^{15}NO_3$) on plots in the field with a small watering device (**a**), and soil incubation in a climate chamber in suitable incubation jars (**b**)

However, the application rate should be slow enough to avoid by-pass in cracks and fissures down the soil profile because this would cause uneven distribution of the ^{15}N.

The time of labelling should be carefully noted because the difference between ^{15}N application and soil analysis during the experiment provides the time after N supply which is required for the model analysis. If both, soil extractions and gaseous measurements are planned then in the field an area for the gas sampling and an adjacent soil sampling area should be setup (in Plate 7.1, gas measurement are in the forefront, and the area for soil sampling is at the top). In laboratory incubations usually one set of jars is reserved for gaseous measurements (which will be extracted at the end of the experiment) while for each analysis day, separate sets of jars are prepared for destructive sampling. The question arises for how long we need to carry out a typical incubation study. Since the application of N may cause a stimulation of microbial activity resulting in faster gross N rates shortly after N application, the duration of a typical tracing study should be continued until after this initial stimulation has subsided. A typical duration of such a study is approximately 14 days. To characterise the non-linear dynamics of the gross N rates over time it is necessary to determine the N pool sizes and their ^{15}N enrichment at least 5 times throughout that period. Gas analysis should be carried out more frequently but at least at the times when soils are extracted.

Soil incubations have typically been carried out under controlled conditions at a pre-defined moisture content (set to a certain water filled pore space (WFPS) or water holding capacity (WHC)) and temperatures in a climate chamber (Plate 7.1).

Case study

To investigate the effect of a nitrification inhibitor in two soils on gross N transformations the following setup is realistic (using a triple ^{15}N labelling approach, numbers in brackets refer to the number of entities):

Soils (2) × Inhibitors (2) × ^{15}N labels (3) × Replicates (3) × Time of soil extraction (5) = 180 jars.

Thus, a total of 180 jars (i.e. 36 jars per extraction day) need to be prepared. The label needs to be applied with minimal disturbance while providing an equal distribution in the soil. This can be done using a long needle with side ports. If 150 g of dry soil equivalent should be used per jar, then approximately 14 kg of soil is required from each soil.

The extraction times should be timed in such a way that the first extraction happens as soon as possible after ^{15}N labelling (typically after 2 h), then on day 1, 3, 7 and 15. Note, soils can react quite differently, therefore, the times and duration of the experiment should be adjusted accordingly.

7.5.4 Analyses of Experimental Data

7.5.4.1 Soil Extraction

If nitrite concentrations should be investigated it is recommendable to carry out the blending procedure of Stevens and Laughlin (1995). They discovered that nitrite is chemically reduced to N_2 in the KCl extract at pH below 5.5. They recommended a soil extraction at pH 7 and fast soil extraction. The blending procedure of Stevens and Laughlin (1995) is typically carried out at a soil: solution ratio of 1:1 in a blender for 90 s (Plate 7.2).

Immediately after the blending, the soil suspension needs to be centrifuged at 2000 × g for 5 min, and the supernatant filtered sequentially through a GF/D and GF/F (Plate 7.2).

The extracts have to be analysed for NO_3^- and NH_4^+ and possibly for NO_2. Based on the concentration a certain μmol of N of each N species will then be converted to ^{15}N-N_2O or via a diffusion approach.

7.5.4.2 Chemical Conversion of Mineral N to ^{15}N-N_2O

A precise method to determine the ^{15}N content of ammonium, NO_3^- and nitrite is via a method that converts the N species to nitrous oxide (N_2O). The reduction of NO_3^- to N_2O is described by Stevens and Laughlin (1994). For this, sulphamic acid (2.5 ml of 0.2 M solution) is added to 50 ml soil extract and shaken by hand for 5 s to ensure conversion of NO_2^- to N_2. Then 5 ml of 1 M sodium acetate- 1 M acetic

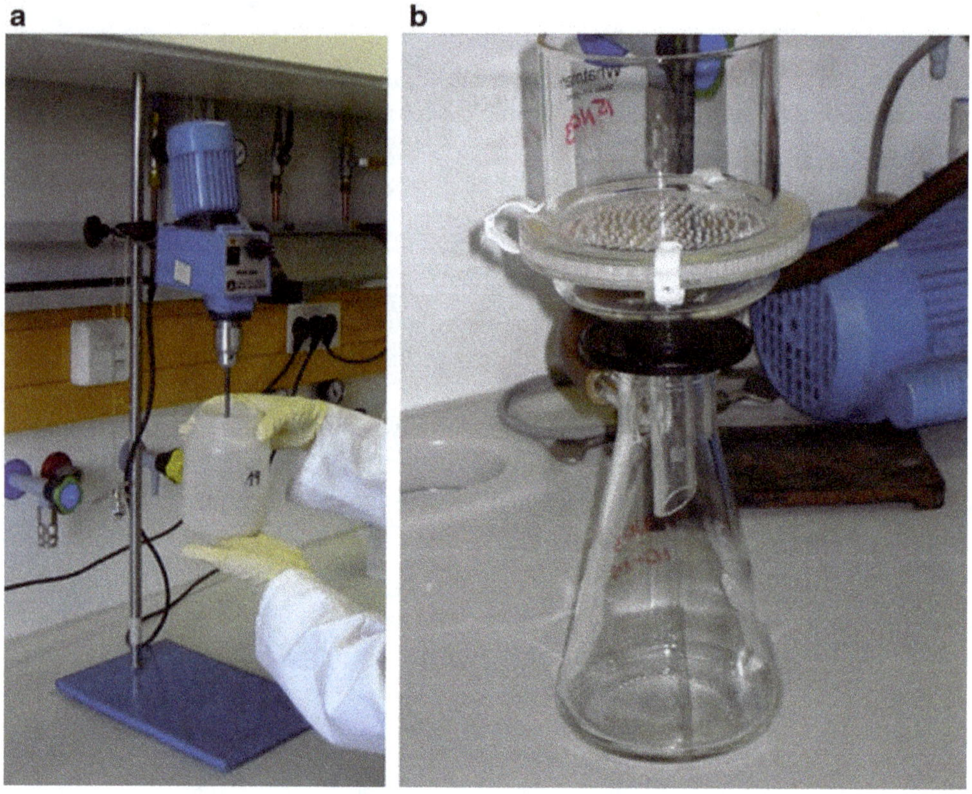

Plate 7.2 Extraction procedure for quick soil extraction (**a**) and glass filter unit for glass fibre filter papers (**b**)

acid buffer has to be added to increase the pH to 4.7. Then a CD-Cu reductor has to be placed in the bottle (Plate 7.3).

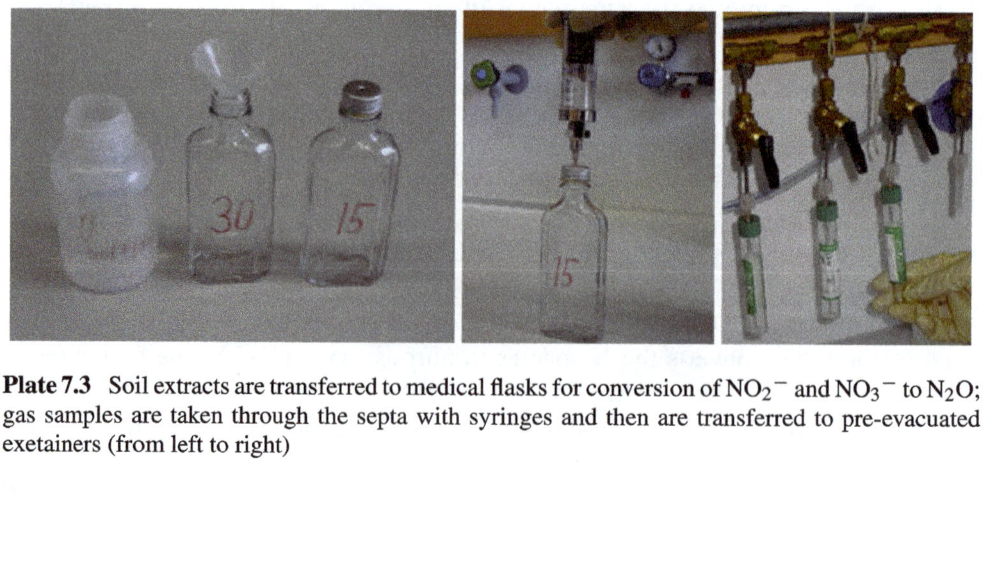

Plate 7.3 Soil extracts are transferred to medical flasks for conversion of NO_2^- and NO_3^- to N_2O; gas samples are taken through the septa with syringes and then are transferred to pre-evacuated exetainers (from left to right)

The flask, capped, has to be laid flat in an orbital incubator at 20°C and shaken at 120 rpm with an orbit diameter of 50 mm for 2 h. A gas sample of the headspace is analysed with an IRMS for the ^{15}N content of the N_2O. The ^{15}N content of the NO_3^- is considered to be the same as that of the N_2O.

The production of N_2O from NH_4^+ is described in Laughlin et al. (1997). Firstly, the ammonium must be diffused into $(NH_4)_2SO_4$. For this, 50 ml of the soil extract has to be pipetted in the diffusion unit (Plate 7.4). Above this liquid, a small flask containing 3 ml of H_2SO_4 has to be placed. Before the diffusion jar is closed, 0.2 g of heavy MgO must be added. The MgO has to be brought into suspension by gentle swirling for 30 s. After this, the diffusion jar has to be left for 4 days. After diffusion of the NH_3, the $(NH_4)_2SO_4$-H_2SO_4 has to be poured into a 12 ml glass exetainer and evaporated to dryness in a 150 °C oven, before cooling it in a desiccator and sealing it with a septum and cap. Then the vial has to be evacuated and filled with He to atmospheric pressure. One ml of NaOBr, with the molarity of NaOH adjusted to 10 M has to be injected through the septum. The vial has to be tilted and the solution gently swirled to ensure that the NaOBr reacts with as much of the $(NH_4)_2SO_4$ as possible. The concentration and ^{15}N content of the N_2O in each vial has to be determined by an IRMS system.

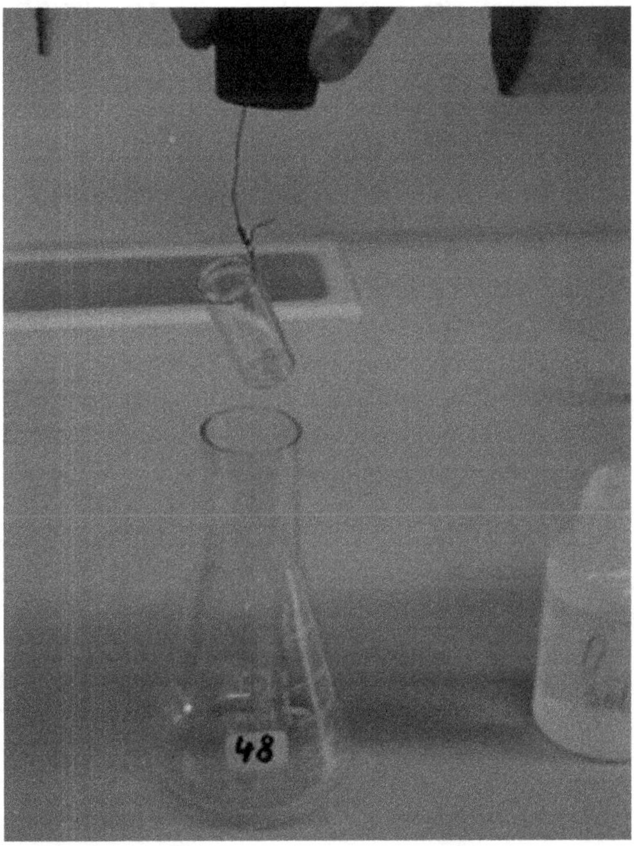

Plate 7.4 Glass equipment used for the conversion of NH_4^+ to NH_3 which is trapped in the acid contained in the small hanging flask

For the *Ntrace*$_{Nitrite}$ model, data on NO_2^- concentration and ^{15}N content are also necessary. The NO_2^- concentrations can be determined by a manual photometer method.

The ^{15}N content of the NO_2^- extracts can also be determined by a method based on conversion to N_2O as described by Stevens and Laughlin (1994). For this 50 ml of the soil extract has to be pipetted into a bottle. One ml of 1 M HCl and 0.5 ml of 0.04 M NH_2OH has to be added to the bottle. The bottle should then be capped and laid flat in an orbital incubator and shaken at 120 rpm with an orbit of 50 mm for 16 h. A 12 ml sample of the headspace has to be transferred to an evacuated septum-capped glass vial, and the ^{15}N content of the N_2O in each vial can be determined by IRMS. The atom% excess in ^{15}N in NO_2^- is calculated as two times the ^{15}N atom% excess in N_2O minus the ^{15}N atom% excess in NH_2OH.

The specific steps of the conversion method to N_2O are summarised below.

1. NH_4^+-N: NH_4^+-N is first oxidised to NO_2^--N by BrO^- in a vacuum with N_2O being the by-product (Eq. 7.22). The production of N_2O can be catalysed by Cu^+ at the appropriate pH (Laughlin et al. 1997).

$$NH_4^+ + NaBrO \rightarrow NaBr + H_2O + N_2 \uparrow + N_2O \tag{7.22}$$

2. NO_3^-–N: NO_2^-–N must be removed by NH_2SO_3H before NO_3^--N is reduced. NO_3^-–N is reduced to NO_2^-–N and NH_2OH by copper-plating cadmium grains at a pH of 4.7. Then NO_2^-–N reacts with NH_2OH to produce N_2O, and the production of N_2O is positively correlated to the production of NO_3^--N (Eq. 7.23). The ammonium and N from other sources has no effect on the determination of NO_3^--N; (Stevens and Laughlin 1994).

$$NO_3^- \rightarrow NO_2^- + NH_2OH \rightarrow N_2O + 2H_2O \tag{7.23}$$

3. NO_2^--N: NO_2^--N reacts with NH_2OH to produce N_2O (Eq. 7.24) and the reaction is pH-dependent. When $pH < 4$, the reaction rate increases rapidly, and the reaction time should be at least 16 h. Because N_2O is formatted via an asymmetric intermediate (N-nitroso-hydroxyl-amine) under acidic condition, the reaction requires at least 10 μmol of NH_2OH (Stevens and Laughlin 1994).

$$HNO_2 + NH_2OH \rightarrow N_2O + 2H_2O \tag{7.24}$$

The amount of N_2O produced is about half of the theoretical yield. According to the isotopic distribution, the two N atoms in N_2O are formed from NO_2^--N and NH_2OH, respectively. Hence, the atom% in NO_2^--^{15}N needs to be calculated with Eq. 7.25 (assume the atom% ^{15}N in NH_2OH is 0.365 atom%) (Laughlin et al. 1997):

$$^{15}N\ atom\%\left(NO_2^- - N\right) = 2 \times {}^{15}N\ atom\%(N_2O) - 0.365\ atom\% \tag{7.25}$$

Apparatus

PT-IRMS (purge and trap system coupled to isotope ratio mass spectrometry)

Vacuum pump

50 ml reaction vials

Glass vials with Al caps and septums

Reagents

NH_4^+-N to N_2O method:

MgO: combusted at 450 °C for 4 h

0.01 M H_2SO_4 with 0.5 mM $CuSO_45H_2O$

Basic NaBrO (10 M NaOH)

NO_3^--N to N_2O:

0.2 M NH_2SO_3H

1 M CH_3COOH- CH_3COONa (pH = 4.7)

Copper-plated cadmium granules

NO_2^--N to N_2O:

1 M HCl

0.04 M NH_2OH-HCl

Procedures

1. NH_4^+-N:

   (a) Pipet 15–20 ml (about 20 μg N) soil extract into a semi-micro steam distiller. Carry out steam distillation immediately after adding 0.2 g MgO. The NH_3 is absorbed by 5 ml 0.01 M H_2SO_4. After 5 min of steam distillation, the distillate is concentrated to 2–3 ml. Transfer part of the concentrate into a 50 ml reaction vial, and evaporate to dryness at 90 °C;
   (b) Evacuate the vials and fill them with He. Then inject 1 ml NaBrO together with 10 M NaOH through the septum, and swirl the solution around the sides of the vial to ensure that NaOBr reacts with as much of the $(NH_4)_2SO_4$ as possible;
   (c) Transfer a known amount of sample to an evacuated septum-capped glass vial. The ^{15}N content of the N_2O is then determined by IRMS.

2. NO_3^--N:

   (a) Pipet 20–25 ml (about 20 μg N) of the soil extract into a flask. Add 2.5 ml 0.2 M NH_2SO_3H, and shake the flask for 5 min to ensure conversion of NO_2^--N to N_2;

(b) Place 50 mg copper-plated cadmium granules together with 5 ml of 1 M CH_3COOH- CH_3COONa into a 50 ml reaction bottle. Keep the bottle capped tightly, evacuate and then fill with pure He;
(c) Inject 15–20 ml of the nitrite-free soil extract (about10 μg N) into a reaction bottle. Place the reaction bottle on a shaker at 120 rpm for 2 h;
(d) Transfer a known amount of sample to an evacuated septum-capped glass vial. The ^{15}N content of the N_2O is then determined with mass spectrometry.

3. NO_2^--N:

(a) Pipet 10–15 ml (about 0.5–1.0 μg NO_2^--N) of the soil extract into a 50 ml reaction bottle. Keep the bottle capped tightly, evacuate and then fill with pure He;
(b) Inject 1 ml 1 M HCl and 0.5 ml 0.04 M NH_2OH-HCl into the bottle;
(c) Place the bottle on a shaker running at 120 rpm for 16 h;
(d) Transfer a known amount of sample to an evacuated septum-capped glass vial. The ^{15}N content of the N_2O is then determined with mass spectrometry. Finally, calculate the NO_2^--^{15}N using Eq. 7.25.

7.5.4.3 Inorganic Nitrogen Isotopic Analysis in Soil Extracts via the Diffusion Method

An alternative method for the determination of ^{15}N in NO_3^- and NH_4^+ is the diffusion method. The diffusion method is easier to apply and has the advantage that only a solid analysis on an IRMS is required rather than a gas measurement. These Mass spectrometers are more readily available. However, it should also be pointed out that chemical conversion method described above is quicker and is free from contamination by atmospheric N. It has very low detection limits, which are 20 μg N for NH_4^+-N, 5 μg N for NO_3^--N and 0.5 μg N for NO_2^--N.

Principle

During diffusion, ammonium in the soil samples is converted to ammonia by the use of MgO (Eq. 7.26). Then the ammonia is absorbed by using a filter paper containing a weakly acidic absorbent liquid during the volatilisation process. For determination of NO_3^--N, titrate some alkaline reagent to remove NH_4^+-N in the sample then add some Devarda's alloy to reduce the NO_3^--N into NH_4^+-N.

$$2NH_4^+ + MgO \rightarrow Mg^+ + 2NH_3 \uparrow + H_2O \tag{7.26}$$

Apparatus need are:

EA-IRMS

Shaker

250 ml airtight containers

Perforated silicon films

Perforated filter papers

paper clips

Glass beads

Reagents:

MgO: Combusted at 450 °C for 4 h

Devarda's alloy: crushed to allow passage through a 300-mesh sieve

1 M $H_2C_2O_4$

Procedures

1. Put clips on the perforated silicon film and place it in the cap of a flask. Then place two pieces of 6 mm-diameter filter paper (Whatman #41 ashless filter paper) which are perforated by a needle on the clip;
2. For soil extracts > 2 mg l^{-1} in inorganic N concentration, only 20 ml of soil extract is needed. Put the 20 ml of soil extract into the container and add 3 glass beads before adding the MgO and Devarda's alloy. Onto each piece of filter paper pipette 10 μl of 1 M $H_2C_2O_4$ solution;
3. Add 0.3 g MgO and close the container quickly. Swirl the container carefully for 15 s. Then incubate the sample for 24 h at 25 °C in a shaker running at 140 rpm to complete the diffusion and recovery of NH_4^+-N;
4. To determine the ^{15}N enrichment of NO_3^--N from the same sample, replace the used filter paper with two new pieces also spiked with $H_2C_2O_4$. Incubate the sample in a shaker running at 140 rpm for 48 h to remove the remaining NH_4^+-N. Then replace the used filter paper with two new acid-spiked pieces again, add 0.3 g Devarda's alloy, and incubate it for 24 h to complete the processes of diffusion and recovery of NO_3^--N;
5. Remove the filter papers from the clips by forceps and dry them in a desiccator containing an open container of concentrated H_2SO_4 (to remove traces of NH_3) and silica gel. Then wrap the filter papers in tin capsules and analyse them for ^{15}N enrichment by using a coupled elemental analyser-isotope ratio mass spectrometer (EA-IRMS);
6. Use the amount of N measured in diffusion blanks to calculate the corrected ^{15}N enrichment of the sample (Eq. 7.27):

$$E_s = \frac{E_m M_{s+b} - E_b M_b}{M_{s+b} - M_b} \tag{7.27}$$

where E_s is the corrected abundance ^{15}N enrichment of the sample, E_m is the enrichment of the sample + blank measured by mass spectrometry, M_{s+b} is the mass of N (sample + blank) recovered in the acid trap, M_b is the mass of N recovered in the acid trap from the diffusion blank, and E_b is the enrichment in the blank (assumed to be 0.3663 atom%).

Note, for soil extracts < 2 mg l^{-1} in inorganic N concentration, 50 ml of extract is needed to ensure accurate determination. When the abundances of NH_4^+-N and NO_3^--N are very different, it is better to diffuse NH_4^+-N and NO_3^--N separately (do not use the same extract).

7.5.4.4 Inorganic Nitrogen Isotopic Analysis in Soil Extracts at Natural Abundance

The diffusion method and chemical conversion method described above are both suitable for N at high abundance, but not for N at natural abundance. They both have a high demand for N and high levels of background N can interfere with the reaction. There are two modified chemical methods for N isotopic analysis at natural abundance. These simplify the preparation procedures, reduce the preparation time and do not require large amounts of N. The chemical method for ammonium requires 2.5 μg N in a 4 ml sample volume for analysis, and its accuracy of $\delta^{15}N$ measurements is less than 0.3‰. The method for NO_3^- needs only 4.5 μg N, and its accuracy of $\delta^{15}N$ and $\delta^{18}O$ reaches 0.31‰ and 0.55‰, respectively.

Conversion of ammonium at natural abundance

Principle

The method is to oxidise NH_4^+-N to NO_2^--N by BrO^- instead of extraction of NH_4^+-N in solution. Subsequently, the NO_2^--N is reduced to N_2O by NH_2OH-HCl, thus replacing HN_3 (Liu et al. 2014; Stedman 1959) (Eq. 7.28).

$$NH_4^+ + NaBrO \rightarrow NO_2^- + NH_2OH \rightarrow N_2O + H_2O \tag{7.28}$$

Apparatus

PT-IRMS

Shaker

20 ml headspace glass vials: Acid rinsed and combusted at 450 °C for 4 h

Reagents

10 M NaOH: Evaporate 100 ml of 5 M NaOH to 50 ml of 10 M NaOH

NaBrO:

(a) Bromate–bromide stock solution: Mix 0.6 g $NaBrO_3$ and BrNa into 250 ml DIW (deionised water) (can be stored ≥ 6 months);

(b) Take 1 ml bromate/bromide stock solution into 50 ml water, and place the solution in the dark with 3 ml 6 M HCl added for 5 min to produce Br_2;
(c) Add 50 ml of 10 M NaOH quickly to produce BrO^-.

$NaAsO_2$: Mix5.1 g $NaAsO_2$ and 100 ml DIW

NH_2OH, HCl:

(a) NH_2OH-HCl stock solution: Add 0.2778 g NH_2OH-HCl in 100 ml DIW (can be stored $\leq$ 7 d)
(b) Take 3 ml NH_2OH-HCl stock solution and dilute in 500 ml DIW

6 M HCl
5 M NaOH

Procedures

1. Take samples (1:10 (v:v) sample to NaBrO, e.g. 4 ml) and place it into 20 ml headspace glass vials. Dilute the sample to 10-20 μM to maximise oxidation yield. NO_2^--N must be removed by NH_2SO_3H earlier to ensure the accurate determination of NH_4^+-N;
2. Add NaBrO (e.g. 0.4 ml) into the vial, shake the vial vigorously, and then let it stand for 30 min;
3. Pipet 0.05 ml $NaAsO_2$ to remove excess BrO^- and terminate oxidation;
4. Add 6 M HCl to lower pH (pH < 1) and seal the vials;
5. Inject NH_2OH-HCl with a gas-tight syringe ($n(NH_4^+)$: $n(NH_2OH) = 1:2$). Put the samples in a shaker running at 120 rpm at 37 °C for 16 h;
6. Inject 0.5 ml 5 M NaOH to absorb CO_2 in the vials and terminate the reaction;
7. Transfer a known amount of gas to a PT-IRMS for analysis;
8. Treat 3 international NH_4^+-N standards (IAEA N1, +0.4‰; USGS 25, −30.4‰; USGS 26, +53.7‰) using the same protocol for calibration (Eq. 7.29):

$$\delta^{15}N_{NH_4^+ sample} = (\delta^{15}N_{N_2O sample} - \text{intercept})/\text{slope} \tag{7.29}$$

where the intercept and slope are obtained from the linear regression of the $\delta^{15}N$ measured and the $\delta^{15}N$ assigned from N_2O produced by the standards.

Conversion of NO_3^- at natural abundance

Principle

The NO_3^--N is reduced to NO_2^--N by copper-plated cadmium granules in a weakly alkaline environment (Eq. 7.30):

$$NO_3^- + Cd + H_2O \leftrightarrow NO_2^- + Cd^{2+} + 2OH^- \tag{7.30}$$

NO_2^--N is converted into N_2O by N_3^- in a weakly acid buffer. When pH > 2, the reaction will be (Eqs. 7.31–7.33):

$$NO_2^- + H^+ \leftrightarrow HNO_2 \quad (7.31)$$

$$HNO_2 + H_2O \leftrightarrow H_2NO_2^+ + OH^- \quad (7.32)$$

$$H_2NO_2^+ + N_3^- \rightarrow N_3NO + H_2O \rightarrow N_2O + N_2 \quad (7.33)$$

When there is a large number of halogen ions present the reaction will be accelerated (Stedman 1959) (Eqs. 7.34, 7.35):

$$H_2NO_2^+ + Cl^- \rightarrow NOCl + H_2O \quad (7.34)$$

$$NOCl + N_3^- \rightarrow N_3NO + Cl^- \rightarrow N_2O + N_2 \quad (7.35)$$

N_2O is an asymmetric molecule with a molecular structure of N-N-O. The $\delta^{15}N_{Air}$ of N_2O is the mean value of $\delta^{15}N_{Air}$ in two N atoms (Eq. 7.36):

$$\delta^{15}N_{Air}‰(N_2O) = \frac{\delta^{15}N_{Air}‰(^{15}N - N - O) + \delta^{15}N_{Air}‰(N -^{15}N - O)}{2} \quad (7.36)$$

and the N_2O produced is composed of a N atom and an oxygen atom provided by the NO_2^--N and a N atom provided by N_3^-, a N source. The isotope ratio of N and oxygen of the NO_2^- is identical with that of NO_3^- in the original solution. So, the following relationships hold (Eqs. 7.37, 7.38):

$$\delta^{15}N_{Air}‰(N_2O) = \frac{\delta^{15}N_{Air}‰(N_3^-) + \delta^{15}N_{Air}‰(NO_2^-)}{2} = \frac{\delta^{15}N_{Air}‰(N_3^-) + \delta^{15}N_{Air}‰(NO}{2} \quad (7.37)$$

$$\delta^{18}O_{SMOW}‰(N_2O) = \delta^{18}O_{SMOW}‰(NO_2^-) + \delta^{18}O_{SMOW}‰(NO_3^-) \quad (7.38)$$

Therefore, when the isotope ratio of N_3^- is constant, the N and oxygen isotope ratios of N_2O produced is linear with the N and oxygen isotope ratios of NO_3^-, and the theoretical slopes of their correlation curves are 0.5 (N) and 1.0 (O), respectively.

Apparatus

PT-IRMS

Shaker

pH metre

Peristaltic pump: flow rate ≥ 5 ml min^{-1}

Water-thermostat

Fume hood

Filter papers

Headspace glass vials

Reagents

0.5 M NaCl

0.5 M HCl

1 M $C_3H_4N_2$

NaN_3- CH_3COOH:

(a) Mix 15 ml of 20% CH_3COOH with 15 ml of 2 M NaN_3 in a fume hood;
(b) Ultrasonic surge the mixing solution for 15 min;
(c) Blow the solution with He (280 ml min^{-1}) for 30 min if there are impurities;
(d) The remaining acid is neutralised with a strong base (NaOH).

6 M NaOH

Copper-plated cadmium granules

Cadmium reduction column

Procedures

1. Place at least 40 ml of sample into the vials to ensure that there is still 16 ml sample with 4.5 μg N for the reaction after any loss. If the concentration of NO_3^--N in sample is above 20 μM, dilute the sample with 0.5 M NaCl. If the concentration of Cl^- is below 0.5 M, add solid NaCl to ensure that the concentration of Cl^- is 0.5 M.
2. The blank must be analysed during each analysis to test the seal of vials and the reagent blank. The signal value of blank should be below 0.6 nA.
3. Add 0.5 M HCl into the samples to adjust pH to 2–3. The blank will only need one drop of 0.5 M HCl. Then add 1 M $C_3H_4N_2$ to adjust the pH to 7.8–8.0.
4. (a) Connect the cadmium reduction column to a peristaltic pump of which the flow rate is 5 ml min^{-1}. Plug the end of the column with foam sponge. After filling the column with copper-plated cadmium granules, also plug the other end. Rinse the pipeline with 0.5 M NaCl (pH = 8) to activate the column. Then transfer 20 ml of adjusted sample into a 25 ml beaker, and place the inflow and outflow ends of the column in the beaker. After 80 min of continuous reduction, rinse the pipeline with 40 ml 0.5 M NaCl again. When moving the column air must not enter the column to prevent oxidation of the cadmium column.
(b) Copper-plated cadmium granules can be added directly to the samples. Place the samples in a shaker running at 200 rpm at 30 °C for 3 h to reduce NO_3^--N. Filter the sample into another vial.
5. Take a 16 ml sample and place it into a 50 ml headspace glass vial with the cap sealed. Evacuate the vial and fill it with He gas.

6. Inject 0.8 ml NaN_3- CH_3COOH into the vial (pH = 4.5) and shake the solution vigorously for 1.0 min. Then place the samples in a water-thermostat at 30 °C for 30 min, or in a shaker running at 200 rpm at 35 °C for 30 min.
7. Inject 0.5 ml 6 M NaOH (pH $\geq$ 10) to terminate the reaction.
8. Transfer a known amount of the gas to a PT-IRMS for determination of $\delta^{15}N$ and $\delta^{18}O$ of N_2O in the sample.
9. Mix 2 international NO_3^--N standards (USGS 32, $\delta^{15}N_{Air}‰ = 180‰$, $\delta^{18}O_{SMOW}‰ = 25.7‰$; USGS 34, $\delta^{15}N_{Air}‰ = -1.8‰$, $\delta^{18}O_{SMOW}‰ = -27.9‰$) in different proportions (e.g. 6:0, 4:2, 0:6), then treat them with the same protocol for calibration. The calibration equation shown below is

$$\delta^{15}N_{AirNO_3^-} = (\delta^{15}N_{AirN_2O} - \text{intercept})/\text{slope} \tag{7.39}$$

$$\delta^{15}N_{(180)WNO_3^-} = (\delta^{15}N_{SMOWN_2O} - \text{intercept})/\text{slope} \tag{7.40}$$

where the intercept and slope are obtained from the linear regression of the $\delta^{15}N$ and $\delta^{18}O$ measured from N_2O produced by the standards and the $\delta^{15}N$ and $\delta^{18}O$ are assigned from the standards.

7.5.5 *^{15}N Tracing Model Analyses via* Ntrace

7.5.5.1 Data Requirements for the *Ntrace* Model

Data obtained through the ^{15}N tracing experiment will be further analysed by the *Ntrace* model to quantify gross N transformation rates and pathway-specific N_2O emissions. The various *Ntrace* model versions differ in their data requirements. The *Ntrace*$_{Basic}$ model has the fewest data requirements. The other models require more data on top of the data required for the *Ntrace*$_{Basic}$ model. The *Ntrace*$_{Basic}$ requires the fertiliser application rate (in μmol N g^{-1}) and its ^{15}N excess (in atom%). It also requires the average NO_3^- and NH_4^+ concentration and ^{15}N excess (in atom% excess) including their standard deviations for each data point in time. The NO_3^- and NH_4^+ concentrations should be given in the same unit as the fertiliser application rate. Next to this, a one-time measurement of total organic N (in %) is required. This measurement can be taken from basic soil characteristics. The *Ntrace*$_{Plant}$ model also requires plant N and plant ^{15}N data, and total plant biomass data, at each time when destructive sampling was carried out, i.e. preferably the same time points when NO_3^- and NH_4^+ were determined. Ideally, above- and belowground biomass (roots) should be determined. The *Ntrace*$_{Urea}$ model requires the urea application rate and its excess, and if plants are included, it also has the additional requirements of the *Ntrace*$_{Plant}$ model. The *Ntrace*$_{Nitrite}$ model requires measurements at multiple time points (preferably the same as for NO_3^- and NH_4^+) of the average NO_2^- concentration and its ^{15}N excess (in atom%) including standard deviations.

Transformation can follow zero-order, first-order, or Michaelis–Menten kinetics. The type of kinetics used needs to be specified for each transformation separately. If the transformation uses N from a large pool, it is generally appropriate to use zero-order kinetics. For N transformations coming from pools that change rapidly it is generally more realistic to use first-order or Michaelis–Menten kinetics. Especially the transformations associated with the NH_4^+ consumption (e.g. nitrification) may be most realistically represented by Michaelis–Menten kinetics. However, under conditions when microbial activities may be affected by conditions other than substrate, the N transformation rate may also follow first-order kinetics. For example, when the rate is governed by temperature or soil moisture.

7.5.5.2 The *Ntrace* Model System

The ^{15}N tracing model *Ntrace* described by Müller et al. (2007) is a tool to quantify gross soil N transformations; this model considers five N pools and twelve simultaneously occurring N transformations (Fig. 7.12). The five soil N pools are ammonium (NH_4^+) adsorbed NH_4^+ ($NH_4^+{}_{ads}$); labile soil organic N (N_{lab}); NO_3^-, nitrate stored ($NO_3^-{}_{sto}$) and recalcitrant soil organic N (N_{rec}). The model considered twelve gross N transformations from these five N pools: mineralisation of recalcitrant organic-N to NH_4^+ (M_{Nrec}); mineralisation of labile N to NH_4^+ (M_{Nlab}); immobilisation of NH_4^+ into recalcitrant organic-N (I_{NH4_Nrec}); immobilisation of NH_4^+ into labile organic-N (I_{NH4_Nlab}); oxidation of NH_4^+ to NO_3^- (O_{NH4}); oxidation of recalcitrant organic-N to NO_3^- (O_{Nrec}); immobilisation of NO_3^- to recalcitrant organic-N (I_{NO3}); dissimilatory NO_3^- reduction to NH_4^+ (D_{NO3}); adsorption of NH_4^+ on cation exchange sites (A_{NH4}); and release of adsorbed ammonia (R_{NH4a}), adsorption and release of NO_3^- on/from stored NO_3^-, i.e. A_{NO3s} and R_{NO3s}, respectively. *Ntrace* is a family of ^{15}N tracing models to quantify gross N transformations in soils and sediments. The model consists of a N transformation model that is programmed in Simulink, a graphical programming language associated to Matlab, and a parameter optimisation routine based on a Markov Chain Monte Carlo routine in combination with the Metropolis algorithm (Müller et al. 2007).

Several extensions exist of the *Ntrace*$_{\text{Basic}}$ model, namely *Ntrace*$_{\text{Plant}}$ (Fig. 7.13a), *Ntrace*$_{\text{Urea}}$ (Fig. 7.13b), *Ntrace*$_{\text{Nitrite}}$ (Fig. 7.14) and *Ntrace*$_{\text{Gas}}$ (Fig. 7.15). The boxes represent the different N pools, and the transformations are represented by the arrows between the boxes. For each model all transformations are quantified simultaneously (Fig. 7.12).

The Matlab-Simulink files (m-files and mdl-files) alongside their description that are part of the *Ntrace* model are presented in Table 7.7.

Currently, a new optimisation routine for the *Ntrace* model is being implemented. This will further improve optimisation speed, and more importantly be quicker to find a global minimum as opposed to a local one. The method used for determining optimal parameters will be Matlab's GlobalSearch algorithm (Ugray et al., 2007).

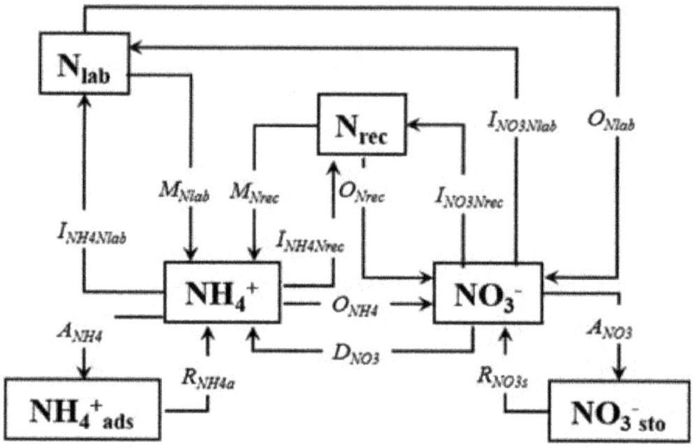

Fig. 7.12 Conceptual *Ntrace*$_{Basic}$ model an extended version of the model published by Müller et al. (2007), extended by additional exchange processes between N_{lab} and NO_3^- (b)

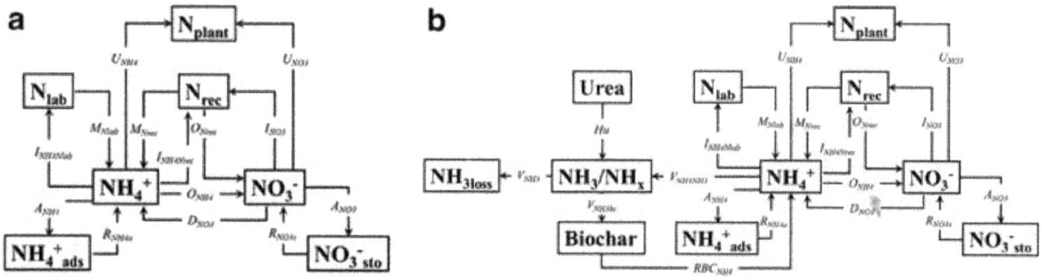

Fig. 7.13 **a** *Ntrace* $_{Plant}$ model, based on Inselsbacher et al. (2013) **b** *Ntrace*$_{Urea}$ model

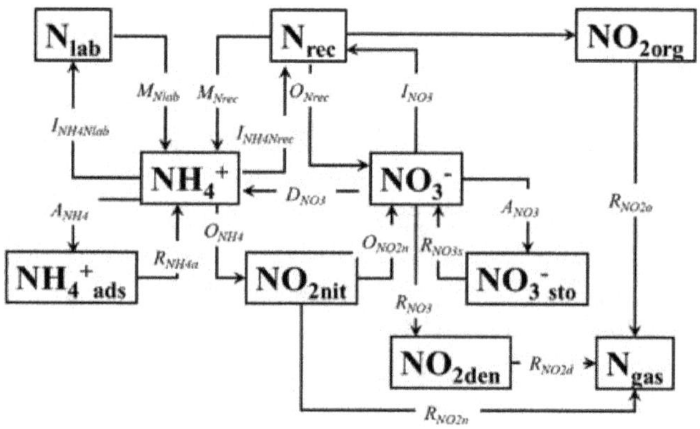

Fig. 7.14 *Ntrace*$_{Nitrite}$ model, based on Müller et al. (2006) and Rütting and Müller (2008)

Standard deviations of the optimised parameters will be determined as described by Gavin (2019).

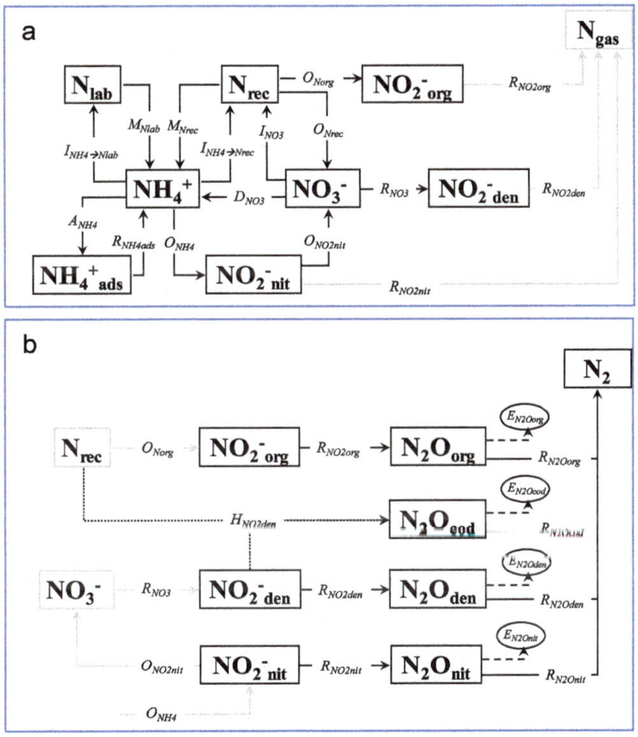

Fig. 7.15 *Ntrace*Gas model an extension to *Ntrace*Nitrite as described by Müller et al. (2014)

7.5.6 *Parameter Optimisation with* Ntrace

7.5.6.1 Setup

After filling out an input file for the model (DataNtrace.xlsx) that contains all the required data specified in data requirements including kinetics for each transformation, initial parameters and minimum and maximum values for the parameters, the model can be run (Fig. 7.16).

7.5.6.2 Procedure

The first step of the optimisation is generally done by hand. The model is run with the initial parameter set, and graphs of modelled versus measured data are inspected. From this, parameters are adjusted until a visually reasonable fit is obtained. Thereafter, all parameters are optimised simultaneously using a Markov chain Monte Carlo (MCMC) method, and this is explained in more detail by Müller et al. (2007). For this, the parameters are slightly adjusted, and the model is run. At the end of the run, the misfit is calculated based on the difference between the modelled and measured values. If the misfit is lower compared to the previous run, i.e. a better fit between modelled and measured data, the new parameter set is accepted, and it will start again by adjusting the newly accepted parameter set and execute the next iteration. If the

Table 7.7 The different MatLab (m-files) and Simulink file (mdl-files) that are part of the *Ntrace* tracing system

File name	Description	Model			
		Basic	Nitrite	Plant/urea	Gaseous N
Ntrace.m	This is the main file. From here the other files are called	x	x	x	x
Ntrace_biomassFit.m	Creates a function for plant growth based on measurements in DataNtrace.xls			x	x
Ntrace_dataAnalysis.m	In this file some basic statistics are determined. And the final run of the model is performed with a fixed time step	x	x	x	x
Ntrace_endOptimizationStep.m	This file is called to show the optimisation is done	x	x	x	x
Ntrace_graphs.m	In this, the graphs are created	x	x	x	x
Ntrace_initialPoolSizes.m	In this file the initial pool sizes calculated. Only the initial separate nitrite pool sizes are calculated somewhere else as the splitfactor for this is optimised as well	x	x	x	x
Ntrace_inputData.m	In this file, all data from the DataNtrace.xls input file are obtained	x	x	x	x
Ntrace_model InputAdjustment.m	If optimisation is done in 2 steps, input parameters are adjusted in this file, so that first the basic model can be run		x	x	x
Ntrace_optimization.m	This is the actual optimisation routine	x	x	x	x
Ntrace_runSimulink.m	This is called from Ntrace_optimization.m to facilitate the optimisation of the split factors. In here the initial Nitrite pool sizes (for $Ntrace_{\mathrm{Nitrite}}$) are determined and also the ^{14}N and ^{15}N amount in each pool based on the abundance and concentration. From this file the actual Simulink model is called	x	x	x	x
Ntrace_stats.m	Optional file to use after optimisation to write correlation coefficients into Ntrace_stats.xls and show statistics	.	.	.	

(continued)

Table 7.7 (continued)

File name	Description	Model			
		Basic	Nitrite	Plant/urea	Gaseous N
Ntrace_values_out.m	Optional file for after optimisation to write pool sizes and transformation rates for every time step into Modelvalues_out.xls	.	.	.	
Ntrace_writeOutput.m	In this the modelled output is written and excel file called data_out, and if requested into the specific excel results file	x	x	x	x
xlsPasteTo.m	This file is used in Ntrace_writeOutput.m to paste the measured vs modelled graph in the result file	x	x	x	x
DataNTrace.xls	Input file, that contains all the settings for the model, initial parameters, and measured data	x	x	x	x
Ntrace_basic_d.mdl	Simulink model for the $Ntrace_{\text{Basic}}$	x			
Ntrace_nitrite_d.mdl	Simulink model for the $Ntrace_{\text{Nitrite}}$		x		
Ntrace_Urea_simPlant.mdl	Simulink model for $Ntrace_{\text{urea}}$ if N applied is via Urea, also exchange with Biochar can be simulated			x	
Ntrace_Urea_simB4d_4.mdl	Simulink model for $Ntrace_{\text{Plant}}$ and $Ntrace_{\text{urea}}$ if plant growth model is used			x	
Ntrace_gas_d	Simulink model for $Ntrace_{\text{Gas}}$				x

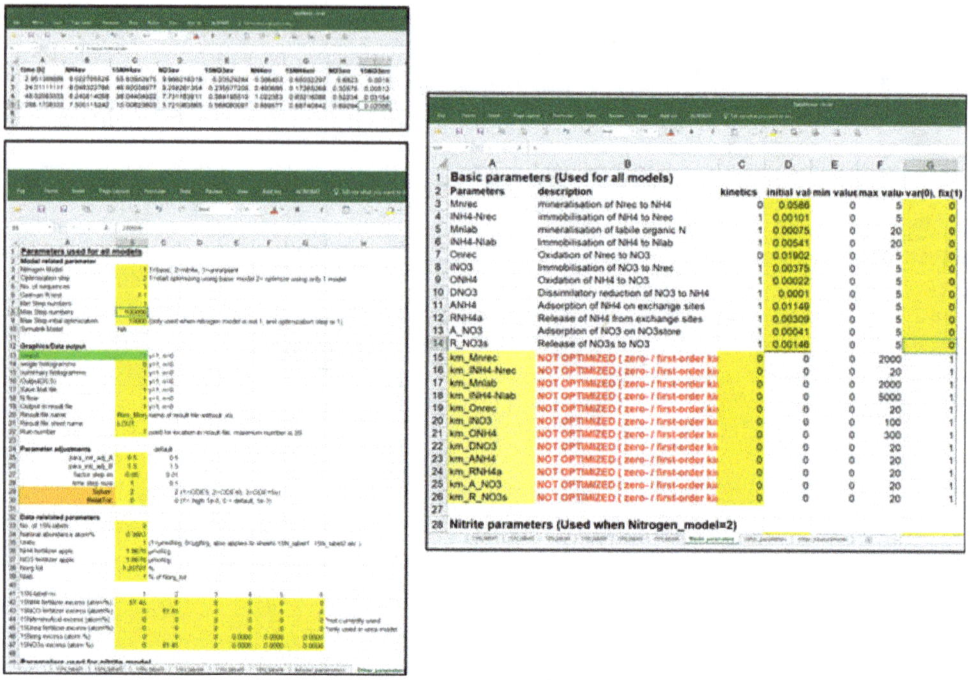

Fig. 7.16 Data *Ntrace* Excel file that is used to setup *Ntrace* model runs. The sheets show: above-left: input sheet of experimental data, below-left: setup of model run with input of basic parameters, e.g. Norg content, N application and the ^{15}N enrichment, right: parameter sheet. Note: yellow areas are input areas

misfit is higher, so a worse fit, there is also a chance the parameter set is accepted via a likelihood function. By this, it is possible that the algorithm moves out of a local minimum and enters the global minimum. If a new parameter set is not accepted, the last accepted parameter set will be used for the parameter adjustment. This iterative procedure of parameter adjustment and running the model should go on till the probability density functions (PDFs) are well characterised for all parameters. If the initial parameter set is fairly close to the optimal set, PDFs are generally well characterised after 50,000 to 100,000 iterations. So, for the final run, the maximum number of iterations is generally set between 50,000 and 100,000. During the optimisation, generally three parallel sequences, each with different starting parameters, are calculated. The number of parallel sequences should be defined before running the model, but three is generally appropriate. From these parallel sequences, a reduction factor is determined, which determines the accuracy of the sampling (Gelman et al. 2003). If the reduction factor is below a pre-specified number (default 1.1) for all parameters, the optimisation will also be stopped, regardless of reaching the maximum number of iterations. The reduction factor near 1 indicates that all parallel sequences resulted in statistically the same parameter set.

Inspection of the PDFs can show that for certain parameters the peak is close to zero. This would indicate that this particular parameter can be neglected. The parameter can then be set to zero, and excluded from optimisation when the model is re-run.

7.5.6.3 *Ntrace* Model Output

At the end of the optimisation, the model output will be exported to an Excel file. This output contains the initial parameter value, the optimised parameter value, the standard deviation of the optimised parameter, the average N flow for each transformation, the overall R^2 and the AIC. The standard deviation (SD) of the transformation rate is based on the SD and average (AVG) parameter values as shown in Eqs. 7.41 to 7.43:

$$SD_{\text{transformation rate}} = \frac{AVG_{\text{transformation rate}} \cdot SD_{\text{parameter}}}{AVG_{\text{parameter}}} \tag{7.41}$$

Response ratios between transformation rates can be calculated by McGeough et al. (2016):

$$\mathrm{R} = \ln\left(\frac{\overline{\mathrm{X}}_{\mathrm{E}}}{\overline{\mathrm{X}}_{\mathrm{C}}}\right) \tag{7.42}$$

With the associated standard deviation:

$$sd_R = \sqrt{\frac{sd_E^2}{n_E \overline{X}_E^2} + \frac{sd_C^2}{n_C \overline{X}_C^2}} \tag{7.43}$$

where $\overline{X}_E$ and $\overline{X}_C$ are the average N transformations of the elevated and control group, sd_E and sd_C the associated standard deviations and n_E and n_C the repetitions.

To compare the effect of different treatments on transformation rates, the individual treatments have to be run separately. After this, the rates can be compared using a one-way ANOVA based on the averages and standard deviations. Pairwise comparisons can be calculated with the Holm-Šídák test.

Another way to compare gross N transformation is via the determination of least significant difference (LSD) as described by Müller et al. (2011).

Output of the correlation matrix can be used to find parameters that tend to be strongly constrained together. A correlation value of above 0.8 indicates that it is constrained. There is also another output file that gives the pool sizes and transformation rates for each time step. This output can be used to create graphs.

7.5.7 *Determination of N_2O Pathways*

7.5.7.1 *Ntrace* Approach to Quantify N_2O Pathways

Nitrous oxide (N_2O) can be emitted via a number of pathways including inorganic and also organic pathways, involving a range of microbes (e.g. bacteria, fungi) (Butterbach-Bahl et al. 2013). The *Ntrace*$_{Gas}$ model can be applied to quantify N_2O pathways based on the underlying N transformations and especially based on the nitrite dynamics (*Ntrace*$_{Nitrite}$). To accurately estimate N_2O pathways via *Ntrace*$_{Gas}$ also the N_2 production is calculated. The two predominant biological processes for N_2O production in soil are traditionally considered to be autotrophic nitrification and heterotrophic denitrification (Ambus 1998; Wrage-Mönnig et al. 2018; Wrage et al. 2001). In both processes NO_2^- is the key precursor to N_2O production. In nitrifier nitrification, it is rather NH_2OH or at least something before nitrite. In nitrifier denitrification, it is nitrite.

Assuming that the N_2O production is derived from a single NO_2^- pool, the ^{15}N enrichments of the N_2O and the NO_2^- should be similar. However, experimental data show that the enrichment of the N_2O is deviating from the theoretical 1:1 line (Fig. 7.17) leading to the conclusion that the N_2O originated from various NO_2^- sub-pools and also from sources which were at or close to natural abundance. Based on the experimental setup, the only common unlabeled N pool in all ^{15}N treatments is organic N. Therefore, two possible processes were included in the *Ntrace*$_{Gas}$ model to account for such a dilution effect:

(a) reduction of NO_2^- originating from oxidation of organic N derived ($NO_2^-{}_{org}$) and

(b) hybrid-reaction for N_2O production whereby one atom of the N_2O is derived from an enriched NO_2^- pool and another from organic N at natural abundance.

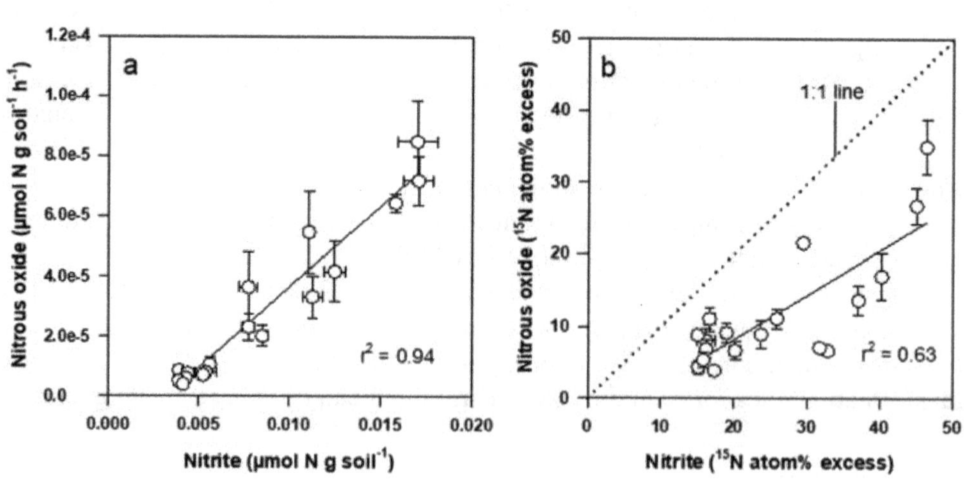

Fig. 7.17 Relationships between nitrite and nitrous oxide **a** as well as the ^{15}N enrichment **b** (Müller et al. 2014)

Heterotrophic nitrifiers can also denitrify (Blagodatsky et al. 2006; Papen et al. 1989) and it is, therefore, possible that $NO_2^-{}_{org}$ in the *Ntrace*$_{Nitrite}$ model (Rütting and Müller 2008) originating from N_{org} oxidation could be further reduced to gaseous N products. A hybrid-reaction between NO_2^- and organic N is also possible which occurs, for instance, in the fungus *Fusarium oxysporum* (Kurakov et al. 2000; Tanimoto et al. 1992) and possibly in other heterotrophic organisms (Kumon et al. 2002).

Based on the above considerations the *Ntrace*$_{Gas}$ model analyses four N_2O processes. The entire model includes all the previous ^{15}N tracing models (see *Ntrace* family above). In *Ntrace*$_{Gas}$ NO_2^- sub-pools are reduced to associated N_2O pools which may further be reduced to N_2 (Eq. 7.44).

$$2NO_2^-{}_x \rightarrow N_2O_x \rightarrow N_2 \tag{7.44}$$

(x = nit, den or org)

In addition, a hybrid-reaction between denitrification derived NO_2^- ($NO_2^-{}_{den}$) and recalcitrant organic N (N_{rec}) was introduced (Eq. 7.45).

$$NO_2^-{}_{den} + N_{rec} \rightarrow N_2O_{cod} \rightarrow N_2 \tag{7.45}$$

Each soil N_2O sub-pool can be further reduced to N_2 via specific N_2O reduction rates or emitted to the atmosphere, which is governed by gas diffusion parameters. For N_2O emission a first-order notation has been implemented (Cho and Mills 1979; p. 97 in Müller 2000).

The total N_2O emission is calculated (Eq. 7.46) by

$$\sum E_x = \sum k_x \cdot N_2O_x \tag{7.46}$$

where E_x is the emission rate (μmol N g^{-1} h^{-1}), k_x is the emission rate constant (h^{-1}) and N_2O_x the soil N_2O pool concentration (μmol N g^{-1}). The symbol x stands for the process specific pools, i.e. nit, den, org and cod.

In the following section, two simplified approaches are presented that are based on the abundance of NH_4^+, NO_3^-, N_{org} and N_2O. The methods are based on the assumption that N_2O is derived from three uniformly labelled pools, i.e. NH_4^+, NO_3^- and N_{org}. and bases the analysis on the ^{15}N enrichments of the different N species.

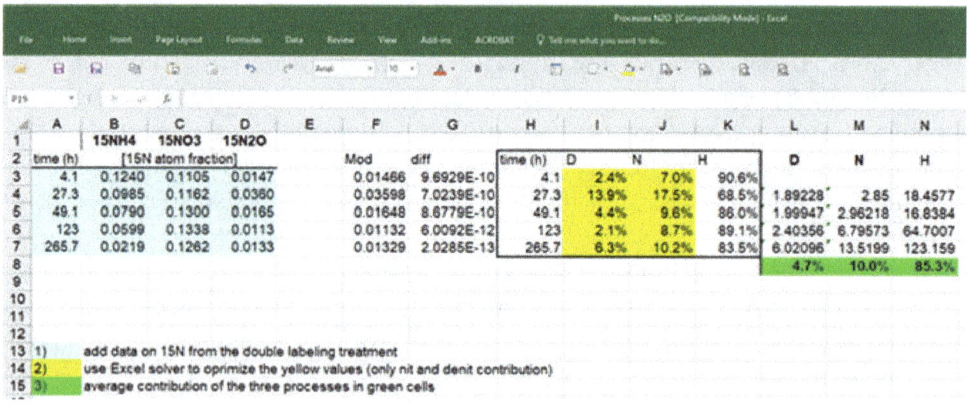

	A	B	C	D	E	F	G	H	I	J	K	L	M	N
1		15NH4	15NO3	15N2O										
2	time (h)	[15N atom fraction]				Mod	diff	time (h)	D	N	H	D	N	H
3	4.1	0.1240	0.1105	0.0147		0.01466	9.6929E-10	4.1	2.4%	7.0%	90.6%			
4	27.3	0.0985	0.1162	0.0360		0.03598	7.0239E-10	27.3	13.9%	17.5%	68.5%	1.89228	2.85	18.4577
5	49.1	0.0790	0.1300	0.0165		0.01648	8.6779E-10	49.1	4.4%	9.6%	86.0%	1.99947	2.96218	16.8384
6	123	0.0599	0.1338	0.0113		0.01132	6.0092E-12	123	2.1%	8.7%	89.1%	2.40356	6.79573	64.7007
7	265.7	0.0219	0.1262	0.0133		0.01329	2.0285E-13	265.7	6.3%	10.2%	83.5%	6.02096	13.5199	123.159
8												4.7%	10.0%	85.3%
13	1)	add data on 15N from the double labeling treatment												
14	2)	use Excel solver to oprimize the yellow values (only nit and denit contribution)												
15	3)	average contribution of the three processes in green cells												

Fig. 7.18 Three source model for determination of N_2O pathways based on Rütting et al. (2010), see text for details

7.5.8 Source Partitioning to Quantify N_2O Pathways

7.5.8.1 Three-Pool Model

Based on the two-pool source-partitioning model by Stevens et al. (1997) a three-pool solver method (Rütting et al. 2010) was developed. The solver method (Microsoft Excel 2007) calculates the N_2O fractions associated with NH_4^+ (n) and NO_3^- (d) by minimisation of the absolute difference between observed and calculated ^{15}N enrichments of N_2O according to the equation:

$$a_{N2O} = d * a_d + n * a_n + (1 - d - n) * a_o \tag{7.47}$$

where *n* and *d* are the fractions related to the NH_4^+ and NO_3^- pools, respectively, and a_d, a_n and a_o represent the ^{15}N abundance of the NO_3^-, NH_4^+ and N_{org} (assumed to be at natural abundance) respectively. The data are setup in an Excel spreadsheet and the Excel Solver is used to minimise the difference between measured and calculated ^{15}N N_2O enrichments (Rütting et al. 2010) (Fig. 7.18).

With this method, it is possible to subdivide total N_2O emission into the three sources, autotrophic nitrification, denitrification and heterotrophic nitrification.

7.5.8.2 Four-Pool Model

The three-pool model has been developed further by Jansen-Willems et al. (2016) to analyse four simultaneous processes (nitrification, denitrification, co-denitrification and oxidation of organic matter). The assumption is that isotopic discrimination is negligible. The conceptual model for this approach is illustrated in Fig. 7.19.

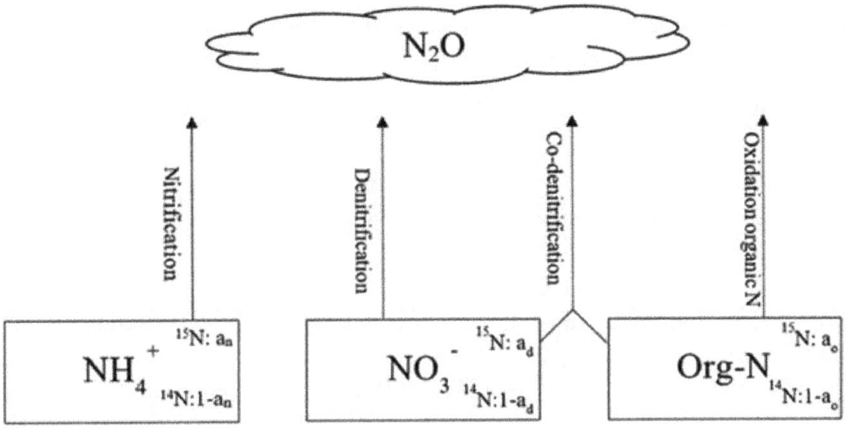

Fig. 7.19 Conceptual model to analyse N_2O pathways according to Jansen-Willems et al. (2016)

Background and development of the four-pool model

Each N pool contains both ^{15}N and ^{14}N atoms. If one N atom would be randomly selected from a pool, the chance it would be a ^{15}N atom is equal to the ^{15}N atom fraction of that pool. So for the NH_4^+ pool this would be a_n, for the NO_3^- pool a_d, and for the organic-N pool a_o. The chance it would be a ^{14}N atom equals 1 minus the ^{15}N atom fraction of that pool. So, for the NH_4^+ pool it would be $1\text{-}a_n$, for the NO_3^- pool $1\text{-}a_d$, and for the organic-N pool it would be $1\text{-}a_o$.

N_2O consists of two N atoms. For nitrification, denitrification and oxidation of organic N, both N atoms come from the same pool. For these three processes:

- The chance that N_2O contains no ^{15}N atoms is the chance that the first atom is a ^{14}N atom multiplied by the chance that the second atom is a ^{14}N atom (Eq. 7.48).
- The chance that N_2O contains one ^{15}N atom is the chance that the first atom is a ^{15}N atom, multiplied by the chance that the second atom is a ^{14}N atom, plus the chance that the first atom is a ^{14}N atom and the second is a ^{15}N atom (Eq. 7.49)
- The chance that N_2O contains two ^{15}N atoms is the chance that the first atom is a ^{15}N atom, multiplied by the chance that the second atom is a ^{15}N atom (Eq. 7.50)

In Eqs. 7.48 to 7.50, a_x would be a_n for nitrification, a_d for denitrification and a_o for oxidation of organic N.

$$\text{Chance of zero}\,^{15}\text{N atoms}: (1-a_x)(1-a_x) = (1-a_x)^2 \tag{7.48}$$

$$\text{Chance of one}\,^{15}\text{N atom}: a_x(1-a_x) + (1-a_x)a_x = 2(1-a_x)a_x \tag{7.49}$$

$$\text{Chance of two}\,^{15}\text{N atoms}: a_x a_x = a_x^2 \tag{7.50}$$

For co-denitrification, one atom comes from the NO_3^-, and one comes from the organic N pool. So, for co-denitrification, the chance that N_2O contains

- No ^{15}N atoms, is the chance of a ^{14}N atom from the NO_3^- pool, multiplied by the chance of a ^{14}N atom from the organic N pool (Eq. 7.51).
- One ^{15}N atom, is the chance of a ^{15}N atom from the NO_3^- pool, multiplied by the chance of a ^{14}N atom from the organic N pool plus the chance of a ^{14}N atom from the NO_3- pool multiplied by the chance of a ^{15}N atom from the organic N pool (Eq. 7.52).
- Two ^{15}N atoms is the chance of a ^{15}N atom from the NO_3^- pool, multiplied by the chance of a ^{15}N atom from the organic N pool (Eq. 7.53).

$$\text{Chance of zero}^{15}\text{N atoms: } (1-a_d)(1-a_o) \tag{7.51}$$

$$\text{Chance of one}^{15}\text{N atom: } a_d(1-a_o) + (1-a_d)a_0 \tag{7.52}$$

$$\text{Chance of two}^{15}\text{N atoms: } a_d a_o \tag{7.53}$$

The N_2O in the gas sample is assumed to come from one of four processes. The fraction that comes from nitrification is written as *n*, the fraction that comes from denitrification is written as *d* and the fraction that comes from oxidation of organic matter is written as *o*. The fraction that comes from co-denitrification is written as *c*. As these are the only four processes considered, the four fractions should add up to one. Therefore, the following two equations apply:

$$a + d + o + c = 1 \tag{7.54}$$

$$c = 1 - a - d - o \tag{7.55}$$

The fraction of N_2O in the gas sample that is expected to contain zero ^{15}N atoms can be calculated by multiplying the fraction of that sample from a specific process by the chance that the N_2O from that process contains zero ^{15}N atoms. So for nitrification, this would be $n(1–a_n)^2$, and for co-denitrification this would be $(1–n–d–o)$ $(1–a_d)$ $(1–a_0)$. This should be done for all four processes, and then should be added together (Eq. 7.56). The fraction of the N_2O in the gas sample that is expected to contain one ^{15}N atom is calculated in the same way (Eq. 7.57), and the expected fraction containing two ^{15}N atoms as well (Eq. 7.58)

Chance of zero ^{15}N atoms:

$$n(1 - a_n)^2 + d(1 - a_d)^2 + o(1 - a_o)^2 + (1 - n - d - o)(1 - a_d)(1 - a_0) \tag{7.56}$$

Chance of one ^{15}N atom:

$$2n(1-a_n)a_n + 2d(1-a_d)a_d + 2o(1-a_o)a_o + (1-n-d-o)(a_d(1-a_0) + a_0(1-a_d)) \quad (7.57)$$

Chance of two ^{15}N atoms:

$$naN_2 + da_d^2 + oaN_2 + (1-n-d-o)a_d a_0 \quad (7.58)$$

7.5.8.3 Mass Spectrometer Measurements and Calculation of Fractions

To determine the fractions of the different processes ^{45}R and ^{46}R measurements are needed. These need to be corrected for the presence of ^{18}O. Therefore, this means that ^{45}R is the fraction of N_2O molecules containing one ^{15}N atom divided by the fraction of N_2O molecules containing zero ^{15}N atoms, and ^{46}R is the fraction of N_2O molecules containing two ^{15}N atoms divided by the fraction of N_2O molecules containing zero ^{15}N atoms.

The expected fractions for N_2O containing zero, one or two ^{15}N atoms are given in Eqs. 7.56−7.58. In the study published by Jansen-Willems et al. (2016), a_0 was set to 0.003663 (natural abundance), and a_n and a_d were considered to be the ^{15}N abundance of NH_4^+ and NO_3^-. Using these values, *n, d* and *o* were quantified using the *fminsearchbnd* function in MatLab (The MathWorks Inc, Natick, MA). Thus, from this, *c* could be calculated according to Eq. 7.55.

References

Addy K, Kellogg DQ, Gold AJ, Groffman PM, Ferendo G, Sawyer C (2002) In Situ Push-Pull method to determine ground water denitrification in Riparian zones. J Environ Qual 31:1017–1024

Ambus P (1998) Nitrous oxide production by denitrification and nitrification in temperate forest, grassland and agricultural soils. Eur J Soil Sci 49(3):495–502

Arah JRM, Smith KA, Crighton IJ, Li HS (1991) Nitrous oxide production and denitrification in Scottish arable soils. J Soil Sci 42:351–367

Arah JRM (1992) New formulae for mass spectrometric analysis of nitrous oxide and dinitrogen emissions. Soil Sci Soc Am J 56:795–800

Arah JRM, Crichton IJ, Smith KA (1993) Denitrification measured directly using a single-inlet mass spectrometer and by acetylene inhibition. Soil Biol Biochem 25:233–238

Aulakh MS, Doran JW, Mosier AR (1991) Field-evaluation of 4 methods for measuring denitrification. Soil Sci Soc Am J 55:1332–1338

Bai E, Houlton BZ (2009) Coupled isotopic and process-based modelling of gaseous nitrogen losses from tropical rain forests. Global Biogeo Cycl 23(2):1–10

Baily A, Watson CJ, Laughlin R, Matthews D, McGeough K, Jordan P (2012) Use of the ^{15}N gas flux method to measure the source and level of N_2O and N_2 emissions from grazed grassland. Nutr Cycl Agro 94:287–298

Barford CC, Montoya JP, Altabet MA, Mitchell R (1999) Steady-state nitrogen isotope effects of N_2 and N_2O production in Paracoccus denitrificans. Appl Environ Microb 65:989–994

Barraclough D, Puri G (1995) The use of ^{15}N pool dilution and enrichment to separate the heterotrophic and autotrophic pathways of nitrification. Soil Biol Biochem 27(1):17–22

Bergsma TT, Ostrom NE, Emmons M, Robertson GP (2001) Measuring simultaneous fluxes from soil of N2O and N2 in the field using the 15 N-gas "Nonequilibrium" technique. Environ. Sci. Tech. 35:4307–4312

Blagodatsky SA, Kesik M, Papen H, Butterbach-Bahl K (2006) Production of NO and N_2O by the heterotrophic nitrifier Alcaligenes faecalis parafaecalis under varying conditions of oxygen saturation. Geomicrobiol J 23:165–176

Bollmann A, Conrad R (1997) Enhancement by acetylene of the decomposition of nitric oxide in soil. Soil Biol Biochem 29(7):1057–1066

Brand WA, Huang L, Muskai H, Chivulescu A, Richter J, Rothe M (2009) How well do we know VVPDB variability of ^{13}C and ^{18}O in CO_2 generated from NBS19-calcite. Rapid Commun Mass Spectrom 23:915–926

Brenninkmeijer CAM, Röckmann T (1999) Mass spectrometry of the intramolecular nitrogen isotope distribution of environmental nitrous oxide using fragment-ion analysis. Rapid Commun Mass Spectrom 13:2028–2033

Brewer P, Kim J, Lee S, Tarasova O, Viallon J, Flores E et al (2019) Advances in reference materials and measurement techniques for greenhouse gas atmospheric observations. Metrologia 56(3):034006

Buchen C, Lewicka-Szczebak D, Flessa H, Well R (2018) Estimating N_2O processes during grassland renewal and grassland conversion to maize cropping using N_2O isotopocules. Rapid Commun Mass Spectrom 32(13):1053–1067

Buchen C, Lewicka-Szczebak D, Fuß R, Helfrich M, Flessa H, Well R (2016) Fluxes of N_2 and N_2O and contributing processes in summer after grassland renewal and grassland conversion to maize cropping on a Plaggic Anthrosol and a Histic Gleysol. Soil Biol Biochem 101:6–19

Butterbach-Bahl K, Breuer L, Gasche R, Willibald G, Papen H (2002) Exchange of trace gases between soils and the atmosphere in Scots pine forest ecosystems of the northeastern German lowlands 1. Fluxes of N_2O, NO/NO_2 and CH_4 at forest sites with different N-deposition. For Ecol Manage 167 (1–3): 123–134

Butterbach-Bahl K, Baggs EM, Dannenmann M, Kiese R, Zechmeister-Boltenstern S (2013) Nitrous oxide emissions from soils, how well do we understand the processes and their controls. Philos T R Soc B 368:16–21

Cabrera ML, Kissel DE (1989) Review and simplifications or calculations in ^{15}N tracer studies. Fertil Res 20:11–15

Cao YC, Zhong M, Gong H et al (2013) Determining ^{15}N abundance in ammonium, nitrate and nitrite in soil by measuring nitrous oxide produced (In Chinese). Acta Pedol Sinica 50(1):113–119

Cárdenas LM, Hawkins JMB, Chadwick D, Scholefield D (2003) Biogenic gas emissions from soils measured using a new automated laboratory incubation system. Soil Biol Biochem 35:867–870

Cardenas L, Bol R, Lewicka-Szczebak D, Gregory AS, Matthews GP, Whalley WR, Misselbrook R, Scholefield D, Well R (2017) Effect of soil saturation on denitrification in a grassland soil. Biogeosc 14:4691–4710

Cho CM, Mills JG (1979) Kinetic formulation of the denitrification process in soil. Can J Soil Sci 59:249–257

Clough T, Condron L (2010) Biochar and the nitrogen cycle: introduction. J Environ Qual 39:1218–1223

Clough T, Stevens R, Laughlin R, Sherlock R, Cameron K (2001) Transformations of inorganic-N in soil leachate under differing storage conditions. Soil Biol Biochem 33(11):1473–1480

Coplen TB (2011) Guidelines and recommended terms for expression of stable-isotope-ratio and gas-ratio measurement results. Rapid Commun Mass Spectrom 25:2538–2560

Cox GM, Gibbons JM, Wood ATA, Craigon J, Ramsden SJ, Crout NMJ (2006) Towards the systematic simplification of mechanistic models. Ecol Model 198:240–246

Decock C, Six J (2013) How reliable is the intramolecular distribution of ^{15}N in N_2O to source partition N_2O emitted from soil? Soil Biol Biochem 65:114–127

Denk TRA, Mohn J, Decock C, Lewicka-Szczebak D, Harris E, Butterbach-Bahl K, Kiese R, Wolf B (2017) The nitrogen cycle: A review of isotope effects and isotope modelling approaches. Soil Biol Biochem 105:121–137

Denk TRA, Kraus D, Kiese R, Butterbach-Bahl K, Wolf B (2019) Constraining N cycling in the ecosystem model LandscapeDNDC with the stable isotope model SIMONE. Ecology 100:e02675

Deppe M, Giesemann A, Well R, Spott O, Flessa H (2017) Soil N_2O fluxes and related processes in laboratory incubations simulating ammonium fertilizer depots. Soil Biol Biochem 104:68–80

De Klein CAM, Harvey M (eds) (2012) Nitrous oxide chamber methodology guidlines, Ministry of Primary Industries, Wellington.

Enrich-Prast A, Santoro AL, Coutinho RS, Nielsen LP, Esteves FA (2015) Sediment Denitrification in Two Contrasting Tropical Shallow Lagoons. Est Coast 39:657–663

Erler D, Duncan T, Murray R, Maher D, Santos I, Gatland J et al (2015) Applying cavity ring-down spectroscopy for the measurement of dissolved nitrous oxide concentrations and bulk nitrogen isotopic composition in aquatic systems: Correcting for interferences and field application. Limnol Oceanogr-Meth 13(8):391–401

Eschenbach W, Well R (2012) Predicting long-term denitrification capacity of sandy aquifers from incubation experiments and sediment properties. Biogeosciences Discuss. 9:1–52

Eschenbach W, Well R, Walther W (2015) Predicting the denitrification capacity of sandy aquifers from in situ measurements using push-pull N-15 tracer tests. Biogeosciences 12:2327–2346

Felber R, Conen F, Flechard CR, Neftel A (2012) Theoretical and practical limitations of the acetylene inhibition technique to determine total denitrification losses. Biogeosc 9(10):4125–4138

Firestone MK, Davidson EA (1989) Microbiological Basis of NO and N_2O Production and Consumption in Soils. In: Andreae MO, Schimel DS (eds) Exchange of trace gases between terrestrial ecosystems and the atmosphere. Willey, New York, pp 7–21

Florian Stange C, Spott O, Apelt B, Russow R (2007) Automated and rapid online determination of ^{15}N abundance and concentration of ammonium, nitrite, or nitrate in aqueous samples by the SPINMAS technique. Isotopes Environ Health Stud 43(3):227–236

Frame CH, Casciotti KL (2010) Biogeochemical controls and isotopic signatures of nitrous oxide production by a marine ammonia-oxidizing bacterium. Biogeosc 7:2695–2709

Gelman A et al (2003) Bayesian data analysis. Boca Raton, Chapman & Hall/CRC, Boca Raton

Gavin H P (2019). The Levenberg-Marquardt algorithm for nonlinear least squares curve-fitting problems. Department of Civil and Environmental Engineering, Duke University, 1–19

Gordon I, Rothman LS, Hill C, Kochanov RV, Tan Y, Bernath PF, Birk M, Boudon V, Campargue A, Chance KV, Drouin BJ, Flaud J-M, Garmache RR, Hodges JT, Jacquemart D, Perevalov VI, Perrin A, Shine KP, Zak EJ (2017) The HITRAN2016 molecular spectroscopic database. J Quant Spectrosc Radiat Transf 203:3–69

Granger J, Sigman DM, Lehmann MF, Tortell PD (2008) Nitrogen and oxygen isotope fractionation during dissimilatory nitrate reduction by denitrifying bacteria. Limnol Oceanogr 53:2533–2545

Griffith D (2018) Calibration of isotopologue-specific optical trace gas analysers: a practical guide. Atmos Meas Tech. 11(11):6189–6201

Groffman PM, Altabet MA, Böhlke JK, Butterbach-Bahl K, David MB, Firestone MK, Giblin AE, Kana TM, Nielsen LP, Voytek MA (2006) Methods for measuring denitrification diverse approaches to a difficult problem. Ecol Appl 16(6):2091–2122

Harris S, Liisberg J, Xia L, Wei J, Zeyer K, Yu L, Barthel M, Wolf B, Kelly BFJ, Cendón DI, Blunier T, Six J, Mohn J (2020) N_2O isotopocule measurements using laser spectroscopy: analyzer characterization and intercomparison. Atmos Meas Tech 13:2797–2831

Hauck RD, Melsted SW, Yankwick PE (1958) Use of N-isotope distribution in nitrogen gas in the study of denitrification. Soil Sci 86:287–291

He X, Chi Q, Cai Z, Zhang J, Müller C (2020) ^{15}N tracing studies including plant N uptake processes provide new insights on gross N transformations in soil-plant system. Soil Biol Biochem 141:107666

Hedges LV, Gurevitch J, Curtis PS (1999) The meta-analysis of response ratios in experimental ecology. Ecology 80(4):1150–1156

Heil J, Wolf B, Bruggemann N, Emmenegger L, Tuzson B, Vereecken H, Mohn J (2014) Site-specific ^{15}N isotopic signatures of abiotically produced N_2O. Geochim Cosmochim Ac 139:72–82

Hiltbold AE, Bartholomew WV, Werkman CH (1951) The use of trace techniques in the simultaneous measurement of mineralization and immobilization of nitrogen in soil. Soil Sci Soc Am Proc 15:166–173

Ibraim E, Wolf B, Harris E, Gasche R, Wei J, Longfei Y, Kiese R, Eggleston S, Butterbach-Bahl K, Zeeman M, Tuzson B, Emmenegger L, Six J, Henne S, Mohn J (2019) Attribution of N_2O sources in a grassland soil with laser spectroscopy based isotopocule analysis. Biogeosc 16:3247–3266

Inselsbacher E, Wanek W, Strauss J, Zechmeister-Boltenstern S, Müller C (2013) A novel ^{15}N tracer model reveals: Plant nitrate uptake governs nitrogen transformation rates in agricultural soils. Soil Biol Biochem 57:301–310

Istok JD, Humphreya MD, Schrotha MH, Hymanb MR, O'Reilly KT (1997) Single-well, "Push-Pull" test for in situ determination of microbial activities. Ground Water 35:619–631

Jansen Willems AB, Lanigan GJ, Clough TJ, Andresen LC, Müller C (2016) Long-term elevation of temperature affects organic N turnover and associated N_2O emissions in a permanent grassland soil. Soil 2:601–614

Jensen ES (1991) Evaluation of automated analysis of ^{15}N and total N in plant material and soil. Plant Soil 133(1):83–92

Jinuntuya-Nortman M, Sutka RL, Ostrom PH, Gandhi H, Ostrom NE (2008) Isotopologue fractionation during microbial reduction of N_2O within soil mesocosms as a function of water-filled pore space. Soil Biol Biochem 40:2273–2280

Jung MY, Well R, Min D, Giesemann A, Park SJ, Kim JG, Kim SJ, Rhee SK (2014) Isotopic signatures of N_2O produced by ammonia-oxidizing archaea from soils. ISME J 8:1115–1125

Kaiser J, Röckmann T (2008) Correction of mass spectrometric isotope ratio measurements for isobaric isotopologues of O_2, CO, CO_2, N_2O and SO_2. Rapid Commun Mass Spectrom 22:3997–4008

Kato T, Toyoda S, Yoshida N, Tang YH, Wada E (2013) Isotopomer and isotopologue signatures of N_2O produced in alpine ecosystems on the Qinghai-Tibetan Plateau. Rapid Commun Mass Spectrom 27:1517–1526

Kelley KR, Ditsch DC, Alley MM (1991) Diffusion and automated nitrogen-15 analysis of low-mass ammonium samples. Soil Sci Soc Am J 55(4):1016–1020

Kirkham D, Bartholomew WV (1954) Equations for following nutrient transformations in soil, utilizing tracer data. Soil Sci Soc Am Proc 18(1):33–34

Kjeldby M, Eriksen AB, Holtan-Hartwig L (1987) Direct measurement of dinitrogen evolution from soil using nitrogen-15 emission spectrometry. Soil Sci Soc Am J 51:1180–1183

Knorr W, Kattge J (2005) Inversion of terrestrial ecosystem model parameter values against eddy covariance measurements by Monte Carlo sampling. Glob Chang Biol 11:1333–1351

Knowles R (1982) Denitrification. Microbiol Revs 46:43–70

Koba K, Osaka K, Tobari Y, Toyoda S, Ohte N, Katsuyama M (2009) Biogeochemistry of nitrous oxide in groundwater in a forested ecosystem elucidated by nitrous oxide isotopomer measurements. Geochim Cosmochim Acta 73(11):3115–3133

Kool DM, Müller C, Wrage N, Oenema O, van Groenigen JW (2009) Oxygen exchange between nitrogen oxides and H_2O can occur during nitrifier pathways. Soil Biol Biochem 41:1632–1641

Kool DM, Wrage N, Zechmeister-Boltenstern S, Pfeffer M, Brus D, Oenema O, Van Groenigen JW (2010) Nitrifier denitrification can be a source of N_2O from soil: a revised approach to the dual-isotope labelling method. Eur J Soil Sci 61:759–772

Kool DM, van Groenigen JW, Wrage N (2011) Source determination of nitrous oxide based on nitrogen and oxygen isotope tracing: dealing with oxygen exchange. Meth Enzymol 496:139–160

Kool DM, Wrage N, Oenema O, Dolfing J, Van Groenigen JW (2007) Oxygen exchange between (de) nitrification intermediates and H_2O and its implications for source determination of NO_3^- and N_2O: a review. Rapid Commun Mass Spectrom 21:3569–3578

Köster JR, Cardenas LM, Bol R, Lewicka-Szczebak D, Senbayram M, Well R, Giesemann A, Dittert K (2015) Anaerobic digestates lower N_2O emissions compared to cattle slurry by affecting rate and product stoichiometry of denitrification—An N_2O isotopomer case study. Soil Biol Biochem 84:65–74

Köster JR, Well R, Tuzson B, Bol R, Dittert K, Giesemann A, Emmenegger L, Manninen A, Cardenas L, Mohn J (2013) Novel laser spectroscopic technique for continuous analysis of N_2O isotopomers - application and intercomparison with isotope ratio mass spectrometry. Rapid Commun Mass Spectrom 27:216–222

Kulkarni MV, Burgin AJ, Groffman PM, Yavitt JB (2014) Direct flux and ^{15}N tracer methods for measuring denitrification in forest soils. Biogeochem. 117:359–373

Kumon Y, Sasaki Y, Kato I, Takaya N, Shoun H, Beppu T (2002) Codenitrification and denitrification are dual metabolic pathways through which dinitrogen evolves from nitrate in Streptomyces antibioticus. J Bacteriol 184(11):2963–2968

Kurakov AV, Nosikov AN, Skrynnikova EV, L'vov NP (2000) Nitrate reductase and nitrous oxide production by Fusarium oxysporum 11dn1 under aerobic and anaerobic conditions. Curr Microbiol 41: 114–119

Laughlin RJ, Stevens RJ (2002) Evidence for fungal dominance of denitrification and codenitrification in a grassland soil. Soil Sci Soc Am J 66:1540–1548

Laughlin RJ, Stevens RJ, Zhuo S (1997) Determining nitrogen-15 in ammonium by producing nitrous oxide. Soil Sci Soc Am J 61:462–465

Lewicka-Szczebak D, Dyckmanns J, Kaiser J, Marca A, Augustin J, Well R (2016) Oxygen isotope fractionation during N_2O production by soil denitrification. Biogeosc 13:1129–1144

Lewicka-Szczebak D, Well R, Bol R, Gregory A, Matthews P, Misselbrook T, Whalley R, Cardenas L (2015) Isotope fractionation factors controlling isotopocule signatures of soil-emitted N_2O produced by denitrification processes of various rates. Rapid Commun Mass Spectrom 29:269–282

Lewicka-Szczebak D, Well R, Giesemann A, Rohe L, Wolf U (2013) An enhanced technique for automated determination of ^{15}N signatures of N_2, $(N_2 + N_2O)$ and N_2O in gas samples. Rapid Commun Mass Spectrom 27:1548–1558

Lewicka-Szczebak D, Augustin J, Giesemann A, Well R (2017) Quantifying N_2O reduction to N_2 based on N_2O isotopocules–validation with independent methods (helium incubation and ^{15}N gas flux method). Biogeosc 14(3):711–732

Lewicka-Szczebak D, Well R (2020) The ^{15}N gas-flux method to determine N_2 flux: a comparison of different tracer addition approaches. Soil 6:145–152

Lewicka-Szczebak D, Well R, Koster JR, Fuss R, Senbayram M, Dittert K, Flessa H (2014) Experimental determinations of isotopic fractionation factors associated with N_2O production and reduction during denitrification in soils. Geoch Cosmo Ac 134:55–73

Liu D, Fang Y, Tu Y, Pan Y (2014) Chemical method for nitrogen isotopic analysis of ammonium at natural abundance. Anal Chem 86(8):3787–3792

Lu RK (1999) Analytical Methods for Soil and Agricultural Chemistry (In Chinese). China Agric Sci and Tech Press, Beijing, pp 107–108

Maeda K, Spor A, Edel-Hermann V, Heraud C, Breuil MC, Bizouard F, Toyoda S, Yoshida N, Steinberg C, Philippot L (2015) N_2O production, a widespread trait in fungi. Sci Rep-Uk 5:9691–9697

Mandernack KW, Mills CT, Johnson CA, Rahn T, Kinney C (2009) The delta N-15 and delta O-18 values of N_2O produced during the co-oxidation of ammonia by methanotrophic bacteria. Chem Geol 267:96–107

Mariotti A, Germon JC, Hubert P, Kaiser P, Letolle R, Tardieux A, Tardieux P (1981) Experimental determination of nitrogen kinetic isotope fractionation - some principles - illustration for the denitrification and nitrification processes. Plant Soil 62:413–430

Mary B, Recous S, Robin D (1998) A model for calculating nitrogen fluxes in soil using ^{15}N tracing. Soil Biol Biochem 30(14):1963–1979

McGeough KL, Watson CJ, Müller C, Laughlin RJ, Chadwick DR (2016) Evidence that the efficacy of the nitrification inhibitor dicyandiamide (DCD) is affected by soil properties in UK soils. Soil Biol Biochem 94:222–232

McIlvin MR, Altabet MA (2005) Chemical conversion of nitrate and nitrite to nitrous oxide for nitrogen and oxygen isotopic analysis in freshwater and seawater. Anal Chem 77:5589–5595

Menyailo OV, Hungate BA (2006) Stable isotope discrimination during soil denitrification: Production and consumption of nitrous oxide. Global Bioge Cy 20: GB3025

Melin J, Nõmmik H (1983) Denitrification measurements in intact soil cores. Acta Agric Scand 33:145–151

Metropolis N, Rosenbluth AW, Rosenbluth MN, Teller AH (1953) Equation of state calculations by fast computing machines. J Chem Phys 21(6):1087–1092

Meyer A, Bermann J, Butterbach-Bahl K, Brüggemann N (2010) A new ^{15}N tracer method to determine N turnover and denitrification of Pseudomonas stutzeri. Isotopes Environ Health Stud 46(4):409–421

Mohn J, Guggenheim C, Tuzson B, Vollmer MK, Toyoda S, Yoshida N, Emmenegger L (2010) A liquid nitrogen-free preconcentration unit for measurements of ambient N_2O isotopomers by QCLAS. Atmos Meas Tech 3:609–618

Mohn J, Gutjahr W, Toyoda S, Harris E, Ibraim E, Geilmann H, Schleppi P, Kuhn T, Lehmann MF, Decock C, Werner RA, Yoshida N, Brand WA (2016) Reassessment of the NH_4NO_3 thermal decomposition technique for calibration of the N_2O isotopic composition. Rapid Commun Mass Spectrom 30:2487–2496

Mohn J, Tuzson B, Manninen A, Yoshida N, Toyoda S, Brand WA, Emmenegger L (2012) Site selective real-time measurements of atmospheric N_2O isotopomers by laser spectroscopy. Atmos Meas Tech 5:1601–1609

Mohn J, Wolf B, Toyoda S, Lin CT, Liang MC, Bruggemann N, Wissel H, Steiker AE, Dyckmans J, Szwec L, Ostrom NE, Casciotti KL, Forbes M, Giesemann A, Well R, Doucett RR, Yarnes CT, Ridley AR, Kaiser J, Yoshida N (2014) Interlaboratory assessment of nitrous oxide isotopomer analysis by isotope ratio mass spectrometry and laser spectroscopy: current status and perspectives. Rapid Commun Mass Spectrom 28:1995–2007

Moser G, Gorenflo A, Brenzinger K, Keidel L, Braker G, Marhan S, Clough TJ, Müller C (2018) Explaining the doubling of N_2O emissions under elevated CO_2 in the Giessen FACE via in-field ^{15}N tracing. Glob Chang Biol 24(9):3897–3910

Müller C (2000) Modelling soil-biosphere interactions. CAB International, Wallingford

Müller C, Clough TJ (2014) Advances in understanding nitrogen flows and transformations: Gaps and research pathways. J Agric Sci 152(S1):34–44

Müller C, Stevens RJ, Laughlin RJ (2006) Sources of nitrite in a permanent grassland soil. Eur J Soil Sci 57:337–343

Müller C, Rütting T, Kattge J, Laughlin RJ, Stevens RJ (2007) Estimation of parameters in complex ^{15}N tracing models by Monte Carlo sampling. Soil Biol Biochem 39(3):715–726

Müller C, Laughlin RJ, Christie P, Watson CJ (2011) Effects of repeated fertilizer and slurry applications over 38 years on N dynamics in a temperate grassland soil. Soil Biol Biochem 43:1362–1371

Müller C, Laughlin RJ, Spott O, Rütting T (2014) A ^{15}N tracing method to quantify N_2O pathways from terrestrial ecosystems. EGU (ed), p. 1, European Geological Union, Vienna.

Mulvaney RL, Vandenheuvel RM (1988) Evaluation of N-15 tracer techniques for direct measurement of denitrification in soil. 4. Field studies. Soil Sci e Soc Am J 52(5):1332–1337

Mulvaney RL (1984) Determination of ^{15}N labeled dinitrogen and nitrous oxide with triple collector mass spectrometers. Soil Sci Soc Am J 48:690–692

Myrold DD, Tiedje JM (1986) Simultaneous estimation of several nitrogen cycle rates using ^{15}N: theory and application. Soil Biol Biochem 18(6):559–568

Nadeem SM, Shaharoona B, Arshad M, Crowley DE (2012) Population density and functional diversity of plant growth promoting rhizobacteria associated with avocado trees in saline soils. Appl Soil Ecol 62:147–154

Naudé SM (1929a) An isotope of nitrogen, mass 15. Phys Review 34(11):1498–1499

Naudé SM (1929b) The isotopes of nitrogen, mass 15, and oxygen mass 18 and 17, and their abundance. Phys Review 36:333–346

Nielsen LP (1992) Denitrification in sediment determined from nitrogen isotope pairing. FEMS Microb Letters 86:357–362

Nõmmik H (1956) Investigation on denitrification in soil. Acta Agric Scand 6:195–227

Norman AG, Werkman CH (1943) The use of the nitrogen isotope N15 in determining nitrogen recovery from plant materials decomposing in soil. J Am Soc Agron 35:1023–1025

Ostrom NE, Ostrom PH (2011) The isotopomers of nitrous oxide: analytical considerations and application to resolution of microbial production pathways, in: Baskaran M (Ed.) Handbook of Envir Isotope Geoch Springer, pp 453–477

Ostrom NE, Gandhi H, Coplen TB, Toyoda S, Böhlke JK, Brand WA, Casciotti KL, Dyckmans J, Giesemann A, Mohn J, Well R, Yu L, Yoshida N (2018) Preliminary assessment of stable nitrogen and oxygen isotopic composition of USGS51 and USGS52 nitrous oxide reference gases and perspectives on calibration needs. Rapid Commun Mass Spectrom 32:1829–1830

Ostrom NE, Pitt A, Sutka R, Ostrom PH, Grandy AS, Huizinga KM, Robertson GP (2007) Isotopologue effects during N_2O reduction in soils and in pure cultures of denitrifiers. J Geophys Res-Biogeo 112(G02005):02001–02012

Papen H, von Berg R, Hinkel I, Thoene B, Rennenberg H (1989) Heterotrophic nitrification by Alcaligenes faecalis: NO_2^-, NO_3^-, N2O, and NO production in exponentially growing cultures. Appl Environ Microb 55(8):2068–2072

Pataki DE, Ehleringer JR, Flanagan LB, Yakir D, Bowling DR, Still CJ, Buchmann N, Kaplan JO, Berry JA (2003) The application and interpretation of Keeling plots in terrestrial carbon cycle research. Glob Biogeochem Cyc 17, 2001GB001850

Röckmann T, Kaiser J, Brenninkmeijer CAM, Brand WA (2003) Gas chromatography/isotope-ratio mass spectrometry method for high-precision position-dependent ^{15}N and ^{18}O measurements of atmospheric nitrous oxide. Rapid Commun Mass Spectrom 17:1897–1908

Rohe L, Anderson TH, Braker G, Flessa H, Giesemann A, Lewicka-Szczebak D, Wrage-Mönnig N, Well R (2014) Dual isotope and isotopomer signatures of nitrous oxide from fungal denitrification – a pure culture study. Rapid Commun Mass Spectrom 28:1893–1903

Rohe L, Well R, Lewicka-Szczebak D (2017) Use of oxygen isotopes to differentiate between nitrous oxide produced by fungi or bacteria during denitrification. Rapid Commun Mass Spectrom 31:1297–1312

Rütting T, Clough TJ, Müller C, Lieffering M, Newton PCD (2010) Ten years of elevated atmospheric carbon dioxide alters soil nitrogen transformations in a sheep-grazed pasture. Glob Chang Biol 16:2530–2542

Rütting T, Boeckx P, Müller C, Klemedtsson L (2011) Assessment of the importance of dissimilatory nitrate reduction to ammonium for the terrestrial nitrogen cycle. Biogeosc 8:1779–1791

Rütting T, Müller C (2008) Process-specific analysis of nitrite dynamics in a permanent grassland soil by using a Monte Carlo sampling technique. Eur J Soil Sci 59:208–215

Ryabenko E (2013) Stable isotope methods for the study of the nitrogen cycle. In Zambianchi E (Ed) Topics in oceanography. IntechOpen 1–40

Schilman B (2007) Teplyakov N (2007) Detailed protocol for nitrate chemical reduction to nitrous oxide for $\delta^{15}N$ and $\delta^{18}O$ analysis of nitrate in fresh and marine waters. Geolog Survey 15:1–20

Scholefield D, Hawkins J, Jackson S (1997) Use of a flowing helium atmosphere incubation technique to measure the effects of denitrification controls applied to intact cores of a clay soil. Soil Biol Biochem 29:1337–1344

Schorpp Q, Riggers C, Lewicka-Szczebak D, Giesemann A, Well R, Schrader S (2016) Influence of Lumbricus terrestris and Folsomia candida on N_2O formation pathways in two different soils–with particular focus on N_2 emissions. Rapid Commun Mass Spectrom 30:2301–2314

Senbayram M, Well R, Bol R, Chadwick DR, Jones DL, Wu D (2018) Interaction of straw amendment and soil NO_3- content controls fungal denitrification and denitrification product stoichiometry in a sandy soil. Soil Biol Biochem 126:204–212

Sgouridis F, Stot A, Ullah S (2016) Application of the N-15 gas-flux method for measuring in situ N_2 and N_2O fluxes due to denitrification in natural and semi-natural terrestrial ecosystems and comparison with the acetylene inhibition technique. Biogeosc 13:1821–1835

Siegel RS. Hauck, RD, Kurtz, LT (1982) Determination of $^{30}N_2$ and application to measurement of N_2 evolution during denitrification. Soil Sci Soc Am J 46: 68 – 74

Snider D, Venkiteswaran JJ, Schiff SL, Spoelstra J (2013) A new mechanistic model of $d^{18}O$-N_2O formation by denitrification. Geoch Cosmoch Ac 112:102–115

Snider DM, Venkiteswaran JJ, Schiff SL, Spoelstra J (2011) Deciphering the oxygen isotope composition of nitrous oxide produced by nitrification. Global Change Biol 18:356–370

Spott O, Russow R, Stange CF (2011) Formation of hybrid N_2O and hybrid N_2 due to codenitrification: first review of a barely considered process of microbially mediated N-nitrosation. Soil. Biol Biochem 1993–2011

Spott O, Russow R, Apelt B, Stange CF (2006) A ^{15}N-aided artificial atmosphere gas flow technique for online determination of soil N_2 release using the zeolite Köstrolith SX6. Rapid Commun Mass Spectrom 20:3267–3274

Spott O, Stange CF (2007) A new mathematical approach for calculating the contribution of anammox, denitrification and atmosphere to an N_2 mixture based on a ^{15}N tracer technique. Rapid Commun Mass Spectrom 21:2398–2406

Stark JM (2000) Nutrient transformations. Methods in Ecosystem Science. OE Sala, RB Jackson, HA Mooney and RW Howarth. New York, Springer: 215–234

Stark JM, Firestone MK (1995) Isotopic labeling of soil nitrate pools using ^{15}N-nitric oxide gas. Soil Sci Soc Am J 59:844–847

Stark JM, Hart SC (1996) Diffusion technique for preparing salt solutions, kjeldahl digests and persulfate digests for nitrogen-15 analysis. Soil Sci Soc Am J 60(6):1846–1855

Stedman G (1959) Mechanism of the azide-nitrite reaction. Part I. J Chem Soc 9:2943–2949

Stevens RJ, Laughlin RJ (1994) Determining nitrogen-15 nitrite or nitrate by producing nitrous oxide. Soil Sci Soc Am J 58(4):1108–1116

Stevens RJ, Laughlin RJ (1995) Nitrite transformations during soil extraction with potassium chloride. Soil Sci Soc Am J 59(3):933–938

Stevens RJ, Laughlin RJ, Atkins GJ, Prosser SJ (1993) Automated determination of nitrogen-15-labelled dinitrogen and nitrous oxide by mass spectrometry. Soil Sci Soc Am J 57(4):981–988

Stevens RJ, Laughlin RJ, Burns LC, Arah JRM, Hood RC (1997) Measuring the contributions of nitrification and denitrification to the flux of nitrous oxide from soil. Soil Biol Biochem 29(2):139–151

Stevens RJ, Laughlin RJ, Malone JP (1998) Soil pH affects the process reducing nitrate to nitrous oxide and di-nitrogen. Soil Biol Biochem 30(8/9):1119–1126

Stieglmeier M, Mooshammer M, Kitzler B, Wanek W, Zechmeister-Boltenstern S, Richter A, Schleper C (2014) Aerobic nitrous oxide production through N-nitrosating hybrid formation in ammonia-oxidizing archaea. ISME J 8:1135–1146

Sun JF, Bai E, Dai WW, Peng B, Qu G, Jiang P (2014) Improvements of the diffusion method to measure inorganic nitrogen isotope of ^{15}N labeled (In Chinese). Chinese J Ecol 33(9):2574–2580

Sutka RL, Adams GC, Ostrom NE, Ostrom PH (2008) Isotopologue fractionation during N_2O production by fungal denitrification. Rapid Commun Mass Spectrom 22:3989–3996

Sutka RL, Ostrom NE, Ostrom PH, Breznak JA, Gandhi H, Pitt AJ, Li F (2006) Distinguishing nitrous oxide production from nitrification and denitrification on the basis of isotopomer abundances. Appl Environ Microb 72:638–644

Tanimoto T, Hatano K, Kim D, Uchiyama H, Shoun H (1992) Co-denitrification by the denitrifying system of the fungus Fusarium oxysporum. FEMS Microb Lett 93:177–180

Toyoda S, Yoshida N (1999) Determination of nitrogen isotopomers of nitrous oxide on a modified isotope ratio mass spectrometer. Anal Chem 71:4711–4718

Toyoda S, Mutobe H, Yamagishi H, Yoshida N, Tanji Y (2005) Fractionation of N_2O isotopomers during production by denitrifier. Soil Biol Biochem 37:1535–1545
Toyoda S, Yano M, Nishimura S, Akiyama H, Hayakawa A, Koba K, Sudo S, Yagi K, Makabe A, Tobari Y, Ogawa NO, Ohkouchi N, Yamada K, Yoshida N (2011) Characterization and production and consumption processes of N_2O emitted from temperate agricultural soils determined via isotopomer ratio analysis. Global Biogeo Cy 25: GB2008
Toyoda S, Yoshda N, Koba K (2017) Isotopocule analysis of biologically produced nitrous oxide in various environments. Mass Spectrom Rev 36:135–160
Ugray Z, Lasdon L, Plummer J, Glover F, Kelly J, Martí R (2007) Scatter search and local NLP solvers: A multistart framework for global optimization. INFORMS Journal on Computing, 19(3):328–340
Verhoeven E, Barthel M, Yu L, Celi L, Said-Pullicino D, Sleutel S, Lewicka-Szczebak D, Six J, Decock C (2019) Early season N_2O emissions under variable water management in rice systems: source-partitioning emissions using isotope ratios along a depth profile. Biogeosci 16(2):383–408
Wang X, Cao YC, Han Y, Tang H, Wang R, Sun X, Sun Y (2015) Determination of nitrogen and oxygen isotope ratio of nitrate in water with a chemical conversion method (In Chinese). Acta Pedol Sinica 52(3):558–566
Wächter H, Mohn J, Tuzson B, Emmenegger L, Sigrist M (2008) Determination of N_2O isotopomers with quantum cascade laser-based absorption spectroscopy. Opt Express 16(12):9239
Wassenaar L, Douence C, Altabet M, Aggarwal P (2018) N and O isotope (δ15 Nα, δ15 Nβ, δ18O, δ17O) analyses of dissolved NO_3^- and NO_2^- by the Cd-azide reduction method and N_2O laser spectrometry. Rapid Commun Mass Spectrom 32(3):184–194
Well R (1993) Denitrifikation im Wurzelraum unterhalb der Ackerkrume. Meßmethodik und Vergleich der ^{15}N-Bilanz mit der ^{15}N-Gasfreisetzungs-Methode. Diss. Fachber. Agrarwiss., Uni Göttingen. 112 p
Well R, Myrold DD (1999) Laboratory evaluation of a new method for in situ measurement of denitrification in water-saturated soils. Soil Biol Biochem 31:1109–1119
Well R, Becker KW, Meyer B (1993) Equilibrating of ^{15}N-Gases by Electrodeless Discharge: A Method of Indirect Mass Spectrometric Analysis of $^{30}N_2$ for Denitrification Studies in Soils. Isot Environ Health Stud 29(1–2):175–180
Well R, Meyer K (1998) Direct measurement of gaseous denitrification products in hydromorphic soils: concept, method and initial results. Mitt Dtsch Bodenkdl Ges 88:47–50
Well R, Becker KW, Langel R, Meyer B, Reineking A (1998) Continuous flow equilibration for mass spectrometric analysis of dinitrogen emissions. Soil Sci Soc Am J 62:906–910
Well R, Myrold DD (2002) A proposed method for measuring subsoil denitrification in situ. Soil Sci Soc Am J 66:507–518
Well R, Augustin J, Meyer K, Myrold DD (2003) Comparison of field and laboratory measurement of denitrification and N_2O production in the saturated zone of hydromorphic soils. Soil Bio Biochem 35:783–799
Well R, Kurganova I, Lopes de Gerenyu V, Flessa H (2006) Isotopomer signatures of soil emitted N_2O under different moisture conditions – a microcosm study with arable loess soil. Soil Biol Biochem 38:2923–2933
Well R, Flessa H, Xing L, Ju XT, Romheld V (2008) Isotopologue ratios of N_2O emitted from microcosms with NH_4^+ fertilized arable soils under conditions favoring nitrification. Soil Biol Biochem 40:2416–2426
Well R, Flessa H (2009) Isotopologue enrichment factors of N_2O reduction in soils. Rapid Commun Mass Spectrom 23:2996–3002
Well R, Eschenbach W, Flessa H, von der Heide C, Weymann D (2012) Are dual isotope and isotopomer ratios of N_2O useful indicators for N_2O turnover during denitrification in nitrate contaminated aquifers? Geochim Cosmochim Ac 90:265–282
Well R, Buchen C, Deppe M, Eschenbach W, Gattinger A, Giesemann A, Krause H-M, Lewicka-Szczebak D (2015) Non-homogeneity of isotopic labelling in ^{15}N gas flux studies: theory,

some observations and possible lessons, In: EGU (Ed.), EGU General Assembly. EGU EGU2015−11636

Well R, Maier M, Lewicka-Szczebak D, Koster JR, Ruoss N (2019a) Underestimation of denitrification rates from field application of the N-15 gas flux method and its correction by gas diffusion modelling. Biogeosciences 16:2233–2246

Well R, Burkart S, Giesemann A, Grosz B, Köster JR, Lewicka-Szczebak D (2019b) Improvement of the ^{15}N gas flux method for in situ measurement of soil denitrification and its product stoichiometry. Rapid Commun Mass Spectrom 33:437–448

Well R, Weymann D, Flessa H (2005) Recent research progress on the significance of aquatic systems for indirect agricultural N_2O emissions. Environ. Sci. –. J. Integr. Environ. Res. 2:143–152

Werle P, Miicke R, Slemr F (1993) The limits of signal averaging in atmospheric trace-gas monitoring by tunable diode-laser absorption spectroscopy (TDLAS). Appl Phys B Photoph and Laser Chem 57(2):131–139

Wolf B, Merbold L, Decock C, Tuzson B, Harris E, Six J, Emmenegger L, Mohn J (2015) First on-line isotopic characterization of N_2O above intensively managed grassland. Biogeosci 12:2517–2531

Wrage N, Velthof GL, van Beusichem ML, Oenema O (2001) Role of nitrifier denitrification in the production of nitrous oxide. Soil Biol Biochem 33:1723–1732

Wrage-Mönnig N, Horn MA, Well R, Müller C, Velthof G, Oenema O (2018) The role of nitrifier denitrification in the production of nitrous oxide revisited. Soil Biol Biochem 123:3–16

Wu H, Dannenmann M, Wolf B, Han XG, Zheng X, Butterbach-Bahl K (2012) Seasonality of soil microbial nitrogen turnover in continental steppe soils of Inner Mongolia. Ecosphere 3

Wu D, Well R, Cárdenas LM, Fuß R, Lewicka-Szczebak D, Köster JR, Brüggemann N, Bol R (2019) Quantifying N_2O reduction to N_2 during denitrification in soils via isotopic mapping approach: Model evaluation and uncertainty analysis. Environ Res 108806

Yoshida N (1988) N-15-Depleted N_2O as a Product of Nitrification. Nature 335:528–529

Yu L, Harris E, Lewicka-Szczebak D, Barthel M, Blomberg MRA, Harris SJ, Johnson MS, Lehmann MF, Liisberg J, Müller C, Ostrom NE, Six J, Toyoda S, Yoshida N, Mohn J (2020). What can we learn from N2O isotope data? - Analytics, processes and modelling. Rapid Commun Mass Spectrom 34:e8858

Zaman M, Nguyen ML, Matheson F, Blennerhassett JD, Quin BF (2007) Can soil amendments (zeolite or lime) shift the balance between nitrous oxide and dinitrogen emissions from pasture and wetland soils receiving urine or urea-N? Aust J Soil Res 45:543–553

Zaman M, Nguyen ML, Gold AJ, Groffman PM, Kellogg DQ, Wilcock RJ (2008a) Nitrous oxide generation, denitrification and nitrate removal in a seepage wetland intercepting surface and subsurface flows from a grazed dairy catchment. Aust J Soil Res 46:565–577

Zaman M, Nguyen ML, Saggar S (2008b) N_2O and N_2 emissions from pasture and wetland soils with and without amendments of nitrate, lime and zeolite under laboratory condition. Aust J Soil Res 46:526–534

Zaman M, Nguyen ML, Šimek M, Nawaz S, Khan MJ, Babar MN, Zaman S (2012) Emissions of nitrous oxide (N_2O) and di-nitrogen (N_2) from agricultural landscape, sources, sinks, and factors affecting N_2O and N_2 ratios. In: Greenhouse Gases - Emission, Measurement and Management. (Ed. Guoxiang Liu). Intech Croatia. 1–32

Zhang PY, Wen T, Zhang JB (2017) On Improving the Diffusion Method for Determination of $\delta^{15}N\text{-}NH_4^+$ and $\delta^{15}N\text{-}NO_3$- in Soil Extracts (In Chinese). Acta Pedol Sinica 54(4):948–957

Zou Y, Hirono Y, Yanai Y, Hattori S, Toyoda S, Yoshida N (2014) Isotopomer analysis of nitrous oxide accumulated in soil cultivated with tea (Camellia sinensis) in Shizuoka, central Japan. Soil Biol Biochem 77:276–291

8

CSA Practices for Reducing GHG Emissions

M. Zaman, K. Kleineidam, L. Bakken, J. Berendt, C. Bracken, K. Butterbach-Bahl, Z. Cai, S. X. Chang, T. Clough, K. Dawar, W. X. Ding, P. Dörsch, M. dos Reis Martins, C. Eckhardt, S. Fiedler, T. Frosch, J. Goopy, C.-M. Görres, A. Gupta, S. Henjes, M. E. G. Hofmann, M. A. Horn, M. M. R. Jahangir, A. Jansen-Willems, K. Lenhart, L. Heng, D. Lewicka-Szczebak, G. Lucic, L. Merbold, J. Mohn, L. Molstad, G. Moser, P. Murphy, A. Sanz-Cobena, M. Šimek, S. Urquiaga, R. Well, N. Wrage-Mönnig, S. Zaman, J. Zhang, and C. Müller

Abstract Agricultural lands make up approximately 37% of the global land surface, and agriculture is a significant source of greenhouse gas (GHG) emissions, including carbon dioxide (CO_2), methane (CH_4) and nitrous oxide (N_2O). Those GHGs are

M. Zaman (✉) · L. Heng
Soil and Water Management & Crop Nutrition (SWMCN) Section, Joint FAO/IAEA Division of Nuclear Techniques in Food and Agriculture, International Atomic Energy Agency (IAEA), Vienna, Austria
e-mail: m.zaman@iaea.org; zamanm_99@yahoo.com

K. Kleineidam · C. Eckhardt · A. Jansen-Willems · G. Moser · C. Müller
Institute of Plant Ecology, Justus Liebig University Giessen, Giessen, Germany

L. Bakken
Norwegian University of Life Sciences (NMBU), Aas, Norway

J. Berendt · S. Fiedler · N. Wrage-Mönnig
University of Rostock, Rostock, Germany

C. Bracken
School of Agriculture and Food Science and Earth Institute, University College Dublin, Dublin, Ireland

K. Butterbach-Bahl
Institute of Meteorology and Climate Research, Atmospheric Environmental Research (IMK-IFU), Karlsruhe Institute of Technology, Karlsruhe, Germany

Z. Cai
School of Geography Sciences, Nanjing Normal University, Jiangsu, China

S. X. Chang
Department of Renewable Resources, University of Alberta, Edmonton, AB T6G 2E3, Canada

T. Clough
Department of Soil & Physical Sciences, Faculty of Agriculture & Life Sciences, Lincoln University, Lincoln, New Zealand

responsible for the majority of the anthropogenic global warming effect. Agricultural GHG emissions are associated with agricultural soil management (e.g. tillage), use of both synthetic and organic fertilisers, livestock management, burning of fossil fuel for agricultural operations, and burning of agricultural residues and land use change. When natural ecosystems such as grasslands are converted to agricultural production, 20–40% of the soil organic carbon (SOC) is lost over time, following cultivation. We thus need to develop management practices that can maintain or even increase SOC storage in and reduce GHG emissions from agricultural ecosystems. We need to design systematic approaches and agricultural strategies that can ensure sustainable food production under predicted climate change scenarios, approaches that are being called climate-smart agriculture (CSA). Climate-smart agricultural management practices, including conservation tillage, use of cover crops and biochar application to agricultural fields, and strategic application of synthetic and organic fertilisers have been considered a way to reduce GHG emission from agriculture. Agricultural management practices can be improved to decreasing disturbance to the soil by decreasing the frequency and extent of cultivation as a way to minimise soil C loss and/or to increase soil C storage. Fertiliser nitrogen (N) use efficiency can be improved to reduce fertilizer N application and N loss. Management measures can also be taken to minimise agricultural biomass burning. This chapter reviews the current literature on CSA practices that are available to reduce GHG emissions and

K. Dawar
Department of Soil and environmental Sciences, University of Agriculture, Peshawar, Pakistan

W. X. Ding
Institute of Soil Science, Chinese Academy of Sciences, Nanjing, China

P. Dörsch · L. Molstad
Faculty of Environmental Sciences and Natural Resource Management,
Norwegian University of Life Sciences (NMBU), Aas, Norway

T. Frosch
Leibniz Institute of Photonic Technology, Technical University
Darmstadt, Darmstadt, Germany

J. Goopy
International Livestock Research Institute (ILRI), Nairobi, Kenya

C.-M. Görres
Department of Soil Science and Plant Nutrition/Department of Applied Ecology, Hochschule Geisenheim University, Geisenheim, Germany

A. Gupta
Independent Consultant India, Mumbai, India

S. Henjes · M. A. Horn
Institute of Microbiology, Leibniz University Hannover, Hannover, Germany

M. E. G. Hofmann
Picarro B.V., 's-Hertogenbosch, The Netherlands

M. M. R. Jahangir
Department of Soil Science, Bangladesh Agricultural University, Mymensingh, Bangladesh

increase soil C sequestration and develops a guideline on best management practices to reduce GHG emissions, increase C sequestration, and enhance crop productivity in agricultural production systems.

Keywords Agriculture · Carbon dioxide · Climate-smart agriculture · C sequestration · GHG · Methane · Mitigation · Nitrous oxide

8.1 Introduction on Climate-Smart Agriculture Practices and Greenhouse Gas Emissions

Agriculture is a major source of greenhouse gases (GHGs) that affect climate change and is also itself a victim of climate change. Agricultural lands make up 37.6% of the global land surface, and agriculture is a significant source of GHG emissions (IPCC 2014; Smith et al. 2008), where CO_2, CH_4 and N_2O are the major forms of trace gases that are responsible for the majority of the global warming effect. Agricultural GHG emissions are associated with agricultural soil management (e.g. tillage), use of both synthetic and organic fertilisers, livestock management, burning of fossil fuel for agricultural operations and burning of agricultural residues. In particular, agriculture can be the source for 52% and 84%

D. Lewicka-Szczebak
Laboratory of Isotope Geology and Geoecology, Institute of Geological Sciences, University of Wrocław, Wrocław, Poland

G. Lucic
Picarro Inc., Santa Clara, CA, USA

L. Merbold
Mazingira Centre, International Livestock Research Institute (ILRI), Nairobi, Kenya

J. Mohn
Laboratory for Air Pollution & Environmental Technology, Empa Dübendorf, Dübendorf, Switzerland

P. Murphy
Environment & Sustainable Resource Management Section, School of Agriculture & Food Science, and UCD Earth Institute, University College, Dublin, Ireland

A. Sanz-Cobena
Research Center for the Management of Environmental and Agricultural Risks (CEIGRAM), ETSIAAB, Universidad Politécnica de Madrid, Madrid, Spain

M. Šimek
Institute of Soil Biology, Biology Centre of the Czech Academy of Sciences, and Faculty of Science, University of South Bohemia, České Budějovice, Czech Republic

M. dos Reis Martins · S. Urquiaga
EMBRAPA Agrobiologia Seropédica, Brazilian Agricultural Research Corporation, Seropedica, RJ, Brazil

R. Well
Thünen Institute of Climate-Smart Agriculture, Braunschweig, Germany

of global anthropogenic emissions of CH_4 and N_2O, respectively (Smith et al. 2008). Since the global warming potentials of CH_4 and N_2O are much higher than that of CO_2 based on per unit mass and a 100-year time frame (IPCC 2014), advanced concepts are required to reduce agricultural emissions of CH_4 and N_2O.

In addition to causing increased levels of GHG emission, human settlement in previously unpopulated areas means that natural ecosystems are converted to agricultural production, with 20–40% of the SOC lost following cultivation and with most of that loss occurring in the first a few years (Davidson and Ackerman 1993). A recent estimate indicates that 133 billion tonnes of SOC, which is about 8% of the total global SOC stock, had been lost from the top 2 metres of soil on a global scale since agriculture started about 12,000 years ago, with the rate of loss dramatically increased since the industrial revolution (Sanderman et al. 2017). The Sanderman et al. (2017) study also indicated that the percentage of SOC loss was greater on cropland but the total amount of SOC loss was slightly higher on grazing land as more than twice as much land is grazed. This indicates that there is a greater potential to improve the % SOC gain in cropland but there is a greater potential to increase total SOC storage in grazing land. One of the key aspects of SOC is that the soil and vegetation stores about three times the organic C of the atmosphere (Plate 8.1; FAO 2004), and thus small changes in the organic C stock in the soil and vegetation can cause a large effect on atmospheric CO_2 concentration; therefore, great efforts must be made to increase SOC storage in and to reduce GHG emissions from terrestrial ecosystems. In managed systems, SOC storage can be increased by management practices such as avoiding the burning of crop residues after harvest, and the application of composts and biochar and animal manure to increase organic C input to the soil.

There is significant potential for the agriculture sector to contribute to the reduction of anthropogenic sources of GHG emissions since agriculture is a large source of GHG and to increase soil C storage since much soil C has been lost through cultivation (Plate 8.2; top panel). For example, agricultural management practices can be improved to reduce disturbance to the soil by decreasing the frequency and extent of cultivation as a way to minimise soil C loss and/or to increase soil C storage; if permanent vegetation can be maintained, soil C storage can increase, benefiting from the C cycle becoming more closed in the system and the soil being able to trap more C (Plate 8.2; bottom panel). Fertiliser N use efficiency (NUE) can be improved through strategic application of fertiliser so as to reduce N loss, whether it is through

S. Zaman
University of Canterbury, Christchurch, New Zealand

J. Zhang
School of Geography, Nanjing Normal University, Nanjing, China

K. Lenhart
Bingen University of Applied Sciences, Berlinstr. 109, Bingen 55411, Germany

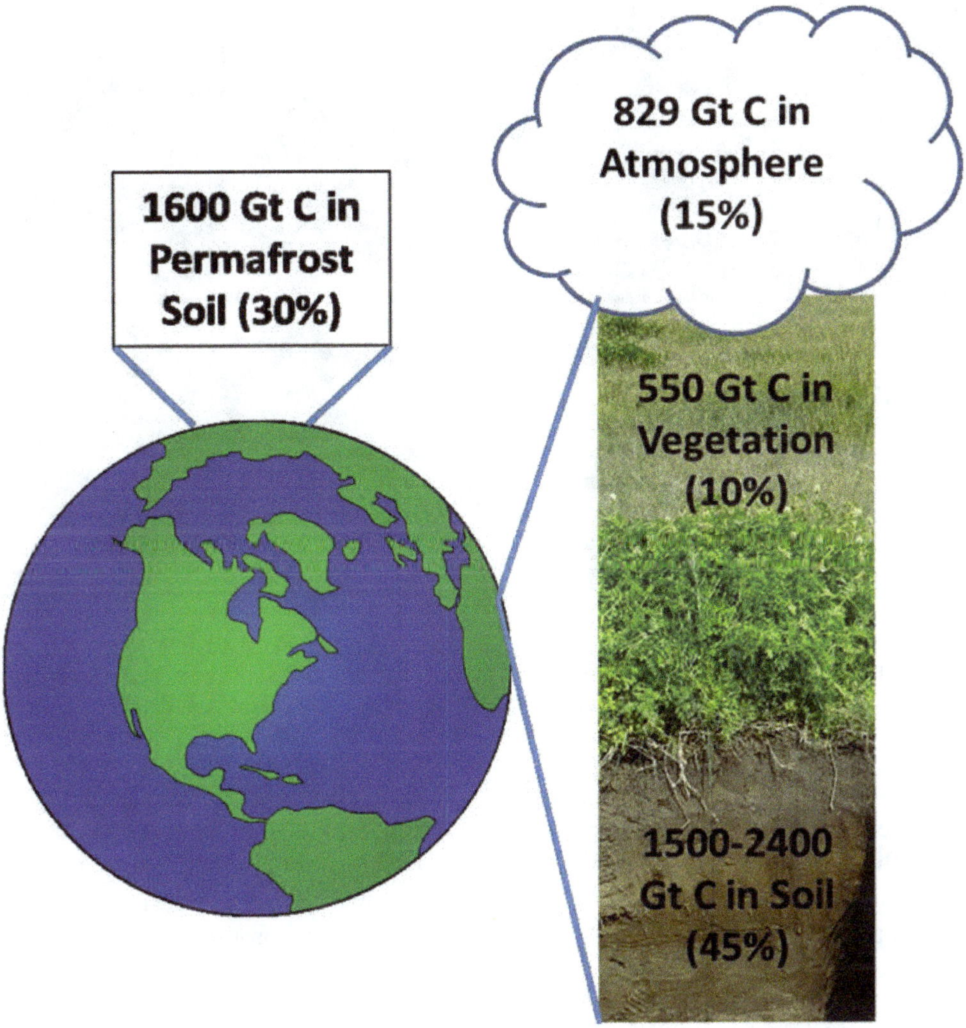

Plate 8.1 An illustration of the distribution of carbon in terrestrial ecosystems, including the atmosphere. (*Source* Schwartz 2014)

leaching or gaseous form of N loss. Management measures can also be taken to minimise agricultural biomass burning. Climate-smart agriculture (CSA) management practices, including strategic use of synthetic and organic fertilisers and water, conservation tillage, use of cover crops, and the use of amendments such as biochar, nitrification inhibitors and lime to agricultural fields, have been considered a way to reduce GHG emissions from agriculture (Bai et al. 2019; Lipper et al. 2014; Zaman et al. 2008a and 2009; Zaman and Blennerhassett 2010). The FAO defines CSA as a systematic approach for developing agricultural strategies that can ensure sustainable food security under predicted climate change scenarios (FAO 2013). Based on this definition, a range of agricultural practices can be developed to help improve food security and environmental quality simultaneously in the context of global change.

Plate 8.2 An illustration of a disturbance to the soil causes an increased release of CO_2 and other greenhouse gases (and leading to the soil being leaky, open treasure chest) and in soils with undisturbed vegetation will cause the carbon cycle to be more closed (*Source* FAO 2013)

Since the soil can act as a sink or source for CO_2 and affect climate change, if we can enhance the C sink strength and remove more CO_2 from the atmosphere by adopting CSA, then we can be in a win–win situation by not only combating the negative effects of climate change but also improving soil quality and health, including the retention of nutrients and water, and increased agricultural productivity.

8.2 Climate-Smart Agricultural Technology to Reduce GHG Emissions

Climate-smart agriculture emphasises on improving risk management, enhancing information flows and promoting local institutions to increase the adaptive capacity of communities to climate change (Campbell et al. 2014), as such CSA plays a pivotal role in maintainable development. Climate-smart agricultural practices such as the use of cover crops, amendments and tillage management play key roles in reducing agricultural GHG emissions. Take N_2O, a potent GHG, as an example, the emission of N_2O is affected by many factors, such as the use of N process inhibitors (urease and nitrification inhibitors), the strategic application of synthetic fertilisers (the right type, the right amount that is based on crop requirement and soil tests, at the right plant growth stage, and the right method (e.g. even spread)), avoiding the application of N fertilisers to very wet or very dry soils, maintaining soil pH above

6 by adding lime to lower the N_2O:N_2 ratio, co-application of animal manure and chemical fertilisers to provide a more balanced nutrient supply, optimising animal stocking rate to avoid over-grazing, keeping animals off the pasture especially in the wet season to minimise N input and avoid soil compaction, and minimising the excessive use of farm machinery. Many of those fall under CSA practices. Below, key CSA practices and their effect on GHG emissions from agriculture are discussed in detail.

8.2.1 Nitrogen Process Inhibitors and Greenhouse Gas Emissions

Multiple microbial soil processes contribute to GHG emissions. One of the major processes contributing to GHG emissions is the mineralization of organic matter by microbial organisms, a process also called soil heterotrophic respiration, where organic C is converted to CO_2 and released to the atmosphere. Nitrification and denitrification are key processes contributing to N_2O emissions from the soil, where ammonia-oxidation and successive nitrifier denitrification or denitrification, among other processes, can be important pathways for N_2O emissions (Fig. 7.8) (Guo et al. 2018). In the ammonia oxidation process, N_2O can be produced by the chemical decomposition of hydroxylamine (NH_2OH). One potential way to mitigate N_2O emissions is to use nitrification and urease inhibitors to slow down the rate of nitrification and reduce the availability of the substrate (NH_4^+) for nitrification.

8.2.1.1 Synthetic Nitrification Inhibitors

Nitrification is the process where ammonium or ammonia is converted to nitrate, via nitrite as an intermediate N species. Since the formation of nitrite is usually the rate-limiting step in the nitrification process, nitrite usually does not accumulate in the soil, unless the soil has a high pH where nitrite oxidisers are inhibited (Rodgers 1986). During the nitrification process, N_2O can be formed and emitted from the soil. The final product of nitrification, i.e nitrate, is subject to leaching losses, and when anaerobic conditions develop, nitrate is denitrified thus leading to the formation of N_2O. Both nitrate leaching and denitrification are major pathways for N losses from the soil. Reducing the rate of nitrification can both conserve N in the soil and reduce N_2O emissions (Abalos et al. 2012). In this respect, the application of nitrification inhibitors can effectively reduce nitrification rates and the buildup of nitrate which can further be denitrified. Nitrification inhibitors are a group of chemical compounds that slow down the conversion rate of ammonium or ammonia to nitrate by inhibiting the ammonia monooxygenase activity, through disrupting the activity of *Nitrosomonas* and *Nitrobacter* bacteria (Abalos et al. 2014). Nitrification inhibitors are applied with ammonium-, ammonia- or urea-based fertilisers as the application of such fertilisers substantially increases the rate of nitrification. The need for nitrification inhibitors in non-fertilised soils is rare as nitrification rates in such soils are low. The ammonium or ammonia can come from urine, manures, composts or crop residues as they decompose, or fertilisers such as ammonium sulphate or urea (Rodgers 1986). Under suitable conditions (e.g. warm soil temperature and a moisture content near field capacity, when there is still ample O_2 available), nitrification occurs within days or weeks of the application of ammonium-based fertilisers and thus if nitrification can be effectively reduced within that timeframe, N loss from the system can be substantially reduced (Sanz-Cobena et al. 2017).

There are at least eight compounds commercially available as nitrification inhibitors but the most commonly used ones are 2-chloro-6-(trichloromethyl)-pyridine (nitrapyrin), dicyandiamide (DCD), 3,4-dimethylpyrazole phosphate (DMPP, e.g. ENTEC®) (IPNI undated) and pronitradine (Centuro™). Nitrapyrin (commercial product: N-Serve™ and Instinct™) can be applied to the soil in different ways: if anhydrous ammonia is the fertiliser being applied, the nitrification inhibitor can be injected with the fertiliser, if solid N fertilisers are being applied, the inhibitor can be coated onto the fertiliser, and if manure is used as an organic fertiliser, the inhibitor can be mixed with the manure before manure application. Nitrapyrin is usually effective for less than 30 days after being applied to the soil and is volatile, therefore, the best way to apply this inhibitor is to incorporate it into the soil; DMPP can be effective for reducing nitrification rates for 25–70 days and is usually pre-blended with fertilisers; DCD can last 25–55 days and can be coated on solid fertilisers (Sanz-Cobena et al. 2012), or surface applied to soils that have been applied with manure or on grazing land to reduce nitrification from urine patches (IPNI undated). While DMPP is somewhat immobile, DCD can be relatively easily leached from the soil; those behaviours of the nitrification inhibitors should be considered when they are applied in the field.

DCD and DMPP have been found to be equally effective in changing soil inorganic N content, leaching of dissolved inorganic N (DIN) and emissions of N_2O in a recent meta-analysis of field trials reported in the literature (Yang et al. 2016). Their cost-benefit analysis (CBA) showed that the economic benefit was about seven times greater with DCD than with DMPP when applied with ammonium-based fertilisers to reduce nitrification (Yang et al. 2016). Those two nitrification inhibitors are among the most commonly used. DCD is cheaper and less volatile, but the application rate of DMPP is typically one-tenth that of DCD, and DMPP has a lower eco-toxicological effect for plants as is summarised in Yang et al. (2016). In an Australian study, DMPP application (as ENTEC®) decreased N_2O emissions by 15% in a subtropical pasture in Queensland (Lam et al. 2018).

The effectiveness of the nitrification inhibitors can be affected by soil properties such as soil water content (Barrena et al. 2017), and soil organic matter and clay content (Zhu et al. 2019). Zhu et al. (2019) reported that the efficiency of DMPP in reducing nitrification and N_2O emissions was lower in soils with high organic matter and clay contents, likely due to the high rates of adsorption of DMPP by soil organic matter and clay. The effectiveness of DCD in reducing N_2O emissions from urine patches in New Zealand is highly season-specific, with reductions of 52, 39 and 16% in autumn, spring and summer, respectively, but DCD application increased NH_3 emissions by 56, 9 and 17% in the respective seasons (Zaman et al. 2009). Management practices can also affect the efficiency of nitrification inhibitors. For example, biochar application to the soil has been shown to decrease the efficiency of DMPP both at 40 and 80% of water-filled pore space (WFPS) in a laboratory incubation study (Fuertes-Mendizábal et al. 2019). The use of nitrification inhibitors increases crop yield and NUE, but the effectiveness was greatest when they are used in coarse-textured soils, irrigated systems and/or crops receiving high rates of N fertiliser input (Abalos et al. 2014).

8.2.1.2 Synthetic Urease Inhibitors

Urease inhibitors retard the activity of urease, which exists in the soil and plant residues. Urease is involved in the conversion of urea to ammonium in a process called hydrolysis. Unfortunately, the urea hydrolysis process increases the pH of the soil and causes a large proportion of the formed ammonium to be volatilized as ammonia. Urease inhibitors would slow down the rate of hydrolysis or the rate of release of ammonium, reduce the loss of N as ammonia through volatilization, and increase the NUE of urea fertilisers applied to the soil. One of the main reasons for the improved NUE is for more urea to be washed into the soil over time (when the rate of hydrolysis is suppressed), as urea is highly soluble in water. The better contact with the soil increases the chances of the released ammonium to be adsorbed by the cation exchange sites.

The N-(n-butyl) thiophosphoric triamide (NBPT) and N-(n-propyl) thiophosphoric triamide (NPPT) are two chemicals that have been shown to be effective in inhibiting urease activities (Rodriguez et al. 2019; Lam et al. 2018; Zaman et al. 2009; Sanz-Cobena et al. 2008). Products containing those urease inhibitors include Agrotain™ (that contains NBPT) and Limus™ (that contains both NBPT and NPPT). The application of NBPT prior to urine deposition was more effective in reducing ammonia volatilization loss (17.5–27.6% reduction) as compared with applying the NBPT after the urine deposition (0.6–2.9% reduction) in pastureland in New Zealand (Rodriguez et al. 2019). The effectiveness of NBPT is highly season-specific for reducing NH_3 volatilization loss from urine patches in New Zealand, with reductions of 29, 93 and 31% in autumn, spring and summer, respectively (Zaman et al. 2009). An Australian study reached similar conclusions that urea applied with NBPT (as Green UreaNV®) was effective in decreasing NH_3 volatilization (by 44%) in a subtropical pasture in Queensland (Lam et al. 2018). The effectiveness of NBPT was greater in alkaline soils ($pH \geq 8$) (Abalos et al. 2014). It has also been observed, under laboratory conditions, that application of urease inhibitors in soils where nitrification is the main pathway in the production of N_2O (i.e. WFPS < 50%), could be an effective way to mitigate these N losses (Sanz-Cobena et al. 2014).

The combined application of urease inhibitor + the nitrification inhibitor, DCD, (e.g. Agrotain Plus) to inhibit both the hydrolysis of urea and the nitrification processes to minimise the N loss showed best results in maximising NUE. When urea and ammonium nitrate (UAN) was applied at 150 mg N kg^{-1} in a sandy loam soil in the United States with Agrotain Plus, N_2O emissions were reduced by 78% compared to the control (Cai et al. 2018). The use of double inhibitors containing both NBPT and DCD (3:7) in a New Zealand study reduced NH_3 volatilization by 14, 78 and 9% in autumn, spring and summer, respectively, and N_2O emissions by 37, 67 and 28%, respectively, from urine patches (Zaman et al. 2009). However, another study showed that adding DCD with NBPT did not further reduce NH_3 volatilization loss, but in fact enhanced the volatilization loss by maintaining a higher soil NH_4^+ concentration and pH for a longer period of time after urea application, indicating that DCD co-applied with NBPT and urea could offset the effect of NBPT in reducing volatilization losses (Soares et al. 2012).

8.2.1.3 Biological Nitrification Inhibitors

Some plant species can release secondary metabolites through root exudation and/or from leaf litter. Such metabolites have the ability to suppress microbial nitrification (Souri and Neumann 2018). In some earlier studies, detectable biological nitrification inhibition (BNI) was found in root exudates of sorghum (*Sorghum bicolor* (L.)), pearl millet (*Pennisetum glaucum* (L.) R. Br.) and groundnut (*Arachis hypogaea* (L.)) among tested cereal and legume crops, while *Brachiaria humidicola* (Rendle) Schweick and *B. decumbens* Stapf had the highest BNI capacity among pasture grass species tested (Subbarao et al. 2007). In addition, when BNI compounds from root exudates were applied to the soil, their inhibitory effects on NO_3^- formation lasted for more than 50 days (Subbarao et al. 2007). Linoleic acid, α-linolenic acid, and methyl linoleate, a fatty acid methyl ester of linoleic acid, are some of the example compounds that are effective in biologically inhibiting nitrification (Subbarao et al. 2008). Subbarao et al. (2007) suggested that some level of BNI is likely a widespread phenomenon in tropical pasture grass species and those properties could be used to suppress nitrification in natural or managed systems. The production in and release from plants of BNIs are triggered by the presence of NH_4^+ in the rhizosphere of plants, which means that BNIs are released where the majority of the nitrifier populations reside (Subbarao et al. 2013a). The pH in the rhizosphere will affect the release of BNIs from roots; for example, sorghum plants release BNIs from their roots in the presence of NH_4^+ when the rhizosphere pH is between 5.0 and 6.0 (Subbarao et al. 2013b), indicating the usefulness of BNI to reduce nitrification in alkaline soils will be non-existent or very low. More research is needed to understand and take advantage of BNI in agricultural production systems to reduce N loss and improve NUE.

8.2.2 *Soil Amendments and Greenhouse Gas Emissions*

8.2.2.1 Mulch

Addition of mulch to the soil will change the availability of carbon (C) and other nutrients to microbial populations and will thus affect soil GHG emissions. Existence of a litter layer (mulch) can induce microbial N immobilisation in the litter layer, and result in reduced available N and reduced plant growth rate; however, the litter layer may benefit plant growth by conserving soil moisture (Matsushima and Chang 2006) but may reduce soil temperature and N mineralization rates (Matsushima and Chang 2007). Addition of mulch can immobilise mineral N in the soil and reduce the availability of NH_4^+ for nitrification and NO_3^- for denitrification, and thus reduce N_2O emissions as compared to no mulch addition (Wu et al. 2013). Using wood bark mulch reduced the nitrate concentration in the soil and cut soil N_2O emissions by up to 28% in a grape (*Vitus vinifera* L. cv. Merlot) yard on a sandy loam soil in British Columbia, Canada, when measured over a two-year period (Fentabil et al. 2016). However, overloading of straw to the soil surface can delay seed germination and result in the need for additional fertiliser supply to compensate for the N that may be immobilised in the critical period of the early growing season (Procházková et al. 2003). As far as CO_2 emissions are concerned, mulching will usually result in increased CO_2 emissions due to the addition of labile C in the mulch, with the rate of CO_2 emissions increasing with the increased rate of mulch addition (Wu et al. 2013). Major anthropogenic sources of methane emissions include fossil fuel production, landfills and livestock farming, but some agricultural soils can be an anthropogenic source of methane emissions as well (IPCC 2007). In rice paddy systems, straw application has been shown to increase CH_4 emissions (Bossio et al. 1999; Ma et al. 2008); however, straw addition significantly reduced CH_4 emissions under an aerobic condition in a laboratory incubation experiment, indicating that under upland conditions, straw application increased the soil's ability to take up CH_4 (Tate et al. 2007).

8.2.2.2 Biochar

Biochar has been widely studied for its effects on GHG emissions. Biochar addition to the soil can change a range of soil properties, including the cycling of C and N. Biochar application has been widely reported to reduce N_2O emissions (Wu et al. 2013; Cayuela et al. 2013; Chang et al. 2016; Hüppi et al. 2016). The application of biochar to the soil has been shown to reduce denitrification and decrease N_2O emissions by 10–90% when tested on 14 different agricultural soils, where a consistent reduction of the $N_2O/(N_2 + N_2O)$ ratio was observed, indicating that biochar reduces N_2O emissions by facilitating the last step of the denitrification process and producing more N_2 rather than N_2O (Cayuela et al. 2013). However, in some soils, biochar application can stimulate nitrification and increase N_2O emissions; therefore, the effect of biochar application on N_2O emissions is related to the dominant N_2O formation pathway that operates in a soil (Sánchez-García et al. 2014).

Biochar application has been reported to reduce CH_4 emissions from paddy soils, one of the largest anthropogenic sources of CH_4 emissions on a global scale (Feng et al. 2012). The reduction in paddy CH_4 emissions by biochar application was not a result of the inhibition of methanogenic archaea, but resulted from increased methanotrophic proteobacterial abundances and decreased ratios of methanogenic to methanotrophic abundances (Feng et al. 2012).

In agricultural production systems, a large quantity of crop residues are produced and the return of crop residues in the raw form vs after the crop residue is converted to biochar can have dramatic effects on the emissions of all three trace gases (Wu et al. 2013), and most research suggests that there are substantial beneficial effects to be gained in mitigating climate change by converting crop residues to biochar and applying biochar to the soil instead. The effect of biochar on GHG emissions itself is highly complex as many factors alter the function of biochar on C and N transformation processes and thus GHG emissions. When the effect of the stable C input in the form of biochar is considered, the application of biochar is considered an effective technique to mitigate climate change due to its negative emissions potential (0.7 Gt Ceq. Yr^{-1}) and its lower impact on land, water use, albedo, energy requirement and cost as compared to other negative emissions technologies such as direct air capture, increased weathering that takes up CO_2 from the air, bioenergy projects with C capture and storage, and afforestation/deforestation (Smith 2016).

8.2.2.3 Liming to Shift the Balance Between N_2O and N_2 Emissions

Soil pH is one of the key regulators of microbiological processes that affect N_2O and N_2 production and their ratio. The soil pH threshold for nitrification is 5; however, nitrification can occur even below pH 5 as some nitrifier strains are adapted to acidic conditions (Bouwman 1990). Denitrification has been reported to occur over a wide range of soil pH (5–8) (Flessa et al. 1998); however, laboratory experiments with artificially adjusted soil pH suggest that, under optimised conditions (very low pO_2, NO_3^- and with glucose amendment), denitrification can proceed even at pHs below 4 or above 10 (Šimek et al. 2002). Numerous laboratory and field studies have shown that soil pH affects N_2O and N_2 emissions and thus the ratio of these gases (e.g. Stevens and Laughlin 1998). In experiments conducted under controlled environmental conditions, raising soil pH to 7 through lime application has been found to significantly increase N_2 emissions from pasture and wetland soils treated with cow urine, urea and KNO_3 at 200 kg N ha^{-1} (Zaman et al. 2007, 2008b). Similar trends of enhanced N_2 emissions after raising soil pH to 7 was observed in pasture soils treated with urea/urine in a field experiment (Zaman and Nguyen 2010). In another study, a site with the greatest animal impact, the ratio of N_2 to N_2O produced during denitrifying enzyme activity (DEA) measurements was fivefold higher, and the pH was 2 units higher, than a site with the least animal impact, indicating that soil conditions were favourable for production of N_2 rather than N_2O in the area with intense excretal returns and treading (Hynšt et al. 2007).

Most researchers attribute high N_2O and low N_2 emissions in acidic conditions to the suppression of N_2O-reductase (inhibition starts at soil pH 4.5) (Daum and Schenk 1998; Flessa et al. 1998; Stevens and Laughlin 1998; Zaman et al. 2007). It is also possible that denitrifying enzymes are susceptible at low soil pH and produce N_2O from intermediate products (Nägele and Conrad 1990). However, the lower rates of N_2 emissions and higher N_2O:N_2 ratio at low soil pH could be due to lower amounts of soil organic C and mineral N available to the denitrifying population rather than a direct effect of low pH on denitrification enzymes (Šimek and Cooper 2002). Regardless of the biochemical mechanism for soil pH effects on N_2 emissions, raising soil pH through the application of amendments such as lime appears a viable approach to mitigate N_2O emissions (Šimek et al. 2002; Zaman and Nguyen 2010; Zaman et al. 2007, 2008b).

8.2.3 Fertiliser Type and Management and Greenhouse Gas Emissions

The use of different fertiliser types and the management of fertiliser applications can have marked effects on nitrification, denitrification and GHG emission rates (Mosier et al. 2006; Wang et al. 2018). As discussed in the earlier section, the type of fertiliser (some fertilisers are acid-forming while others raise soil pH when applied to the soil) applied can affect the total amount of GHG emitted as well as the N_2O:N ratio in the emissions. A number of agronomic practices have been widely tested to minimise N losses from agricultural production systems, for example, alteration of the rate or timing of fertiliser application, such as autumn vs. spring, basal vs. broadcast, deep vs. surface applications, point injection placement of solutions, foliar applications of urea (Subbarao et al. 2013b), and split application. Fertilisers such as polythene-coated urea (PCU) that releases N slowly in the soil have been demonstrated to reduce nitrification (Zvomuya et al. 2003); however, the use of such fertilisers can be limited by the high cost for purchasing such fertilisers (Subbarao et al. 2013b).

8.2.4 Cropping Systems and Greenhouse Gas Emissions

8.2.4.1 Agroecosystems

The type of cropping system used has a significant effect on GHG emissions as cropping systems will differ in their fertilisation regime, crop productivity (and thus the amount of organic matter input to and retention in the soil), crop species being used (N-fixing vs. non-N-fixing crop species), water management practices, and tillage management, among others (Snyder et al. 2009). On an extreme case, GHG emission rates and timing will be dramatically different between rice cultivation in flooded fields vs. wheat production in well-drained upland sites in a rice-wheat rotation, where the rice-wheat belt makes up 24–27 million ha in South and East Asia (Wassmann et al. 2004).

Rice cultivation in paddy fields is a unique cropping system in Southeast Asia. In Vietnam, for example, rice production is the largest source of agricultural GHG, with 37.4 Tg CO_2 equivalent of total emissions, that account for 58% of agricultural GHG emissions in that country (United Nations 2013). Many factors, such as the management of fertiliser applications, animal manure and crop residue management, water regime used during rice production, and use of urease and/or nitrification inhibitors, will affect the emissions of GHGs from paddy fields. One of the biggest concerns of GHG emissions from paddy fields is the emission of CH_4 as paddy fields are mostly submerged in water during the growing season and anaerobic condition caused substantial CH_4 emission to occur. The CH_4 emissions are the balance of CH_4 production and oxidation in the soil, and are affected by factors influencing the transportation of CH_4 from the anoxic soil/free-standing water to the atmosphere (Aulakh et al. 2001). Up to 80% of the CH_4 produced in paddy soils is oxidised in the rhizosphere or microsites that are less anaerobic before it is released to the atmosphere (Sass et al. 1991; Holzapfel-Pschorn et al. 1985) and this process helps to dramatically reduce the rate of CH_4 emissions from paddy soils.

Proper water management can substantially reduce CH_4 emissions from paddy fields; midseason drainage has been reported to reduce CH_4 emissions by 44% and alternating wetting–drying cycles at ten-day intervals by 61% as compared to continuously flooded management in southeast China (Lu et al. 2000). Application of urea can increase CH_4 emissions from paddy fields as ammonium can inhibit CH_4 oxidation (Conrad and Rothfuss 1991). However, others found that application of ammonium-based fertilisers can reduce CH_4 emissions as ammonium enhances methanotrophic bacteria activities in the rhizosphere of rice plants (Bodelier et al. 2000). The effect of ammonium on CH_4 oxidation is dependent on the CH_4 concentration: inhibition at low initial CH_4 concentration (500 $\mu l\, l^{-1}$) but stimulation at high initial CH_4 concentration (2000 $\mu l\, l^{-1}$) (Cai and Mosier 2000). However, ammonium sulphate has been found to be a promising fertiliser to use, as opposed to urea, to mitigate CH_4 emissions as sulphate enhances sulphate-reducing bacteria activities, which decrease the availability of substrates for methanogens in submerged soils (Yagi et al. 1997). Reductions in CH_4 emissions after ammonium sulphate application has been reported to range from 10 to 67% (Schütz et al. 1989; Wassmann et al. 2000). Ammonium is the preferred N form for rice, therefore, applying ammonium form of N that does not enhance CH_4 emissions would be beneficial for the environment. Application of sulphate-containing phosphorus fertilisers (Achtnich et al. 1995) and gypsum ($CaSO_4$) (Lindau und Bollich 1993), a common soil amendment for sodic and/or alkaline soil reclamation, have been shown to reduce CH_4 emissions from paddy soils. Therefore, choice of fertiliser is important in minimising CH_4 emissions from rice paddies, in this case, the use of sulphate-based N or phosphorus fertilisers would be preferred in flooded rice fields.

Applying organic fertilisers such as animal manure and returning crop residue to the soil are necessary to maintain and enhance the sustainability of rice production. However, addition of organic materials to rice fields would increase CH_4 emissions, and organic materials application to the soil can lower the soil redox potential and supply C to methanogens that are responsible for CH_4 production in paddy soils. The production of CH_4 in paddy soils is markedly influenced by the quality and quantity of organic materials added to the soil (Minasny et al. 2017). Therefore, proper manure and crop residue management strategies need to be developed to achieve environment-friendly rice cultivation. One potential alternative to the application of crop residues to paddy soil is to convert crop residues to biochar that is slow to decompose after soil application (Ippolito et al. 2012); biochar application to soil can increase soil aeration and soil C content but mitigate CH_4 emissions as compared with the conventional crop residue application (Feng et al. 2012; Karhu et al. 2011; Liu et al. 2011; Xie et al. 2013). However, other studies reported increased CH_4 emissions from rice fields after biochar application, which may be related to increased substrate supply and enhanced environment for methanogenic activity, and increased rice growth (Knoblauch et al. 2011; Lehmann et al. 2011). The effects of biochar application on CH_4 emissions is thus dependent on soil type, agricultural management practices used, and the type of biochar applied (Lehmann et al. 2011; Waters et al. 2011). Site-specific research should be conducted before any recommendation on agricultural management practices is made for farmers to adopt.

In rice paddies, the flooded condition is conducive for denitrification to occur. Even though less attention has been paid to N_2O as compared to CH_4 emissions from paddy fields, N loss in the form of NH_3 volatilisation and N_2O emissions and NO_3^- leaching affect GHG emissions as well as NUE in rice production systems. Therefore, urease inhibitors can be used to slow down the rate of urea hydrolysis when urea is used as the main N fertiliser (Rogers et al. 2015), while the use of nitrification inhibitors can reduce nitrate leaching loss (Li et al. 2008) and N_2O emissions (Majumdar et al. 2000; Kumar et al. 2000).

8.2.4.2 Organic Farming

Organic farming has long been considered a viable agricultural practice to improve soil health, reduce the resource use intensity, reduce the environmental impact of agriculture and improve food quality (Squalli and Adamkiewicz 2018). The comprehensive study of Squalli and Adamkiewicz (2018), based on longitudinal state-level data in the United States collected between 1997 and 2010, demonstrates that a 1% increase in organic farming acreage can result in a 0.049% reduction in GHG emissions; however, they showed that the net effect of organic farming on GHG emissions is dependent on the contribution of transportation (fuel burning) on methane and nitrous oxide emissions, even though their calculation indicates that the negative environmental effect of transportation output associated with organic food production is small relative to the environmental benefits of organic farming. A study in Switzerland showed a 40% reduction in N_2O emissions for organic compared to conventional systems; however, yield-scaled N_2O emissions under silage maize were not different between organic and conventional systems (Skinner et al. 2019), indicating that even if we consider the lower yield in organic systems, organic farming does not have a negative effect on GHG emissions. In contrast, Aguilera et al. (2015) and Cayuela et al. (2017) showed in two Meta-analyses that fertilisation with solid organic manures (the most used form of N fertilisers in organic agroecosystems) led to the lowest N_2O emission factor in Mediterranean cropping systems worldwide. On the other hand, we must recognise that organic farming does have a lower crop yield (c. 25% on average) which may require a larger area of land to be cultivated to produce the same amount of food compared to conventional farming (Kniss et al. 2016).

8.2.4.3 Row, Intercropping and Crop Rotation

When dealing with a cropping system that involves rice production (e.g. in Southeast Asia), it is often difficult to strike a balance in the mitigation of different GHGs. Using a rice-wheat crop rotation system as an example, the two crops markedly differ in their nature and intensity of GHG fluxes, where CH_4 emissions are a major contributor to GHG emissions from rice paddies. In rice production systems, water regimes, rice cultivars and soil properties all markedly affect CH_4 emissions (Cai et al. 2003). In addition, N_2O is also emitted in large quantities from rice production systems following aerobic-anaerobic cycles; on the other hand, N_2O is emitted in short-term pulses after fertilisation, heavy rainfall or irrigation events and is the main GHG emissions of concern in upland wheat production systems (Wassmann et al. 2004). It is often difficult to balance emissions between CH_4 and N_2O when designing GHG mitigation strategies in a rice-wheat system, as measures to reduce CH_4 emissions often intensify N_2O emissions (Wassmann et al. 2004).

In evaluating the effect of intercropping on GHG emissions, Ricord (2018) studied GHG emissions from a sole maize crop, a sole soybean crop and a maize–soybean intercrop and found that the cereal–legume intercropping system effectively reduced N_2O emissions. In a similar study on the North China Plain, N_2O fluxes were lower from maize–soybean intercropping than a maize monoculture system in three growing seasons (2013–2015), when all cropping systems were applied with 240 kg N ha^{-1} as urea in two split applications (Shen et al. 2018). Shen et al. (2018) showed that the fertiliser N loss as N_2O was lower in the maize-soybean intercropping (1.6%) and soybean monoculture (1.7%) than in the maize monoculture (2.3%), concluding that maize–soybean intercropping should be recommended as a climate-smart cropping systems for use on the North China Plain. A maize–wheat intercropping system coupled with reduced tillage and stubble mulching can increase grain production and decrease C emissions in an arid area in northwest China (Hu et al. 2015; Yin et al. 2018). Therefore, choice of a cropping system to use and the associated management practices are important decisions to make to minimise GHG emissions.

8.3 Climate-Smart Agriculture (CSA) Practices and C Sequestration

Climate-smart agriculture has three key objectives: (1) to increase agricultural production per unit land area so as to increase income, food security and community development, (2) to improve the adaptive capacity at multiple levels, i.e. from the farm to the national level and (3) to reduce GHG emissions and to enhance C sinks in ecosystems (Campbell et al. 2014). Climate-smart agricultural management practices, including conservation tillage, use of cover crops, and biochar application to agricultural fields, has been widely considered a way to reduce GHG emissions from agriculture.

Conventional tillage has been identified as one of the causes of widespread land degradation problems, such as deterioration of soil structure, soil erosion and decreased soil fertility, affecting the long-term sustainability of agricultural production (Barber et al. 1996). Many climate-smart agricultural technologies have been tested to improve SOC storage in the agricultural landscape, and many of those have been demonstrated to be effective. Climate-smart agricultural technologies such as the use of cover crops, use of perennial crops, application of manure and biochar, reduced/minimum tillage or zero tillage, and crop rotation have all been shown to increase SOC storage.

Field experiments in Australia on light-textured soils in southern Australia indicate that conservation tillage (3–19 years in duration) was effective in increasing SOC levels as compared with conventional tillage, but only in areas with >500 mm annual precipitation and in the top 2.5–10.0 cm of the soil; the lack of conservation tillage effects on SOC levels in other climatic condition or soil layers was mainly attributed to low crop yield related to low rainfall, partial removal of stubble due to grazing and the high decomposition rate in areas with high air temperature (Chan et al. 2004).

8.4 Life Cycle Assessment (LCA) for Estimating the C Footprint of Agro-Food Systems

The proposal of effective GHG mitigation strategies in the agri-food sector needs to be based on a whole-system approach. This means that not only direct emissions but also indirect GHG losses (both upstream and downstream from the production systems) must be considered. For that purpose, the calculation and use of the "C footprint" (CF) or "C budget" of agri-food entities, from products to systems, is primordial (highly important) (Plate 1.1).

The calculation of the CF of the agri-food system requires the accounting for GHG emissions that occur in each of the phases of food and feed production, including not only those that take place in the agricultural sector itself, that is, in crop fields and farms, but also during the manufacture of agricultural inputs, or those derived from the distribution, marketing and consumption of food, using an LCA approach.

The LCA adopts a "bottom-up" approach, that is, it records in detail the emissions generated along the product supply chain, using information on production technologies of the different goods and services. To this end, a "product system" has to be defined, which includes both the different phases of the supply chain (i.e. the "life cycle" of the product) and the exchanges that occur with the environment (i.e. GHG emissions); and a "functional unit" for each food (e.g. 1 kg of product). As a result of the application of an LCA, the "emission coefficients" (i.e. the amount of GHG emissions in kg CO_2 eq./kg of product) are obtained, which can be applied to both intermediate and final products, whether domestic or imported.

In estimating the CF of the Spanish Agri-food sector, Aguilera et al. (2015) used the following information as a source (Plate 8.3): (i) inventory analysis based on previous work of the Laboratory of History of Agroecosystems (UPO), based on official data (Yearbook of Agricultural Statistics, FAOSTAT, National Emissions Inventory, etc.); (ii) industrial input emission factors based on "embodied energy" (Aguilera et al. 2015); (iii) Mediterranean N_2O emission factors–meta-analysis (Cayuela et al. 2017); (iv) C sequestration with HSOC model (Aguilera et al. 2018); and (v) calculation of emissions associated with irrigation (Aguilera et al. 2019).

National inventories of atmospheric emissions, prepared by the signatory countries to verify compliance with the Kyoto Protocol, is the IPCC Tier 1 approach that is based on global emission factors. However, there is growing evidence that the

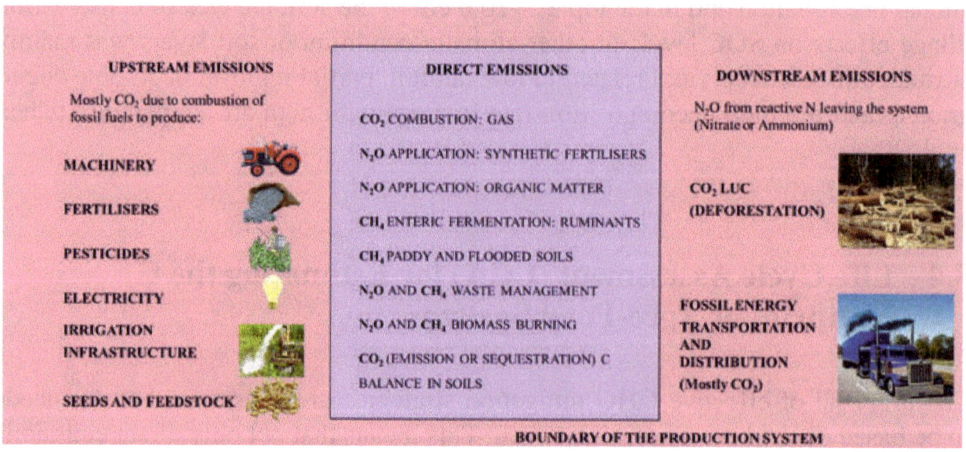

Plate 8.3 An example of processes considered, and the main sources used to estimate the carbon footprint of the Spanish agri-food sector. Reproduced courtesy of Aguilera (2015)

factors differ depending on the type of climate, soil conditions and management, so it is necessary to use more specific factors for a more accurate assessment of emissions. The balance of C in the soil is a crucial process in the CF of agricultural products, since as we have seen it can compensate a large part of the emissions, when the soil gains organic matter and behaves as a sink; otherwise, the soil would contribute more to GHG emissions, when the soil loses organic matter. Despite this, it has been shown that, particularly under arid and semi-arid conditions, C in the soil is very sensitive to changes in management regimes, with changes that can range from the loss of half a ton of C per year in unfertilised soils, to the gain of more than one ton of C in soils in which organic amendments are applied.

The emissions derived from the production of agricultural inputs have decreased significantly due to improvements in efficiency in the industry; therefore, it is necessary to use temporal dynamic factors for the evaluation of historical GHG emissions. Quantitative reviews have been published that analyse precisely these factors for conditions comparable to those in European countries (Aguilera et al. 2015).

The sources of GHG emissions include the construction and maintenance of the agricultural infrastructure, the direct and indirect emissions associated with the use of energy, including traction animals, power generation and fuel use, and CH_4 emissions from water bodies (reservoirs, rafts, ditches and canals) (Aguilera et al. 2019). Emissions related to the use of energy could be estimated considering the changes in the country's electric mix, in the energy efficiency of electric generation, and in fossil fuel extraction techniques, including associated methane emissions. The GHG emissions associated with water bodies should include CH_4, CO_2 and N_2O.

Given the challenges that we face in reducing GHG emissions at the agri-food system level, it is necessary to advance our knowledge about effective mitigation strategies that are adapted to the soil-climatic conditions in each region, for example, by synthesising the existing relevant information regarding the main agricultural management practices and their impact on the mitigation of GHG emissions, C sequestration, other polluting compounds, as well as potential barriers and opportunities for the implementation of these strategies.

8.5 Conclusions

We conclude that CSA practices, with an emphasis on climate change adaptation and mitigation, can take many different forms. The CSA practices have many roles to play in agricultural sustainability and in reducing GHG emissions, as well as in increasing soil C sequestration. Practices such as the use of nitrification and urease inhibitors, mulching, application of biochar to the soil, fertilisation management and use of intercropping and crop rotations are all options available to landowners to effectively adapt to and mitigate regional to global climate change. The reader is, however, cautioned that the best CSA practice to be applied to a specific system or location depends on many different factors. Region- or site-specific research is often needed prior to their application to determine if any of the CSAs might produce a positive result on climate change adaptation and mitigation.

References

Abalos D, Jeffery S, Sanz-Cobena A, Guardia G, Vallejo A (2014) Meta-analysis of the effect of urease and nitrification inhibitors on crop productivity and nitrogen use efficiency. Agric Ecosyst Environ 189:136–144

Abalos D, Sanz-Cobena A, Misselbrook T, Vallejo A (2012) Effectiveness of urease inhibition on the abatement of ammonia, nitrous oxide and nitric oxide emissions in a non-irrigated Mediterranean barley field. Chemosphere 89:310–318

Achtinich C, Bak F, Conrad R (1995) Competition for electron donors among nitrate reducers, ferric iron reducers, sulfate producers and methanogens in anoxic paddy soil. Biol Fertil Soils 19:65–72

Aguilera E (2016) The influence of management practices on the greenhouse gas balance of Mediterranean cropping systems: identifying the climate change mitigation potential through quantitative review and life cycle assessment. PhD Thesis

Aguilera E, Guzmán G, Alonso A (2015) Greenhouse gas emissions from conventional and organic cropping systems in Spain. II. Fruit tree orchards. Agron Sustain Dev 35:725–737

Aguilera E, Guzmán GI, Álvaro-Fuentes J, Infante-Amate J, García-Ruiz R, Carranza-Gallego G, Soto D, González de Molina M (2018) A historical perspective on soil organic carbon in Mediterranean cropland (Spain, 1900–2008). Sci Total Environ 621:634–648

Aguilera E, Vila-Traver J, Deemer BR, Infante-Amate J, Guzmán GI, González De Molina M (2019) Methane emissions from artificial waterbodies dominate the carbon footprint of irrigation: A study of transitions in the food-energy-water-climate nexus (Spain, 1900-2014). Environ Sci Tech 53:5091–5101

Aulakh MS, Wassmann R, Rennenberg H (2001) Methane emissions from rice fields quantification, mechanisms, role of management, and mitigation options. Adv Agron 70:193–260

Bai XX, Huang YW, Ren W, Coyne M, Jacinthe PA, Tao B, Hui DF, Yang J, Matocha C (2019) Responses of soil carbon sequestration to climate-smart agriculture practices: A meta-analysis. Glob Chang Biol 25:2591–2606

Barber RG, Orellana M, Navarro F, Diaz O, Soruco MA (1996) Effects of conservation and conventional tillage systems after land clearing on soil properties and crop yield in Santa Cruz, Bolivia. Soil Tillage Res 38:133–152

Barrena I, Menendez S, Correa-Galeote D, Vega-Mas I, Bedmar EJ, Gonzalez-Murua C, Estavillo JM (2017) Soil water content modulates the effect of the nitrification inhibitor 3,4-dimethylpyrazole phosphate (DMPP) on nitrifying and denitrifying bacteria. Geoderma 303:1–8

Bodelier PLE, Roslev P, Henckel T, Frenzel P (2000) Stimulation by ammonium-based fertilizers of methane oxidation in soil around rice roots. Nature 403:421–424

Bossio DA, Horwath WR, Mutters RG, van Kessel C (1999) Methane pool and flux dynamics in a rice field following straw incorporation. Soil Biol Biochem 31:1313–1322

Bouwman AF (1990) Soils and the greenhouse effect. In: Proceedings of the international conference soils and the greenhouse effect. International Soil Reference and Information Centre (ISRIC), John Wiley and Sons, New York, pp 575

Cai ZC, Mosier AR (2000) Effect of NH_4Cl addition on methane oxidation by paddy soils. Soil Biol Biochem 32:1537–1545

Cai ZC, Sawamoto T, Li CS, Kang GD, Boonjawat J, Mosier A, Wassmann R, Tsuruta H (2003) Field validation of the DNDC model for greenhouse gas emissions in East Asian cropping systems. Glob Biogeochem Cyc 17:1107

Cai ZJ, Gao SD, Xu MG, Hanson BD (2018) Evaluation of potassium thiosulfate as a nitrification inhibitor to reduce nitrous oxide emissions. Sci Total Environ 618:243–249

Campbell BM, Thornton P, Zougmore R, van Asten P, Lipper L (2014) Sustainable intensification: What is its role in climate smart agriculture? Curr Opin Environ Sustain 8:39–43

Cayuela ML, Aguilera E, Sanz-Cobena A, Adams DC, Abalos D, Barton L, Ryals R, Silver WL, Alfaro MA, Pappa VA et al (2017) Direct nitrous oxide emissions in Mediterranean climate cropping systems: emission factors based on a meta-analysis of available measurement data. Agric Ecosyst Environ 238:25–35

Cayuela ML, Sanchez-Monedero MA, Roig A, Hanley K, Enders A, Lehmann J (2013) Biochar and denitrification in soils when, how much and why does biochar reduce N_2O emissions? Sci Rep 3:1732

Chan KY, Heenan DP, So HB (2004) Sequestration of carbon and changes in soil quality under conservation tillage on light-textured soils in Australia: a review. Aust J Exp Agric 43:325–334

Chang JY, Clay DE, Clay SA, Chintala R, Miller JM, Schumacher T (2016) Biochar reduced nitrous oxide and carbon dioxide emissions from soil with different water and temperature cycles. Agron J 108:2214–2221

Conrad R, Rothfuss F (1991) Methane oxidation in the soil surface layer of a flooded rice field and the effect of ammonium. Biol Fertil Soils 12:28–32

Daum D, Schenk MK (1998) Influence of nutrient solution pH on N_2O and N_2 emissions from a soilless culture system. Plant Soil 203:279–287

Davidson EA, Ackerman IL (1993) Changes in soil carbon inventories following cultivation of previously untilled soils. Biogeochemistry 20:161–193

FAO (2004) Carbon sequestration in dryland soils. World Soil Resources Reports 102. FAO, Rome, Italy, pp 108

FAO (2013) Climate-smart agriculture sourcebook. Food and Agriculture Organization of the United Nations, Rome, Italy, p 570

Feng Y, Xu Y, Yu Y, Xie Z, Lin X (2012) Mechanisms of biochar decreasing methane emission from Chinese paddy soils. Soil Biol Biochem 46:80–88

Fentabil MM, Nichol CF, Neilsen GH, Hannam KD, Neilsen D, Forge TA, Jones MD (2016) Effect of micro-irrigation type, N-source and mulching on nitrous oxide emissions in a semi-arid climate: an assessment across two years in a Merlot grape vineyard. Agric Water Manag 171:49–62

Flessa H, Wild U, Klemisch M, Pfadenhauer J (1998) Nitrous oxide and methane fluxes from organic soils under agriculture. Eur J Soil Sci 49:327–335

Food and Agriculture Organization of the United Nations (2015) Soils: our ally against climate change [Video]

Fuertes-Mendizábal T, Huérfano X, Vega-Mas I, Torralbo F, Menéndez S, Ippolito JA, Kammann C, Wrage-Mönnig N, Cayuela ML, Borchard N, Spokas K, Novak J, González-Moro MB, González-Murua C, Estavillo JM (2019) Biochar reduces the efficiency of nitrification inhibitor 3,4-dimethylpyrazole phosphate (DMPP) mitigating N_2O emissions. Sci Rep 9:2346

Guo L, Wang X, Diao T, Ju X, Niu XG, Zheng L, Zhang X, Han X (2018) N_2O emission contributions by different pathways and associated microbial community dynamics in a typical calcareous vegetable soil. Environ Pollut 242(Pt B):2005–2013

Holzapel-Pschorn A, Conrad R, Seiler W (1985) Production, oxidation and emission of methane in rice paddies. FEMS Microbiol Ecol 31:345–351

Hu F, Chai Q, Yu A, Yin W, Cui HY, Gan YT (2015) Less carbon emissions of wheat-maize intercropping under reduced tillage in arid areas. Agron Sustain Dev 35:701–711

Hüppi R, Neftel A, Lehmann MF, Krauss M, Six J, Leifeld J (2016) N use efficiencies and N_2O emissions in two contrasting, biochar amended soils under winter wheat-cover crop-sorghum rotation. Environ Res Lett 11:084013

Hynšt J, Brůček P, Šimek M (2007) Nitrous oxide emissions from cattle-impacted pasture soil amended with nitrate and glucose. Biol Fertil Soils 43:853–859

IPCC (2007) Summary for policymakers in Climate Change (2007) the physical science basis. In: Contribution of working group I to the fourth assessment report of the intergovernmental panel on climate change. Cambridge University Press, Cambridge, NY, USA, 2007

IPCC (2014) Climate Change (2014) synthesis report. In: Core Writing Team, Pachauri RK, Meyer LA (eds) Contribution of working groups I, II and III to the fifth assessment report of the intergovernmental panel on climate change. IPCC, Geneva, Switzerland, pp 151

IPNI (undated) Nitrification inhibitors. Nutrient source specifics(26)

Ippolito JA, Laird DA, Busscher WJ (2012) Environmental benefits of biochar. J Environ Qual 41(4):967–972

Karhu K, Mattila T, Bergström I, Regina K (2011) Biochar addition to agricultural soil increased CH_4 uptake and water holding capacity—results from a short-term pilot field study. Agric Ecosyst Environ 140(1–2):309–313

Kniss AR, Savage SD, Jabbour R (2016) Commercial crop yields reveal strengths and weaknesses for organic agriculture in the United States. PLoS ONE 11(8)

Knoblauch C, Maarifat A, Pfeiffer E, Haefele SM (2011) Degradability of black carbon and its impact on trace gas fluxes and carbon turnover in paddy soils. Soil Biol Biochem 43:1768–1778

Kumar U, Jain MC, Pathak H, Kumar S, Majumdar D (2000) Nitrous oxide emission from different fertilizers and its mitigation by nitrification inhibitors in irrigated rice. Biol Fertil Soils 32:474–478

Lam SK, Suter H, Bai M, Walker C, Davies R, Mosier AR, Chen DL (2018) Using urease and nitrification inhibitors to decrease ammonia and nitrous oxide emissions and improve productivity in a subtropical pasture. Sci Total Environ 644:1531–1535

Lehmann J, Rillig MC, Thies J, Masiello CA, Hockaday WC, Crowley D (2011) Biochar effects on soil biota–a review. Soil Biol Biochem 43:1812–1836

Li H, Liang XQ, Chen YX, Lian YF, Tian GM, Ni WZ (2008) Effect of nitrification inhibitor DMPP on nitrogen leaching, nitrifying organisms, and enzyme activities in a rice-oilseed rape cropping system. J Environ Sci 20:149–155

Lindau CW, Bollich PK (1993) Methane emissions from Louisiana first and ratoon crop rice. Soil Sci 156:42–48

Lipper L, Thornton P, Campbell BM, Baedeker T, Braimoh A, Bwalya M, Caron P, Cattaneo A, Garrity D, Henry K, Hottle R, Jackson L, Jarvis A, Kossam F, Mann W, McCarthy N, Meybeck A, Neufeldt H, Remington T, Sen PT, Sessa R, Shula R, Tibu A, Torquebiau EF (2014) Climate-smart agriculture for food security. Nat Clim Chang 4:1068–1072

Liu Y, Yang M, Wu Y (2011) Reducing CH_4 and CO_2 emissions from waterlogged paddy soil with biochar. J Soils Sediments 11(6):930–939

Lu WF, Chen W, Duan BW, Guo WM, Lu Y, Lantin RS, Wassmann R, Neue HU (2000) Methane emissions and mitigation options in irrigated rice fields in southeast China. Nutr Cycl Agroecosyst 58:65–73

Ma J, Xu H, Yagi K, Cai Z (2008) Methane emission from paddy soils as affected by wheat straw returning mode. Plant Soil 313:167–174

Majumdar D, Kumar S, Pathak H, Jain MC, Kumar U (2000) Reducing nitrous oxide emission from an irrigated rice field of North India with nitrification inhibitors. Agric Ecosyst Environ 81:163–169

Matsushima M, Chang SX (2006) Vector analysis of understory competition, N fertilization, and litter layer removal effects on white spruce growth and nutrition in a 13-year-old plantation. For Ecol Manag 236:332–341

Matsushima M, Chang SX (2007) Effects of understory removal, N fertilization, and litter layer removal on soil N cycling in a 13-year-old white spruce plantation infested with Canada bluejoint grass. Plant Soil 292:243–258

Minasny B, Malone BP, McBratney AB, Angers DA, Arrouays D, Chambers A, Chaplot V, Chen Z-S, Cheng K, Das BS, Field DJ, Gimona A, Hedley CB, Hong SY, Mandal B, Marchant BP, Martin M, McConkey BG, Mulder VL, O'Rourke SM, Richer-de-Forges AC, Odeh I, Padarian J, Paustian K, Pan G, Poggio L, Savian I, Stolbovoy V, Stockmann U, Sulaeman Y, Tsui C-C, Vågen T-G, Van Wesemael B, Winowiecki L (2017) Soil carbon 4 per mille. Geoderma 292:59–86

Mosier AR, Halvorson AD, Reule CA, Liu XJ (2006) Net global warming potential and greenhouse gas intensity in irrigated cropping systems in northeastern Colorado. J Environ Qual 35:1584–1598

Nägele W, Conrad R (1990) Influence of pH on the release of NO and N_2O from fertilized and unfertilized soil. Biol Fertil Soils 10:139–144

Procházková B, Hrubý J, Dovrtěl J, Dostál O (2003) Effects of different organic amendment on winter wheat yields under long-term continuous cropping. Plant Soil Environ 49:433–438

Ricord M (2018) Evaluating intercropping systems as a sustainable agroecosystem alternative to reduce greenhouse gas emissions. MSc thesis, University of Waterloo, pp 151

Rodgers GA (1986) Nitrification inhibitors in agriculture. J Environ Sci Health A 21:701–722

Rodriguez MJ, Saggar S, Berben P, Palmada T, Lopez-Villalobos N, Pal P (2019) Use of a urease inhibitor to mitigate ammonia emissions from urine patches. Environ Tech

Rogers CW, Norman RJ, Brye KR, Slaton NA, Smartt AD (2015) Comparison of urease inhibitors for use in rice production on a silt-loam soil. Crop Forage Turfgrass Manag 1:1–6

Sánchez-García M, Roig A, Sánchez-Monedero MA, Cayuela ML (2014) Biochar increases soil N_2O emissions produced by nitrification-mediated pathways. Front Environ Sci 2:25

Sanderman J, Hengl T, Fiske GJ (2017) Soil carbon debt of 12,000 years of human land use. PNAS USA 114:9575–9580

Sanz Cobena A, Misselbrook TH, Arce A, Mingot JI, Diez JA, Vallejo A (2008) An inhibitor of urease activity effectively reduces ammonia emissions from soil treated with urea under Mediterranean conditions. Agric Ecosyst Environ 126:243–249

Sanz-Cobena A, Lassaletta L, Aguilera E, Del Prado A, Garnier J, Billen G, Iglesias A, Sánchez B, Guardia G, Abalos D, Plaza-Bonilla D, Puigdueta-Bartomoé I, Moral R, Galán E, Arriaga H, Merino P, Infante-Amate J, Meijide A, Pardo G, Álvaro-Fuentes J, Gilsanz C, Báez D, Doltra J, González-Ubierna S, Cayuela ML, Menéndez S, Díaz-Pinés E, Le-Noë J, Quemada M, Estellés F, Calvet S, van Grinsven HJM, Westhoek H, Sanz MJ, Gimeno BS, Vallejo A, Smith P (2017) Strategies for greenhouse gas emissions mitigation in Mediterranean agriculture: a review. Agric Ecosys Environ 238:5–24

Sanz-Cobena A, Abalos D, Meijide A, Sanchez-Martin L, Vallejo A (2014) Soil moisture determines the effectiveness of two urease inhibitors to decrease N_2O emission. Mitig Adapt Strateg Glob Change

Sanz-Cobena A, Sánchez-Martín L, García-Torres L, Vallejo A (2012) Gaseous emissions of N_2O and NO and NO_3^- leaching from urea applied with urease and nitrification inhibitors to a maize (Zeamays) crop. Agric Ecosyst Environ 149:64–73

Sass RL, Fisher FM, Harcombe PA, Turner FT (1991) Methane emission from rice fields as influenced by solar radiation, temperature and straw incorporation. Glob Biogeochem Cyc 5:335–350

Schütz H, Holzapfel-Pschorn A, Conrad R, Rennenberg H, Seiler W (1989) A three-year continuous record on the influence of daytime season and fertilizer treatment on methane emission rates from an Italian rice paddy field. J Geophys Res 94:16405–16416

Schwartz J (2014) Soil as carbon storehouse: new weapon in climate fight? Yale School of Forestry & Environmental Studies

Shen Y, Sui P, Huang J, Wang D, Whalen JK, Chen YQ (2018) Greenhouse gas emissions from soil under maize–soybean intercrop in the North China Plain. Nutr Cycl Agroecosyst 110:451–465

Šimek M, Cooper JE (2002) The influence of soil pH on denitrification, progress towards the understanding of this interaction over the last 50 years. Eur J Soil Sci 53:345–354

Šimek M, Jíšová L, Hopkins DW (2002) What is the so-called optimum pH for denitrification in soil? Soil Biol Bioch 34:1227–1234

Skinner C, Gattinger A, Krauss M, Krause HM, Mayer J, van der Heijden MGA, Mäder P (2019) The impact of long-term organic farming on soil-derived greenhouse gas emissions. Sci Rep 9:1702

Smith P (2016) Soil carbon sequestration and biochar as negative emission technologies. Glob Chang Biol 22:1315–1324

Smith P, Martino D, Cai ZC et al (2008) Greenhouse gas mitigation in agriculture. Philos Trans R Soc B 363(1492):789–813

Snyder CS, Bruulsema TW, Jensen TL, Fixen PE (2009) Review of greenhouse gas emissions from crop production systems and fertiliser management effects. Agric Ecosyst Environ 133:247–266

Soares JR, Cantarella H, Menegale MLD (2012) Ammonia volatilization losses from surface-applied urea with urease and nitrification inhibitors. Soil Biol Biochem 52:82–89

Souri MK, Neumann G (2018) Indications for passive rather than active release of natural nitrification inhibitors in *Brachiaria humidicola* root exudates. J Plant Nutrition 41:477–486

Squalli J, Adamkiewicz G (2018) Organic farming and greenhouse gas emissions: a longitudinal US state-level study. J Clean Prod 192:30–42

Stevens RJ, Laughlin RJ (1998) Measurement of nitrous oxide and di-nitrogen emissions from agricultural soils. Nutr Cycl Agroecosyst 52:131–139

Subbarao GV, Nakahara K, Ishikawa T, Yoshihashi T, Ito O, Ono H, Ohnishi-Kameyama M, Yoshida M, Kawano N, Berry WL (2008) Free fatty acids from the pasture grass Brachiaria humidicola and one of their methyl esters as inhibitors of nitrification. Plant Soil 313:89–99

Subbarao GV, Rao IM, Nakahara K, Sahrawat KL, Ando Y, Kawashima T (2013a) Potential for biological nitrification inhibition to reduce nitrification and N_2O emissions in pasture crop–livestock systems. Animal 7:322–332

Subbarao GV, Nakahara K, Ishikawa T, Ono H, Yoshida M, Yoshihashi T, Zhu YY, Zakir HAKM, Deshpande SP, Hash CT, Sahrawat KL (2013b) Biological nitrification inhibition (BNI) activity in sorghum and its characterization. Plant Soil 366:243–259

Subbarao GV, Rondon M, Ito O, Ishikawa T, Rao IM, Nakahara K, Lascano C, Berry WL (2007) Biological nitrification inhibition (BNI)—is it a widespread phenomenon? Plant Soil 294:5–18

Tate KR, Ross DJ, Saggar S, Hedley CB, Dando J, Singh BK, Lambie SM (2007) Methane uptake in soils from Pinus radiata plantations, a reverting shrubland and adjacent pastures: effects of landuse change, and soil texture, water and mineral nitrogen. Soil Biol Biochem 39:1437–1449

United Nations (2013) Greenhouse gas emissions and options for mitigation in Viet Nam, and the UNs responses

Wang FH, Chen SM, Wang YY, Zhang YM, Hu CS, Liu BB (2018) Long-term nitrogen fertilization elevates the activity and abundance of nitrifying and denitrifying microbial communities in an upland soil: implications for nitrogen loss from intensive agricultural systems. Front Microbiol 9:2424

Wassmann R, Lantin RS, Neue HU, Buendia LV, Corton TM, Lu Y (2000) Characterization of methane emissions from rice fields in Asia III: mitigation options and future research needs. Nutr Cycl Agroecosyst 58:23–36

Wassmann R, Neue HU, Ladha JK, Aulakh MS (2004) Mitigating greenhouse gas emissions from rice-wheat cropping systems in Asia. In: Wassmann R, Vlek PLG (eds) Tropical agriculture in transition—opportunities for mitigating greenhouse gas emissions?. Springer, Dordrecht

Waters D, Van Zwieten L, Singh BP, Downie A, Cowie AL, Lehmann J (2011) Biochar in soil for climate change mitigation and adaptation. In: Singh BP et al. (eds) Soil health and climate change. Soil biology, vol 29. Springer, Heidelberg, pp 345–368

Wu FP, Jia ZK, Wang SG, Chang SX, Startsev A (2013) Contrasting effects of wheat straw and its biochar on greenhouse gas emissions and enzyme activities in a Chernozemic soil. Biol Fert Soils 49:555–565

Xie Z, Xu Y, Liu G, Liu Q, Zhu J, Tu C, James SE, Amonette JE, Cadisch G, Jean WH, Yong JWH, Hu S (2013) Impact of biochar application on nitrogen nutrition of rice, greenhouse-gas emissions and soil organic carbon dynamics in two paddy soils of China. Plant Soil 370:527–540

Yagi K, Tsuruta H, Minami K (1997) Possible options for mitigating methane emission from rice cultivation. Nutr Cycl Agroecosyst 49:213–220

Yang M, Fang YT, Sun D, Shia YL (2016) Efficiency of two nitrification inhibitors (dicyandiamide and 3, 4-dimethypyrazole phosphate) on soil nitrogen transformations and plant productivity a meta-analysis. Sci Rep 6:22075

Yin W, Guo Y, Hu FL, Fan ZL, Feng FX, Zhao C, Yu AZ, Chai Q (2018) Wheat-maize intercropping with reduced tillage and straw retention: a step towards enhancing economic and environmental benefits in arid areas. Front Plant Sci 9:1328

Zaman M, Blennerhassett JD (2010) Effects of the different rates of urease and nitrification inhibitors on gaseous emissions of ammonia and nitrous oxide, nitrate leaching and pasture production from urine patches in an intensive grazed pasture system. Agric Ecosyst Environ 136:236–246

Zaman M, Nguyen ML (2010) Effect of lime or zeolite on N_2O and N_2 emissions from a pastoral soil treated with urine or nitrate-N fertiliser under field conditions. Agric Ecosyst Environ 136:254–261

Zaman M, Nguyen ML, Saggar S (2008a) N_2O and N_2 emissions from pasture and wetland soils with and without amendments of nitrate, lime and zeolite under laboratory condition. Aust J Soil Res 46:526–534

Zaman M, Nguyen ML, Gold AJ, Groffman PM, Kellogg DQ, Wilcock RJ (2008b) Nitrous oxide generation, denitrification and nitrate removal in a seepage wetland intercepting surface and subsurface flows from a grazed dairy catchment. Aust J Soil Res 46:565–577

Zaman M, Nguyen ML, Matheson F, Blennerhassett JD, Quin BF (2007) Can soil amendments zeolite or lime shift the balance between nitrous oxide and dinitrogen emissions from pasture and wetland soils receiving urine or urea-N? Aust J Soil Res 45:543–553

Zaman M, Saggar S, Blennerhassett JD, Singh J (2009) Effect of urease and nitrification inhibitors on N transformation, gaseous emissions of ammonia and nitrous oxide, pasture yield and N uptake in grazed pasture system. Soil Biol Biochem 41:1270–1280

Zhu GD, Ju XT, Zhang JB, Muller C, Rees RM, Thorman RE, Sylvester-Bradley R (2019) Effects of the nitrification inhibitor DMPP (3,4-dimethylpyrazole phosphate) on gross N transformation rates and N_2O emissions. Biol Fertil Soils 55:603–615

Zvomuya F, Rosen CJ, Russelle MP, Gupta SC (2003) Nitrate leaching and nitrogen recovery following application of polyolefin coated urea to potato. J Environ Qual 32:480–489

Permissions

All chapters in this book were first published in MEAGGDMONRTANTGHGS, by Springer; hereby published with permission under the Creative Commons Attribution License or equivalent. Every chapter published in this book has been scrutinized by our experts. Their significance has been extensively debated. The topics covered herein carry significant information for a comprehensive understanding. They may even be implemented as practical applications or may be referred to as a beginning point for further studies.

The contributors of this book come from diverse backgrounds, making this book a truly international effort. We would like to thank all the contributing authors for lending their expertise to make the book truly unique. They have played a crucial role in the development of this book. Without their invaluable contributions this book wouldn't have been possible. They have made vital efforts to compile up to date information on the varied aspects of this subject to make this book a valuable addition to the collection of many professionals and students.

This book was conceptualized with the vision of imparting up-to-date and integrated information in this field. To ensure the same, a matchless editorial board was set up. Every individual on the board went through rigorous rounds of assessment to prove their worth. After which they invested a large part of their time researching and compiling the most relevant data for our readers.

The editorial board has been involved in producing this book since its inception. They have spent rigorous hours researching and exploring the diverse topics which have resulted in the successful publishing of this book. They have passed on their knowledge of decades through this book. To expedite this challenging task, the publisher supported the team at every step. A small team of assistant editors was also appointed to further simplify the editing procedure and attain best results for the readers.

Apart from the editorial board, the designing team has also invested a significant amount of their time in understanding the subject and creating the most relevant covers. They scrutinized every image to scout for the most suitable representation of the subject and create an appropriate cover for the book.

The publishing team has been an ardent support to the editorial, designing and production team. Their endless efforts to recruit the best for this project, has resulted in the accomplishment of this book. They are a veteran in the field of academics and their pool of knowledge is as vast as their experience in printing. Their expertise and guidance has proved useful at every step. Their uncompromising quality standards have made this book an exceptional effort. Their encouragement from time to time has been an inspiration for everyone.

The publisher and the editorial board hope that this book will prove to be a valuable piece of knowledge for students, practitioners and scholars across the globe.

Index

Printed in the USA
CPSIA information can be obtained
at www.ICGtesting.com
LVHW082159201023
761596LV00009B/1503